New Perspectives on the Theory of Inequalities for Integral and Sum

Nazia Irshad • Asif R. Khan • Faraz Mehmood •
Josip Pečarić

New Perspectives on the Theory of Inequalities for Integral and Sum

Nazia Irshad
Department of Mathematics
Dawood University of Engineering
and Technology
Karachi, Pakistan

Asif R. Khan
Department of Mathematics
University of Karachi
Karachi, Pakistan

Faraz Mehmood
Department of Mathematics
Dawood University of Engineering
and Technology
Karachi, Pakistan

Josip Pečarić
Croatian Academy of Sciences
and Arts
Zagreb, Croatia

ISBN 978-3-030-90562-0   ISBN 978-3-030-90563-7 (eBook)
https://doi.org/10.1007/978-3-030-90563-7

Mathematics Subject Classification: 26A16, 26A42, 26A45, 26A46, 26A51, 39B22, 39B62, 26D07, 26D10, 26D15, 26D20, 26D99

© The Editor(s) (if applicable) and The Author(s), under exclusive license to Springer Nature Switzerland AG 2021

This work is subject to copyright. All rights are solely and exclusively licensed by the Publisher, whether the whole or part of the material is concerned, specifically the rights of translation, reprinting, reuse of illustrations, recitation, broadcasting, reproduction on microfilms or in any other physical way, and transmission or information storage and retrieval, electronic adaptation, computer software, or by similar or dissimilar methodology now known or hereafter developed.

The use of general descriptive names, registered names, trademarks, service marks, etc. in this publication does not imply, even in the absence of a specific statement, that such names are exempt from the relevant protective laws and regulations and therefore free for general use.

The publisher, the authors, and the editors are safe to assume that the advice and information in this book are believed to be true and accurate at the date of publication. Neither the publisher nor the authors or the editors give a warranty, expressed or implied, with respect to the material contained herein or for any errors or omissions that may have been made. The publisher remains neutral with regard to jurisdictional claims in published maps and institutional affiliations.

This book is published under the imprint Birkhäuser, www.birkhauser-science.com by the registered company Springer Nature Switzerland AG.
The registered company address is: Gewerbestrasse 11, 6330 Cham, Switzerland

Preface

The discipline of mathematical inequalities has continued to grow rapidly since the beginning of the twentieth century. Jensen, Cauchy, Schwartz, Hölder, Hadamard, Minkowski and Hardy of that era are mathematicians known for their work in inequalities. Inequalities are essential to study mathematics and many related fields. Various applications of inequalities have been established in the field of differential and integral equations, calculus, probability theory, interpolation theory, optimization theory, control theory, game theory, spectral theory, functional analysis, harmonic analysis, economics, physics and geometry. A number of authors use integral inequalities in studying existence, uniqueness, boundedness, stability and asymptotic behaviour of solutions of ordinary and partial differential equations. There are numerous known inequalities, and the list is ongoing. The database of MathSciNet contains over 23,000 references of inequalities and their applications.

Primarily the most relevant target audience of this book are all mathematicians either from pure or applied side. Secondary audience may consist of physicists, statisticians, engineers, researchers and decision makers working in various research and government departments. Research scholars, students, faculty members and professionals who are working and doing research in different field and areas including (but not limited to) statistics, physics, optimization theory, numerical integration, probability theory, convex analysis, mathematical analysis, integration and measure theory, and linear inequalities may got help from this book.

Our book shows new and latest results and methods with new findings and applications as compared to the book titled *General linear inequalities and positivity: Higher order convexity* [101]. To be more specific Chap. 1 of this book is similar to Chapter 5 of [101]. In Chapter 5 of another book, the authors have studied general linear inequalities of Popoviciu type via interpolation polynomials with and without Green function. In Chap. 1 of our book we have studied the same general linear inequalities via interpolation polynomials using 4 new Green functions. Furthermore, in Chap. 4 of our book, we have studied Popoviciu and Čebyšev-Popoviciu type identities and inequalities using ∇ operator and completely monotonic functions, while in monograph [101], the authors have studied similar results for Δ operator. In [101], the authors have stated some results related to functions with nondecreasing increments, while in our book, we not only state some results for functions with nondecreasing increments but also have compared and linked many important concepts including arithmetic integral mean, Wright convex functions, convex functions, ∇-convex functions, Jensen m-convex functions, m-convex functions, m-∇-convex functions, k-monotonic functions, absolutely monotonic functions, completely monotonic functions, Laplace transform and exponentially convex functions, by using the finite difference operator as different cases of $\Delta_h^m f$. We can find other similarities in both the books as well, for instance both books contain results related to Montgomery identities, Čebyšev, Grüss and Ostrowski inequalities, but our book discusses all these identities and inequalities in greater detail.

Now, we put some light on the chapter-wise organization of the book as follows:

In the first chapter, we would like to discuss linear inequalities via interpolation polynomials with Green functions. To be more specific, our focus would be to obtain Popoviciu type inequalities via

- Extension of the Montgomery identity.
- The Taylor's formula.
- Hermite interpolation polynomials.
- The Fink identity.
- Abel-Gontscharoff's interpolation polynomials.

We would like to discuss all the above-mentioned results with effect of newly established Green's functions of a certain dynamical system. For Popoviciu type inequalities, we would also like to obtain results for:

- n-convex functions at a point.
- New upper bounds for remainders through Čebyšev functional.
- Mean value theorems of Lagrange and Cauchy type.
- n-exponential and logarithmic convexity.
- Construction of generalized means and mixed symmetric means, etc.

Some non-trivial examples in all above-mentioned topics will also be presented.

Regarding applications, in our proposed research, we would obtain mean value theorems which provide good estimates for physical quantities. We would also obtain special means (averages) which play an important role in weather forecasting and stock exchange (in terms of moving averages) and help us to study dynamical systems and all the systems that do not have any specific pattern or are unpredictable or have high level of uncertainty. In all these cases, we mostly deal with averages. As far as we are concerned with the exponential convexity, exponentially convex functions have many nice properties, for example, these functions are analytic in their domain. These functions also provide us positive-semidefinite matrices. Moreover, they play an important role in studying the properties of Stolarsky and Cauchy means, such as monotonicity of these means. Additionally, a number of important inequalities arise from the logarithmic convexity of some functions. Logarithmic convexity plays an important role in various fields like reliability theory and survival analysis, economics, statistics, social sciences, information theory and optimization. Its applications can also be found in applied mathematics.

In the second chapter, our focus is on Ostrowski type inequalities. In 1937, a Ukrainian mathematician Alexander Markowich Ostrowski first presented this interesting and useful inequality [141]. It can be used to find out the absolute deviation of functional value from its integral mean. It also approximates area under the curve of a function by a rectangle. It has great importance because of its number of applications in statistics, probability theory, integral operator theory, numerical integration and special means. We present new extensions and generalized results of Ostrowski type integral inequalities including Grüss, Ostrowski-Grüss and Čebyšev involving parameters with and without weights. We estimate the bounds on the deviation of function values from its mean value for different functions including bounded differentiable functions (also for the case of bounded below only differentiable function and bounded above only differentiable function), functions from L_p spaces for $p \geq 1$, function of bounded variation and absolutely continuous function. We generalize Ostrowski-Grüss inequality via weighted Korkine's identity and Peano kernel approach. We also present new results of fractional Ostrowski inequalities using Riemann Liouville fractional integral. In addition, we modify Čebyšev and Ostrowski inequality for two independent variables involving parameters. In the process of generalizing new inequalities, we achieve some refinements of Montgomery identities. These inequalities have some valuable applications related to some standard (including midpoint, trapezoidal, perturbed trapezoidal and Simpson's type) and non-standard numerical quadrature rules and probability theory. Not only they have generalized and better results but also recapture previously obtained inequalities.

The third chapter deals with functions with nondecreasing increments and related results. Functions with nondecreasing increments were introduced by Brunk in 1964. We extend this for mth order by using finite difference operator with equally spaced interval. With the help of this special definition of function with nondecreasing increments, we get relationship among functions with nondecreasing increments and arithmetic integral mean, Wright convex functions, convex functions, ∇-convex functions, Jensen m-convex functions, m-convex functions, m-∇-convex functions, k-monotonic functions, absolutely monotonic functions, completely monotonic functions, Laplace transform and exponentially convex functions, by using the finite difference operator as different cases of $\Delta_h^m f$. We also discuss some examples in each above-stated relation. Generalizations of the Levinson's-type inequality and Jensen-Mercer's-type inequality by using Jensen-Boas inequality for function with nondecreasing increments of third order are also deduced.

In the final chapter, we introduce a new notion of ∇-convex function and completely monotonic functions in two dimension case. We obtain discrete and integral identities for sequences and functions in two dimensions involving higher-order ∇-convex functions and completely monotonic functions. Further, we discuss the Popoviciu type characterization of positivity of sums and integrals of obtained identities. Further, by recalling higher-order completely monotonic function of one and two variables, we use variety of classes of completely monotonic functions and give examples and applications of completely monotonic functions and also recall some generalized results. In order to prove main results of the current chapter, we use various techniques including double induction and Taylor's series expansion. We also discuss some generalized Lagrange and Cauchy-type's mean value theorems for ∇-convex functions of higher order for two independent variables. Further that we construct some nonnegative functionals to study exponential convexity of special type and discuss some properties and we present examples and applications of completely monotonic, exponentially convex functions by using various classes of functions. We also discuss generalization of discrete Čebyšev and Ky Fan's identities for ∇-convex functions of higher order with two independent variables and discuss similar result for discrete case as well. Last but not least, we also obtain the more generalized Montgomery's identities for differentiable function of higher order with two independent variables. Generalized Montgomery's identities support us for contribution in the generalized Ostrowski and Grüss type inequalities for double weighted integrals. We also provide more generalized Ostrowski and Grüss type inequalities for differentiable functions of higher order with two variables as compared to the existing results related to the subject.

Karachi, Pakistan	Nazia Irshad
Karachi, Pakistan	Asif R. Khan
Karachi, Pakistan	Faraz Mehmood
Zagreb, Croatia	Josip Pečarić

Contents

1 Linear Inequalities via Interpolation Polynomials and Green Functions 1
  1.1 Linear Inequalities and the Extension of Montgomery Identity with New Green Functions 5
      1.1.1 Results Obtained by the Extension of Montgomery Identity and New Green Functions 7
      1.1.2 Inequalities for n-Convex Functions at a Point 13
      1.1.3 Bounds for Remainders and Functionals 18
      1.1.4 Mean Value Theorems 22
  1.2 Linear Inequalities and the Taylor Formula with New Green Functions 24
      1.2.1 Results Obtained by the Taylor Formula and New Green Functions 24
      1.2.2 Inequalities for n-Convex Functions at a Point 30
      1.2.3 Bounds for Remainders and Functionals 32
      1.2.4 Mean Value Theorems and Exponential Convexity 36
      1.2.5 Examples with Applications 42
  1.3 Linear Inequalities and Hermite Interpolation with New Green Functions 44
      1.3.1 Results Obtained by the Hermite Interpolation Polynomial and Green Functions 47
      1.3.2 Inequalities for n-Convex Functions at a Point 61
      1.3.3 Bounds for Remainders and Functionals 62
  1.4 Linear Inequalities and the Fink Identity with New Green Functions 63
      1.4.1 Results Obtained by the Fink identity and New Green functions 64
      1.4.2 Inequalities for n-Convex Functions at a Point 68
      1.4.3 Bounds for Remainders and Functionals 69

1.5 Linear Inequalities and the Abel-Gontscharoff's
    Interpolation Polynomial 71
    1.5.1 Results Obtained by the Abel–Gontscharoff's
          Interpolation 72
    1.5.2 Results Obtained by the Abel–Gontscharoff's
          Interpolation Polynomial and Green Functions 76
    1.5.3 Inequalities for n-Convex Functions at a Point 84
    1.5.4 Bounds for Remainders and Functionals 86

2 Ostrowski Inequality 87
2.1 Generalized Ostrowski Type Inequalities with Parameter 90
    2.1.1 Ostrowski Type Inequality for Bounded
          Differentiable Functions 91
    2.1.2 Ostrowski Type Inequalities for Bounded Below
          Only and Bounded Above Only Differentiable Functions 108
    2.1.3 Applications to Numerical Integration 115
2.2 Generalized Ostrowski Type Inequalities for Functions of
    L_p Spaces and Bounded Variation 118
    2.2.1 Ostrowski Type Inequality for Functions of L_p Spaces 118
    2.2.2 Ostrowski Type Inequality for Functions of Bounded
          Variation 122
    2.2.3 Applications to Numerical Integration 126
2.3 Generalized Weighted Ostrowski Type Inequality with Parameter ... 129
    2.3.1 Weighted Ostrowski Type Inequality with Parameter 131
    2.3.2 Applications to Numerical Integration 141
2.4 Generalized Weighted Ostrowski-Grüss Type Inequality
    with Parameter 145
    2.4.1 Weighted Ostrowski-Grüss Type Inequality with
          Parameter by Using Korkine's Identity 148
    2.4.2 Applications to Probability Theory 157
    2.4.3 Applications to Numerical Integration 159
2.5 Generalized Fractional Ostrwoski Type Inequality
    with Parameter 162
    2.5.1 Fractional Ostrowski Type Inequality Involving
          Parameter 163
2.6 Generalized Inequalities for Functions of L_p Spaces via
    Montgomery Identity with Parameters 172
    2.6.1 Montgomery Identity for Functions of Two
          Variables involving Parameters 173
    2.6.2 Generalized Ostrowski Type Inequality 177
    2.6.3 Generalized Grüss Type Inequalities 183

3 Functions with Nondecreasing Increments 193
3.1 Functions with Nondecreasing Increments in Real Life 197
3.2 Relationship Among Functions with Nondecreasing
    Increments and Many Others 199

Contents xi

 3.3 Functions with Nondecreasing Increments of Order 3 207
   3.3.1 On Levinson Type Inequalities 208
   3.3.2 On Jensen-Mercer Type Inequalities 211

4 Popoviciu and Čebyšev-Popoviciu Type Identities and Inequalities ... 213
 4.1 Linear Inequalities for Higher Order ∇-Convex and
   Completely Monotonic Functions 213
   4.1.1 Discrete Identity for Two Dimensional Sequences............ 216
   4.1.2 Discrete Identity and Inequality for Functions of
      Two Variables 219
   4.1.3 Integral Identity and Inequality for Functions of One
      Variable 226
   4.1.4 Integral Identity and Inequality for Functions of
      Two Variables 227
   4.1.5 Mean Value Theorems and Exponential Convexity 245
 4.2 Generalized Čebyšev and Ky Fan Identities and Inequalities
   for ∇-Convex Functions 251
   4.2.1 Generalized Discrete Čebyšev Identity and Inequality 257
   4.2.2 Generalized Integral Čebyšev Identity and Inequality 263
   4.2.3 Generalized Integral Ky Fan Identity and Inequality 273
 4.3 Weighted Montgomery Identities for Higher Order
   Differentiable Function of Two Variables and Related
   Inequalities 275
   4.3.1 Montgomery Identities for Double Weighted
      Integrals of Higher Order Differentiable Functions........... 279
   4.3.2 Ostrowski Type Inequalities for Double Weighted
      Integrals of Higher Order Differentiable Functions 290
   4.3.3 Grüss Type Inequalities for Double Weighted
      Integrals of Higher Order Differentiable Functions 292

Bibliography 299

Index 307

Notations and Terminologies

$\mathbb{R} = (-\infty, \infty)$	Set of all real numbers
$\mathbb{R}_+ = (0, \infty)$	Set of positive real numbers
$\mathbb{R}_* = [0, \infty)$	Set of nonnegative real numbers
$[a, b]$	A closed interval of $\mathbb{R}$ with end points a and b
(a, b)	An open interval of $\mathbb{R}$ with end points a and b
$[a, b)$ or $(a, b]$	A semi-closed interval of $\mathbb{R}$ with end points a and b
$C[a, b]$	Class of all continuous functions defined on $[a, b]$
I	An interval of $\mathbb{R}$
I^o	The interior of the interval I
$AC[a, b]$	Class of all absolutely continuous real-valued functions defined on $[a, b]$
$C^{(n)}[a, b]$	Class of all continuously differentiable real-valued functions of order n defined on $[a, b]$
$L[a, b]$	Class of integrable functions defined on $[a, b]$
$L_p[a, b]$	Function spaces with L_p norm defined on $[a, b]$
$L_\infty[a, b]$	Function spaces with L_∞ norm defined on $[a, b]$
$\|f\|_p$	The norm on L_p defined as $\left(\int_a^b |f(x)|^p dx\right)^{\frac{1}{p}} < \infty$
$\|f\|_\infty$	The norm on L_∞ defined as $ess \sup_{x \in [a,b]} |f(x)| < \infty$
$BV[a, b]$	Space of real-valued functions of bounded variation defined on $[a, b]$
$\bigvee_a^b (f)$	Total variation of f over $[a, b]$
J_a^γ	Fractional integral operator where $\gamma \geq 0$, a is lower limit of the integral

Chapter 1
Linear Inequalities via Interpolation Polynomials and Green Functions

> *One reason why mathematics enjoy special esteem, above all other sciences, is that its law are absolutely certain and indisputable, while those of other sciences are to some extent debatable and in constant danger of being overthrown by newly discovered facts.*
>
> —Albert Einstein

Linear inequalities are those inequalities which involve linear functionals or linear relations. Its best examples are discrete and integral inequalities which are of great importance. In this book several general linear inequalities involving functions of general nature have been stated and proved. To our eye the term general linear inequalities proves subjective to some extent. In our view a general linear inequality is an inequality which is not confined to a specific function rather it is valid for a class of functions. Moreover, it may have the ability to give birth to many other inequalities by substitution of suitable functions and conditions in it. For further study on the topic we refer the monograph [101].

Some definitions and related results which we would be presently used throughout the chapter will be discussed in the following subsection. We commence with the notion of convex functions.

Convex Functions
Although the systematic study of convex functions was commenced by Jensen [82, 83] and one may find its roots in the works of Hermite [72], Hölder [73], Stolz [181] and Hadamard [67], some authors still believe that it may be traced back to Gibbs [178, p. 287]. The theory of convex functions has a unique place in mathematics due to several reasons: firstly, it is amongst the most important theories per se, which touches almost all fields of mathematics such as optimization theory, control theory, operations research, geometry, differential equations, functional analysis, operator theory, probability theory, numerical integration, information theory, integral operator theory etc. The theory of convex functions plays an important role in other branches of sciences as well, such as physics, statistics, mechanics, economics, finance, engineering and management sciences. Due to its

wide range of applications it has attracted many economists, engineers along with pure mathematicians to be more interested in convex analysis [139]: "Convexity appear like an octopus, tentacles reaching far and wide, its shape and color changing as it roams from one area to the next. It is quite clear that research opportunity abounds".

Secondly, convex functions are closely related to the theory of inequalities and many famous inequalities are consequences of the applications of convex functions. It is no exaggeration to argue that the theory of convex functions has become a special domain of inequality theory with a number of powerful results and numerous applications in many branches of mathematics.

Thirdly, it has high geometric and intuitive content. The subject of convex geometry is well developed, as the geometric approach to convex functions is the one many prefer [165].

Fourthly, the comparison of means stands at the core of the notion of convexity. In fact, nowadays the study of convex functions has evolved into a larger theory about functions, which are adapted to other geometries of the domain and/or they obey other laws of comparison of means [139].

Fifthly, the theory of convex functions permits an easy generalization to an abstract setting.

Finally, the class of convex functions may be characterized in a variety of way as we see in present book.

Due to all the aforementioned reasons, many books have been written on the topic of "convex functions". Here we mention some remarkable works such as "Convex analysis" by Rockafellar [166], "Convex functions" by Robert and Verberg [165] and "Convex functions, partial ordering and statistical applications" by Pečarić et al. [158]. Moreover, many classical books of inequalities like "Inequalities" by Hardy et al. [69, 70], "Inequalities" by Beckenbach and Bellman [21], and "Analytic inequalities" by Mitrinović [131] have treated the topic of convex functions extensively. For a detailed discussion on the books related to the topic we refer the reader to [178].

Now, we recall some useful definitions and significant results about convex functions extracted from [158]. Throughout the section I stands for an interval in $\mathbb{R}$.

Definition 1.0.1 A function $f : I \to \mathbb{R}$ is called *convex in the J-sense* or *J-convex* or *mid-convex* if

$$f\left(\frac{x_1 + x_2}{2}\right) \leq \frac{f(x_1) + f(x_2)}{2}$$

holds for each $x_1, x_2 \in I$.

1 Linear Inequalities via Interpolation Polynomials and Green Functions

Definition 1.0.2 A function $f : I \to \mathbb{R}$ is called *convex* if the inequality

$$f(\lambda x_1 + (1 - \lambda) x_2) \leq \lambda f(x_1) + (1 - \lambda) f(x_2) \tag{1.0.1}$$

holds for each $x_1, x_2 \in I$ and $\lambda \in [0, 1]$.

Remark 1.0.1

(a) If inequality (1.0.1) is strict for each $x_1 \neq x_2$ and $\lambda \in (0, 1)$, then f is called *strictly convex*.
(b) If the inequality in (1.0.1) is reversed, then f is called *concave*. If it is strict for each $x_1 \neq x_2$ and $\lambda \in (0, 1)$, then f is called *strictly concave*.
(c) A J-convex function is convex if it is continuous as well. □

The following result gives us an alternate definition of convex functions [158, p. 2].

Theorem 1.0.1 *A function $f : I \to \mathbb{R}$ is convex if the inequality*

$$(x_3 - x_2) f(x_1) + (x_1 - x_3) f(x_2) + (x_2 - x_1) f(x_3) \geq 0$$

holds for each $x_1, x_2, x_3 \in I$ such that $x_1 < x_2 < x_3$.

The following result can be deduce from Theorem 1.0.1.

Theorem 1.0.2 *If a function $f : I \to \mathbb{R}$ is convex, then the inequality*

$$\frac{f(x_2) - f(x_1)}{x_2 - x_1} \leq \frac{f(y_2) - f(y_1)}{y_2 - y_1}$$

holds for each $x_1, x_2, y_1, y_2 \in I$ such that $x_1 \leq y_1$, $x_2 \leq y_2$, $x_1 \neq x_2$, $y_1 \neq y_2$.

Definition 1.0.3 A sequence (a_i) of real numbers is called convex sequence if $2a_i \leq a_{i-1} + a_{i+1}$, $\forall i \in \{2, \ldots, n-1\}$.

Now we define generalized convex function which can be found in [97], [95] and [158].

Definition 1.0.4 The *nth order divided difference* of a function $f : I \to \mathbb{R}$ at distinct points $x_i, x_{i+1}, \ldots, x_{i+n} \in I$ for some $i \in \mathbb{N} \cup \{0\}$ is defined recursively by:

$$[x_j; f] = f(x_j), \quad j \in \{i, \ldots, i+n\};$$

$$[x_i, \ldots, x_{i+n}; f] = \frac{[x_{i+1}, \ldots, x_{i+n}; f] - [x_i, \ldots, x_{i+n-1}; f]}{x_{i+n} - x_i}.$$

It may easily be verified that

$$[x_i, \ldots, x_{i+n}; f] = \sum_{k=0}^{n} \frac{f(x_{i+k})}{\prod_{j=i, j \neq i+k}^{i+n} (x_{i+k} - x_j)}.$$

Remark 1.0.2 Let us denote $[x_i, \ldots, x_{i+n}; f]$ by $\Delta_{(n)} f(x_i)$. The value $[x_i, \ldots, x_{i+n}; f]$ is independent of the order of the points $x_i, x_{i+1}, \ldots, x_{i+n}$. We can extend this definition by including the cases in which two or more points coincide by taking respective limits. □

Definition 1.0.5 A function $f : I \to \mathbb{R}$ is called *convex of order n* or *n-convex* if for all choices of $(n+1)$ distinct points $x_i, \ldots, x_{i+n}$ we have $\Delta_{(n)} f(x_i) \geq 0$.

If nth order derivative $f^{(n)}$ exists, then f is n-convex iff $f^{(n)} \geq 0$.

Remark 1.0.3 For $n = 2$ and $i = 0$, we get the *second order divided difference* of a function $f : I \to \mathbb{R}$ which is defined recursively by

$$[x_j; f] = f(x_j) \qquad , \quad j \in \{0, 1, 2\},$$

$$[x_j, x_{j+1}; f] = \frac{f(x_{j+1}) - f(x_j)}{x_{j+1} - x_j}, \quad j \in \{0, 1\},$$

$$[x_0, x_1, x_2; f] = \frac{[x_1, x_2; f] - [x_0, x_1; f]}{x_2 - x_0}, \tag{1.0.2}$$

for arbitrary points $x_0, x_1, x_2 \in I$. Now, we discuss some limiting cases as follows: taking the limit as $x_1 \to x_0$ in (1.0.2), we get

$$\lim_{x_1 \to x_0} [x_0, x_1, x_2; f] = [x_0, x_0, x_2; f] = \frac{f(x_2) - f(x_0) - f'(x_0)(x_2 - x_0)}{(x_2 - x_0)^2}$$

$$x_2 \neq x_0,$$

provided that $f'(x_0)$ exists. Furthermore, taking the limits as $x_i \to x_0$, $i \in \{1, 2\}$ in (1.0.2), we obtain

$$\lim_{\substack{x_1 \to x_0 \\ x_2 \to x_0}} [x_0, x_1, x_2; f] = [x_0, x_0, x_0; f] = \frac{f''(x_0)}{2},$$

provided that $f''(x_0)$ exists. □

Popoviciu proved following result [162, 163] (see also [158, 160, 164]).

Theorem 1.0.1 *Let $n \geq 2$, $n \in \mathbb{N}$. Inequality*

$$\sum_{i=1}^{m} p_i f(x_i) \geq 0 \tag{1.0.3}$$

is valid ∀ convex functions of order n, $f : [a, b] \to \mathbb{R}$ iff m-tuples $\mathbf{x} \in [a, b]^m$, $\mathbf{p} \in \mathbb{R}^m$ satisfy

$$\sum_{i=1}^{m} p_i x_i^k = 0, \quad \forall k \in \{1, 2, \ldots, n-1\}; \tag{1.0.4}$$

$$\sum_{i=1}^{m} p_i (x_i - t)_+^{n-1} \geq 0, \quad \forall t \in [a, b], \tag{1.0.5}$$

where

$$(x - t)_+^k = \begin{cases} (x - t)^k, & \text{if } x \geq t; \\ 0, & \text{if } x < t. \end{cases}$$

for $k \in \mathbb{N} \cup \{0\}$.

Remark 1.0.4 Case $n = 2$ was of particular interest and in [152] (see also [158, p. 262]) it is proved that if $n = 2$ conditions (1.0.4) and (1.0.5) can be replaced with

$$\sum_{i=1}^{m} p_i = 0 \quad \text{and} \quad \sum_{i=1}^{m} p_i |x_i - x_k| \geq 0 \text{ for } k \in \{1, 2, \ldots, m\}. \tag{1.0.6}$$

The integral analogue of previous result is stated here as:

Theorem 1.0.2 *Let* $n \geq 2$, $n \in \mathbb{N}$, $p : [\alpha, \beta] \to \mathbb{R}$ *and* $g : [\alpha, \beta] \to [a, b]$. *Then, inequality*

$$\int_{\alpha}^{\beta} p(x) f(g(x)) \, dx \geq 0 \tag{1.0.7}$$

is valid ∀ *convex functions of order* n, $f : [a, b] \to \mathbb{R}$ *iff*

$$\int_{\alpha}^{\beta} p(x) g(x)^k \, dx = 0, \quad \forall k \in \{1, 2, \ldots, n-1\}$$

$$\int_{\alpha}^{\beta} p(x) (g(x) - t)_+^{n-1} \, dx \geq 0, \quad \forall t \in [a, b]. \tag{1.0.8}$$

1.1 Linear Inequalities and the Extension of Montgomery Identity with New Green Functions

In the present section, new general linear (integral and discrete) identities and inequalities have been presented for convex functions of order n via extension of Montgomery identity using new Green functions. We also state positivity conditions of these inequalities. We also study n-convexity at a point for our proposed inequalities. Bounds for reminders in new generalizations of general linear inequalities are given by using Čebyšev functional, Ostrowski and Grüss types inequalities. We would also state Mean value theorems of Cauchy type and Lagrange type. The results of present section are extracted from [93].

Montgomery Identity

We state the well known Montgomery identity from "Inequalities for Functions and Their Integrals and Derivatives" by Mitrinović et al. [134, p. 565].

Theorem 1.1.1 *Let $f: I \to \mathbb{R}$, be an absolutely continuous function. Then*

$$f(x) = \frac{1}{b-a}\int_a^b f(t)dt + \frac{1}{b-a}\int_a^b K(x,t)f'(t)dt, \qquad (1.1.1)$$

where Peano kernal $K(x,t)$ is given as

$$K(x,t) = \begin{cases} t - a, & \text{if } t \in [a,x]; \\ t - b, & \text{if } t \in (x,b]. \end{cases} \qquad (1.1.2)$$

In article [4], following extension of Montgomery identity is given.

Theorem 1.1.2 *Let $f : I \to \mathbb{R}$ be such that $f^{(n-1)} \in AC(I)$ for $n \geq 2$, $n \in \mathbb{N}$, where $I \subset \mathbb{R}$ an open interval; $a,b \in I$ with $a < b$. Then following identity is valid*

$$f(x) = \frac{1}{b-a}\int_a^b f(s)\,ds + \sum_{i=1}^{n-1} \frac{f^{(k+1)}(a)}{k!(k+2)} \frac{(x-a)^{k+2}}{b-a}$$

$$- \sum_{i=1}^{n-1} \frac{f^{(k+1)}(b)}{k!(k+2)} \frac{(x-b)^{k+2}}{b-a} + \frac{1}{(n-1)!}\int_a^b T_n(x,t)f^{(n)}(t)\,dt$$

(1.1.3)

where

$$T_n(x,t) = \begin{cases} -\frac{(x-t)^n}{n(b-a)} + \frac{x-a}{b-a}(x-t)^{n-1}, & a \leq t \leq x; \\ -\frac{(x-t)^n}{n(b-a)} + \frac{x-b}{b-a}(x-t)^{n-1}, & x < t \leq b. \end{cases} \qquad (1.1.4)$$

Note that in case $n = 1$ the sum $\sum_{i=1}^{n-1} \cdots$ is empty, so identity (1.1.3) reduces to Montgomery identity.

1.1.1 Results Obtained by the Extension of Montgomery Identity and New Green Functions

In present subsection we obtain some discrete and integral identities and corresponding linear inequalities using new Green functions and applying Montgomery identity. As a special choice Abel-Gontscharoff polynomial for 'two-point right focal' interpolating polynomial for $n = 2$ can be given as (see [159]):

$$f(x) = f(a) + (x - a)f'(b) + \int_a^b G_1(x, t) f''(t) dt, \quad (1.1.5)$$

where $G_1(s, t)$ is Green's function for 'two-point right focal problem' defined as

$$G_1(s, t) = \begin{cases} a - t, & a \leq t \leq s; \\ a - s, & s \leq t \leq b. \end{cases} \quad (1.1.6)$$

Motivated by Abel-Gontscharoff identity (1.1.5) and related Green's function (1.1.6), we would like to recall some new types of Green functions $G_l : [a, b] \times [a, b] \to \mathbb{R}$, for $l \in \{2, 3, 4\}$ defined as in [29]:

$$G_2(s, t) = \begin{cases} s - b, & a \leq t \leq s; \\ t - b, & s \leq t \leq b; \end{cases} \quad (1.1.7)$$

$$G_3(s, t) = \begin{cases} s - a, & a \leq t \leq s; \\ t - a, & s \leq t \leq b; \end{cases} \quad (1.1.8)$$

$$G_4(s, t) = \begin{cases} b - t, & a \leq t \leq s; \\ b - s, & s \leq t \leq b. \end{cases} \quad (1.1.9)$$

In [29], it is also shown that all four Green functions are symmetric and continuous. These new Green functions enables us to introduce some new identities, stated in form of following lemma:

Lemma 1.1.1 *Let $f : [a, b] \to \mathbb{R}$ be a twice differentiable function and G_l for $l \in \{1, 2, 3, 4\}$ be new Green functions defined above. Then along with (1.1.5) following identities are valid:*

$$f(x) = f(b) + (b - x)f'(a) + \int_a^b G_2(x, t) f''(t) dt, \quad (1.1.10)$$

$$f(x) = f(b) - (b - a)f'(b) + (x - a)f'(a) + \int_a^b G_3(x, t) f''(t) dt, \quad (1.1.11)$$

$$f(x) = f(a) + (b-a)f'(a) - (b-x)f'(b) + \int_a^b G_4(x,t)f''(t)dt. \quad (1.1.12)$$

Proof Consider integral

$$\int_a^b G_l(x,t)f''(t)dt = \int_a^x G_l(x,t)f''(t)dt + \int_x^b G_l(x,t)f''(t)dt.$$

Fix $l \in \{1,2,3,4\}$ and integration by parts for specific value of Green's function, we would obtained identities (1.1.5), (1.1.10), (1.1.11) and (1.1.12) for $l = 1,2,3$ and 4 respectively. $\square$

In [103] we proved various results related to general linear inequalities via extended Montgomery identity using Green's function (see also [101]). Recently, in [29] authors have introduced new Green type functions. Our main objective of present subsection is to further extend results of [103] using new definitions.

Theorem 1.1.3 *Fix* $l \in \{1,2,3,4\}$. *Let* $\mathbf{x} = (x_1, \ldots, x_m) \in [a,b]^m$, $\mathbf{p} = (p_1, \ldots, p_m) \in \mathbb{R}^m$ *satisfy conditions*

$$\sum_{i=1}^m p_i = 0, \qquad \sum_{i=1}^m p_i x_i = 0. \quad (1.1.13)$$

Also let $f : I \to \mathbb{R}$ *be a function such that* $f^{(n-1)} \in AC(I)$ *for* $n \geq 3$, $n \in \mathbb{N}$; *where* $I \subset \mathbb{R}$ *is an open interval* $a, b \in I$ *with* $a < b$, *then* $\forall\, s \in [a,b]$ *we have following identity*

$$\sum_{i=1}^m p_i f(x_i) = \frac{f'(a) - f'(b)}{b-a} \int_a^b \sum_{i=1}^m p_i G_l(x_i, s) ds$$

$$+ \sum_{k=2}^{n-1} \frac{k}{(k-1)!} \int_a^b \sum_{i=1}^m p_i G_l(x_i, s)$$

$$\times \frac{f^k(a)(s-a)^{k-1} - f^k(b)(s-b)^{k-1}}{b-a} ds$$

$$+ \frac{1}{(n-3)!} \int_a^b f^{(n)}(t)$$

$$\times \left(\int_a^b \sum_{i=1}^m p_i G_l(x_i, s) \tilde{T}_{n-2}(s,t) ds \right) dt \quad (1.1.14)$$

1.1 Linear Inequalities and the Extension of Montgomery Identity with New...

where

$$\tilde{T}_{n-2}(s,t) = \begin{cases} \frac{1}{b-a}\left[\frac{(s-t)^{n-2}}{(n-2)} + (s-a)(s-t)^{n-3}\right], & a \leq t \leq s \leq b; \\ \frac{1}{b-a}\left[\frac{(s-t)^{n-2}}{(n-2)} + (s-b)(s-t)^{n-3}\right], & a \leq s < t \leq b, \end{cases} \quad (1.1.15)$$

and G_l are as defined in (1.1.6)–(1.1.9). Moreover, we also obtain following identity

$$\sum_{i=1}^{m} p_i f(x_i) = \frac{f'(a) - f'(b)}{b-a} \int_a^b \sum_{i=1}^{m} p_i G_l(x_i, s) ds$$

$$+ \sum_{k=3}^{n-1} \frac{k-2}{(k-1)!} \int_a^b \sum_{i=1}^{m} p_i G_l(x_i, s)$$

$$\times \frac{f^k(a)(s-a)^{k-1} - f^k(b)(s-b)^{k-1}}{b-a} ds$$

$$+ \frac{1}{(n-3)!} \int_a^b f^{(n)}(t)$$

$$\times \left(\int_a^b \sum_{i=1}^{m} p_i G_l(x_i, s) \tilde{T}_{n-2}(s,t) ds\right) dt, \quad (1.1.16)$$

where T_n is as defined in Theorem 1.1.2.

Proof First consider four identities (1.1.5), (1.1.10), (1.1.11) and (1.1.12) and putting $x = x_i$ in all these identities, multiplying each with p_i and then summing over each identity for $i \in \{1, 2, \ldots, m\}$ and using conditions that $\sum_{i=1}^{m} p_i = 0$, $\sum_{i=1}^{m} p_i x_i = 0$ we get by fixing $l \in \{1, 2, 3, 4\}$

$$\sum_{i=1}^{m} p_i f(x_i) = \int_a^b \left(\sum_{i=1}^{m} p_i G_l(x_i, t)\right) f''(t) dt. \quad (1.1.17)$$

Differentiating (1.1.3) twice with respect to $x = s$, we get

$$f''(s) = \frac{f'(a) - f'(b)}{b-a} + \sum_{k=2}^{n-1} \frac{k}{(k-1)!} \frac{f^k(a)(s-a)^{k-1} - f^k(b)(s-b)^{k-1}}{b-a}$$

$$+ \frac{1}{(n-3)!} \int_a^b \tilde{T}_{n-2}(s,t) f^{(n)}(t) dt. \quad (1.1.18)$$

Now using (1.1.18) in (1.1.17) we get

$$\sum_{i=1}^{m} p_i f(x_i) = \frac{f'(a) - f'(b)}{b - a} \int_a^b \sum_{i=1}^{m} p_i G_l(x_i, s) ds$$

$$+ \sum_{k=2}^{n-1} \frac{k}{(k-1)!} \int_a^b \sum_{i=1}^{m} p_i G_l(x_i, s)$$

$$\times \frac{f^k(a)(s-a)^{k-1} - f^k(b)(s-b)^{k-1}}{b - a} ds$$

$$+ \frac{1}{(n-3)!} \int_a^b \sum_{i=1}^{m} p_i G_l(x_i, s) \left(\int_a^b \tilde{T}_{n-2}(s, t) f^{(n)}(t) dt \right) ds$$

and then using Fubini's theorem in last term to get (1.1.14).

Also, by using formula (1.1.3) on function f'', replacing n by $n - 2$ ($n \geq 3$) and rearranging indices we get

$$f''(s) = \frac{f'(a) - f'(b)}{b - a} + \sum_{k=3}^{n-1} \frac{k-2}{(k-1)!} \frac{f^k(a)(s-a)^{k-1} - f^k(b)(s-b)^{k-1}}{b - a}$$

$$+ \frac{1}{(n-3)!} \int_a^b \tilde{T}_{n-2}(s, t) f^{(n)}(t) dt. \qquad (1.1.19)$$

Similarly, using (1.1.19) in (1.1.17) and applying Fubini's Theorem, we get (1.1.16). □

Theorem 1.1.4 *Let all the assumptions of Theorem 1.1.3 be valid with additional condition*

$$\int_a^b \sum_{i=1}^{m} p_i G_l(x_i, s) \tilde{T}_{n-2}(s, t) ds \geq 0, \quad \forall t \in [a, b]. \qquad (1.1.20)$$

Then $\forall$ n-convex function $f : I \to \mathbb{R}$ following inequality is valid

$$\sum_{i=1}^{m} p_i f(x_i) \geq \frac{f'(a) - f'(b)}{b - a} \int_a^b \sum_{i=1}^{m} p_i G_l(x_i, s) ds$$

$$+ \sum_{k=2}^{n-1} \frac{k}{(k-1)!} \int_a^b \sum_{i=1}^{m} p_i G_l(x_i, s) \frac{f^k(a)(s-a)^{k-1} - f^k(b)(s-b)^{k-1}}{b - a} ds.$$

$$(1.1.21)$$

1.1 Linear Inequalities and the Extension of Montgomery Identity with New...

Proof Since function f is n-convex, therefore $f^{(n)} \geq 0$. Hence using n convexity of function and (1.1.20) in (1.1.14) we obtain our required result. $\square$

Theorem 1.1.5 *Let all the assumptions of Theorem 1.1.3 be valid with additional condition*

$$\int_a^b \sum_{i=1}^m p_i G_l(x_i, s) T_{n-2}(s, t) ds \geq 0, \quad \forall\, t \in [a, b]. \tag{1.1.22}$$

Then $\forall$ n-convex function $f : I \to \mathbb{R}$ following inequality is valid

$$\sum_{i=1}^m p_i f(x_i) \geq \frac{f'(a) - f'(b)}{b - a} \int_a^b \sum_{i=1}^m p_i G_l(x_i, s) ds$$
$$+ \sum_{k=3}^{n-1} \frac{k-2}{(k-1)!} \int_a^b \sum_{i=1}^m p_i G_l(x_i, s) \frac{f^k(a)(s-a)^{k-1} - f^k(b)(s-b)^{k-1}}{b-a} ds. \tag{1.1.23}$$

Proof Since function f is n-convex, therefore $f^{(n)} \geq 0$. Hence using n convexity of function and (1.1.22) in (1.1.16) we obtain what we wanted. $\square$

Now we state important consequences.

Theorem 1.1.6 *Let all the assumptions from Theorem 1.1.3 be valid with*

$$\sum_{i=1}^m p_i = 0 \quad \text{and} \quad \sum_{i=1}^m p_i |x_i - x_k| \geq 0 \text{ for } k \in \{1, 2, \ldots, m\}. \tag{1.1.24}$$

If f is n-convex and n is even, then inequalities (1.1.21) and (1.1.23) are valid.

Proof Since new Green functions $G_l(s, t)$ are convex with respect to t $\forall\, s \in [a, b]$ and for each $l \in \{1, 2, 3, 4\}$ and $\mathbf{x} = (x_1, \ldots, x_m)$ and $\mathbf{p} = (p_1, \ldots, p_m)$ satisfy conditions (1.0.6) from Remark 1.0.4, therefore, we have

$$\sum_{i=1}^m p_i G_l(x_i, s) \geq 0 \quad \text{for} \quad s \in [a, b]. \tag{1.1.25}$$

Also note that $\tilde{T}_{n-2}(s, t)$ and $T_{n-2}(s, t)$ both are non-negative for even values of n. Therefore combining present fact with (1.1.25) we get inequality (1.1.20) and inequality (1.1.22) respectively. As f is n-convex, so results follows from Theorems 1.1.4 and 1.1.5 respectively. $\square$

The integral version of our main results may be stated as follows. Since proofs of these results are of similar nature so we omit the details.

Theorem 1.1.7 *Fix $l \in \{1, 2, 3, 4\}$. Let $g : [\alpha, \beta] \to [a, b]$ be a function and let $p : [\alpha, \beta] \to \mathbb{R}$ be a continuous integrable function such that $\int_\alpha^\beta p(x)dx = 0$ and $\int_\alpha^\beta p(x)g(x)dx = 0$. Let $f : I \to \mathbb{R}$ be a function such that $f^{(n-1)} \in AC(I)$, $I \subset \mathbb{R}$ an open interval, $a, b \in I$, $a < b$, then $\forall\, s \in [a, b]$ we have following identity*

$$\int_\alpha^\beta p(x) f(g(x))\, dx = \frac{f'(a) - f'(b)}{b - a} \int_a^b \int_\alpha^\beta p(x)\, G_l(g(x), s)\, dx\, ds$$

$$+ \sum_{k=2}^{n-1} \frac{k}{(k-1)!} \int_a^b \left(\int_\alpha^\beta p(x)\, G_l(g(x), s)\, dx \right)$$

$$\times \frac{f^k(a)(s-a)^{k-1} - f^k(b)(s-b)^{k-1}}{b - a} ds$$

$$+ \frac{1}{(n-3)!} \int_a^b f^{(n)}(t) \left(\int_a^b \left(\int_\alpha^\beta p(x)\, G_l(g(x), s)\, dx \right) \tilde{T}_{n-2}(s, t)ds \right) dt.$$
(1.1.26)

Moreover, we also obtain following identity

$$\int_\alpha^\beta p(x) f(g(x))\, dx = \frac{f'(a) - f'(b)}{b - a} \int_a^b \int_\alpha^\beta p(x)\, G_l(g(x), s)\, dx\, ds$$

$$+ \sum_{k=3}^{n-1} \frac{k-2}{(k-1)!} \int_a^b \left(\int_\alpha^\beta p(x)\, G_l(g(x), s)\, dx \right)$$

$$\times \frac{f^k(a)(s-a)^{k-1} - f^k(b)(s-b)^{k-1}}{b - a} ds$$

$$+ \frac{1}{(n-3)!} \int_a^b f^{(n)}(t) \left(\int_a^b \left(\int_\alpha^\beta p(x)\, G_l(g(x), s)\, dx \right) T_{n-2}(s, t)ds \right) dt$$
(1.1.27)

where $\tilde{T}_n$, T_n and G_l are as in Theorem 1.1.3.

Theorem 1.1.8 *Let all the assumptions of Theorem 1.1.7 be valid with following condition*

$$\int_a^b \int_\alpha^\beta p(x)\, G_l(g(x), s)\, \tilde{T}_{n-2}(s, t)\, dx\, ds \geq 0, \quad \forall\, t \in [a, b]. \qquad (1.1.28)$$

1.1 Linear Inequalities and the Extension of Montgomery Identity with New...

Then $\forall$ n-convex function $f : I \to \mathbb{R}$ following inequality is valid

$$\int_\alpha^\beta p(x) f(g(x)) dx \geq \frac{f'(a) - f'(b)}{b-a} \int_a^b \int_\alpha^\beta p(x) G_l(g(x), s) dx\, ds$$
$$+ \sum_{k=2}^{n-1} \frac{k}{(k-1)!} \int_a^b \left(\int_\alpha^\beta p(x) G_l(g(x), s) dx \right)$$
$$\times \frac{f^k(a)(s-a)^{k-1} - f^k(b)(s-b)^{k-1}}{b-a} ds. \quad (1.1.29)$$

Theorem 1.1.9 *Let all the assumptions of Theorem 1.1.7 be valid with following condition*

$$\int_a^b \int_\alpha^\beta p(x) G_l(g(x), s) T_{n-2}(s,t) dx\, ds \geq 0, \quad \forall t \in [a,b]. \quad (1.1.30)$$

Then $\forall$ n-convex function $f : I \to \mathbb{R}$ following inequality is valid

$$\int_\alpha^\beta p(x) f(g(x)) dx \geq \frac{f'(a) - f'(b)}{b-a} \int_a^b \int_\alpha^\beta p(x) G_l(g(x), s) dx\, ds$$
$$+ \sum_{k=3}^{n-1} \frac{k-2}{(k-1)!} \int_a^b \left(\int_\alpha^\beta p(x) G_l(g(x), s) dx \right)$$
$$\times \frac{f^k(a)(s-a)^{k-1} - f^k(b)(s-b)^{k-1}}{b-a} ds. \quad (1.1.31)$$

Theorem 1.1.10 *Let all the assumptions of Theorem 1.1.7 be valid with additional assumptions that $p : [\alpha, \beta] \to \mathbb{R}$ and $g : [\alpha, \beta] \to [a,b]$ be such that*

$$\int_\alpha^\beta p(x) g(x)^k dx = 0, \quad \forall k \in \{1, 2, \ldots, n-1\};$$
$$\int_\alpha^\beta p(x) (g(x) - t)_+^{n-1} dx \geq 0, \quad \forall t \in [a,b]. \quad (1.1.32)$$

If f is n-convex and n is even, then inequalities (1.1.29) and (1.1.31) are valid.

1.1.2 Inequalities for n-Convex Functions at a Point

In articles [98] and [160], we can find following definition of convexity at a point. For detailed discussion on the topic one can see [101].

Definition 1.1.1 Let I be an interval in $\mathbb{R}$, $c \in I^\circ$ (interior of I) and $n \in \mathbb{N}$. A function $f : I \to \mathbb{R}$ is said to be n-convex at a point c if $\exists$ a constant K such that the function

$$F(x) = f(x) - \frac{K}{(n-1)!} x^{n-1}$$

is $(n-1)$-concave on $I \cap (-\infty, c]$ and $(n-1)$-convex on $I \cap [c, \infty)$. A function f is said to be n-concave at point c if the function $-f$ is n-convex at point c.

A property that explains the name of the class is the fact that a function is n-convex on an interval iff it is n-convex at every point of the interval (see [160]). For further details on the topic kindly see [160]. Let e_i denote the monomials $e_i(x) = x^i$, $i \in \mathbb{N}_0 = \mathbb{N} \cup \{0\}$. Throughout the subsection we let $c \in \langle a, b \rangle$.

Let $T_n^{[a,c]}$ and $T_n^{[c,b]}$ denote equivalent of (1.1.4) on these intervals, i. e.,

$$T_n^{[a,c]}(s,t) = \begin{cases} -\dfrac{(s-t)^n}{n(c-a)} + \dfrac{\eta - a}{c-a}(s-t)^{n-1}, & a \le t \le s \le c, \\[2mm] -\dfrac{(s-t)^n}{n(c-a)} + \dfrac{\eta - c}{c-a}(s-t)^{n-1}, & a \le s \le t \le c, \end{cases}$$

(1.1.33)

$$T_n^{[c,b]}(s,t) = \begin{cases} -\dfrac{(s-t)^n}{n(b-c)} + \dfrac{\eta - c}{b-c}(s-t)^{n-1}, & c \le t \le s \le b, \\[2mm] -\dfrac{(s-t)^n}{n(b-c)} + \dfrac{\eta - b}{b-c}(s-t)^{n-1}, & c \le s \le t \le b. \end{cases}$$

(1.1.34)

Similarly, $\tilde{T}_n^{[a,c]}$ and $\tilde{T}_n^{[c,b]}$ denote equivalent of (1.1.15) on these intervals, i. e.

$$\tilde{T}_{n-2}^{[a,c]}(s,t) = \begin{cases} \dfrac{1}{c-a}\left[\dfrac{(s-t)^{n-2}}{(n-2)} + (s-a)(s-t)^{n-3}\right], & a \le t \le s \le c, \\[2mm] \dfrac{1}{c-a}\left[\dfrac{(s-t)^{n-2}}{(n-2)} + (s-c)(s-t)^{n-3}\right], & a \le s < t \le c. \end{cases}$$

(1.1.35)

$$\tilde{T}_{n-2}^{[c,b]}(s,t) = \begin{cases} \dfrac{1}{b-c}\left[\dfrac{(s-t)^{n-2}}{(n-2)} + (s-c)(s-t)^{n-3}\right], & c \le t \le s \le b, \\[2mm] \dfrac{1}{b-c}\left[\dfrac{(s-t)^{n-2}}{(n-2)} + (s-b)(s-t)^{n-3}\right], & c \le s < t \le b. \end{cases}$$

(1.1.36)

1.1 Linear Inequalities and the Extension of Montgomery Identity with New...

Let $\mathbf{x} \in [a,c]^{m_1}$, $\mathbf{p} \in \mathbb{R}^{m_1}$, $\mathbf{y} \in [c,b]^{m_2}$ and $\mathbf{q} \in \mathbb{R}^{m_2}$ and denote

$$A_1^{[a,c]}(l, m_1, \mathbf{x}, \mathbf{p}, f) = \sum_{i=1}^{m_1} p_i f(x_i) - \frac{f'(a) - f'(c)}{c-a} \int_a^c \sum_{i=1}^{m_1} p_i G_l(x_i, s) ds$$

$$- \sum_{k=2}^{n-1} \frac{k}{(k-1)!} \int_a^c \sum_{i=1}^{m_1} p_i G_l(x_i, s) \frac{f^k(a)(s-a)^{k-1} - f^k(c)(s-c)^{k-1}}{c-a} ds, \tag{1.1.37}$$

$$A_1^{[c,b]}(l, m_2, \mathbf{y}, \mathbf{q}, f) = \sum_{i=1}^{m_2} q_i f(y_i) - \frac{f'(c) - f'(b)}{c-b} \int_c^b \sum_{i=1}^{m_2} q_i G_l(y_i, s) ds$$

$$- \sum_{k=2}^{n-1} \frac{k}{(k-1)!} \int_c^b \sum_{i=1}^{m_2} q_i G_l(y_i, s) \frac{f^k(c)(s-c)^{k-1} - f^k(c)(s-b)^{k-1}}{b-c} ds. \tag{1.1.38}$$

Notice that, using newly introduced functionals identity (1.1.14) applied to intervals $[a,c]$ and $[c,b]$ can be written as

$$A_1^{[a,c]}(l, m_1, \mathbf{x}, \mathbf{p}, f) = \frac{1}{(n-3)!} \int_a^c f^{(n)}(t)$$

$$\times \left(\int_a^c \sum_{i=1}^{m_1} p_i G_l(x_i, s) \tilde{T}_{n-2}^{[a,c]}(s, t) ds \right) dt, \tag{1.1.39}$$

$$A_1^{[c,b]}(l, m_2, \mathbf{y}, \mathbf{q}, f) = \frac{1}{(n-3)!} \int_c^b f^{(n)}(t)$$

$$\times \left(\int_c^b \sum_{i=1}^{m_2} q_i G_l(y_i, s) \tilde{T}_{n-2}^{[c,b]}(s, t) ds \right) dt. \tag{1.1.40}$$

In same manner we can introduce further functionals namely

$$A_2^{[a,c]}(l, m_1, \mathbf{x}, \mathbf{p}, f) = \frac{1}{(n-3)!} \int_a^c f^{(n)}(t)$$

$$\times \left(\int_a^c \sum_{i=1}^{m_1} p_i G_l(x_i, s) T_{n-2}^{[a,c]}(s, t) ds \right) dt, \tag{1.1.41}$$

$$A_2^{[c,b]}(l, m_2, \mathbf{y}, \mathbf{q}, f) = \frac{1}{(n-3)!} \int_c^b f^{(n)}(t)$$
$$\times \left(\int_c^b \sum_{i=1}^{m_2} q_i G_l(y_i, s) T_{n-2}^{[c,b]}(s,t) ds \right) dt, \quad (1.1.42)$$

$$A_3^{[a,c]}(l, [\alpha, \beta], g, p, f) = \frac{1}{(n-3)!} \int_a^c f^{(n)}(t)$$
$$\times \left(\int_a^c \left(\int_\alpha^\beta p(x) G_l(g(x), s) dx \right) \tilde{T}_{n-2}^{[a,c]}(s,t) ds \right) dt, \quad (1.1.43)$$

$$A_3^{[c,b]}(l, [\gamma, \delta], h, q, f) = \frac{1}{(n-3)!} \int_c^b f^{(n)}(t)$$
$$\times \left(\int_c^b \left(\int_\gamma^\delta q(x) G_l(h(x), s) dx \right) \tilde{T}_{n-2}^{[c,b]}(s,t) ds \right) dt, \quad (1.1.44)$$

$$A_4^{[a,c]}(l, [\alpha, \beta], g, p, f) = \frac{1}{(n-3)!} \int_a^c f^{(n)}(t)$$
$$\times \left(\int_a^c \left(\int_\alpha^\beta p(x) G_l(g(x), s) dx \right) T_{n-2}^{[a,c]}(s,t) ds \right) dt, \quad (1.1.45)$$

$$A_4^{[c,b]}(l, [\gamma, \delta], h, q, f) = \frac{1}{(n-3)!} \int_c^b f^{(n)}(t)$$
$$\times \left(\int_c^b \left(\int_\gamma^\delta q(x) G_l(h(x), s) dx \right) T_{n-2}^{[c,b]}(s,t) ds \right) dt, \quad (1.1.46)$$

where $\alpha \leq \beta, \gamma \leq \delta, a < c < b$, $g : [\alpha, \beta] \to [a, c]$, $p : [\alpha, \beta] \to \mathbb{R}$, $h : [\gamma, \delta] \to [c, b]$, $q : [\gamma, \delta] \to \mathbb{R}$ are integrable functions. For sake of brevity we let $A_l^{[a,b]}(\cdot, \cdot, \cdot, \cdot, f) = A_l^{[a,b]}(f)$.

Theorem 1.1.11 *Let* $\mathbf{x} \in [a, c]^{m_1}$, $\mathbf{p} \in \mathbb{R}^{m_1}$, $\mathbf{y} \in [c, b]^{m_2}$ *and* $\mathbf{q} \in \mathbb{R}^{m_2}$ *be such that*

$$\int_a^c \sum_{i=1}^{m_1} p_i G_l(x_i, s) \tilde{T}_{n-2}^{[a,c]}(s,t) ds \geq 0, \quad \forall t \in [a, c], \quad (1.1.47)$$

1.1 Linear Inequalities and the Extension of Montgomery Identity with New...

$$\int_c^b \sum_{i=1}^{m_2} q_i G_l(y_i, s) \tilde{T}_{n-2}^{[c,b]}(s, t)\, ds \geq 0, \quad \forall t \in [c, b], \tag{1.1.48}$$

$$\int_a^c \int_a^c \sum_{i=1}^{m_1} p_i G_l(\eta_i, s) \tilde{T}_{n-2}^{[a,c]}(s, t)\, ds dt = \int_c^b \int_c^b \sum_{i=1}^{m_2} q_i G_l(y_i, s) \tilde{T}_{n-2}^{[c,b]}(s, t)\, ds dt, \tag{1.1.49}$$

where $\tilde{T}_n^{[a,c]}$, $\tilde{T}_n^{[c,b]}$, $A_1^{[a,c]}(f)$ and $A_1^{[c,b]}(f)$ are given by (1.1.35), (1.1.36), (1.1.37) and (1.1.38) respectively. If $f : [a, b] \to \mathbb{R}$ is $(n+1)$-convex at point c, then

$$A_1^{[a,c]}(f) \leq A_1^{[c,b]}(f). \tag{1.1.50}$$

If inequalities in (1.1.47) and (1.1.48) are reversed, then (1.1.50) is valid with reversed sign of inequality.

Proof Let $F = f - \frac{K}{n!} e_n$ be as in definition of function n-convex at a point, i. e., function F is n-concave on $[a, c]$ and n-convex on $[c, b]$ and e_i denote monomials $e_i(x) = x^i$, $i \in \mathbb{N}_0 = \mathbb{N} \cup \{0\}$. Applying Theorem 1.1.4 to F on interval $[a, c]$ we have

$$0 \geq A_1^{[a,c]}(F) = A_1^{[a,c]}(f) - \frac{K}{n!} A_1^{[a,c]}(e_n) \tag{1.1.51}$$

and applying Theorem 1.1.4 to F on interval $[c, b]$ we have

$$0 \leq A_1^{[c,b]}(F) = A_1^{[c,b]}(f) - \frac{K}{n!} A_1^{[c,b]}(e_n). \tag{1.1.52}$$

Identities (1.1.39) and (1.1.40) applied to function e_n yield

$$A_1^{[a,c]}(e_n) = \frac{n!}{(n-3)!} \int_a^c \left(\sum_{i=1}^m p_i \tilde{T}_{n-2}^{[a,c]}(x_i, s) \right) ds,$$

$$A_1^{[c,b]}(e_n) = \frac{n!}{(n-3)!} \int_c^b \left(\sum_{i=1}^l q_i \tilde{T}_{n-2}^{[c,b]}(y_i, s) \right) ds.$$

Therefore, assumption (1.1.49) is equivalent to $A(e_n) = B(e_n)$. Now, from (1.1.51) and (1.1.52) we obtain stated inequality. □

Remark 1.1.1 In proof of Theorem 1.1.11 we have, actually, shown that

$$A_1^{[a,c]}(f) \leq \frac{K}{n!} A_1^{[a,c]}(e_n) = \frac{K}{n!} A_1^{[c,b]}(e_n) \leq A_1^{[c,b]}(f).$$

In fact, inequality (1.1.50) still is valid if we replace assumption (1.1.49) with weaker assumption that $K \left(A_1^{[c,b]}(e_n) - A_1^{[a,c]}(e_n) \right) \geq 0$.

Here we have another similar result.

Theorem 1.1.12 *Let* $\mathbf{x} \in [a, c]^{m_1}$, $\mathbf{p} \in \mathbb{R}^{m_1}$, $\mathbf{y} \in [c, b]^{m_2}$ *and* $\mathbf{q} \in \mathbb{R}^{m_2}$ *be such that*

$$\int_a^c \sum_{i=1}^{m_1} p_i G_l(x_i, s) T_{n-2}^{[a,c]}(s, t)\, ds \geq 0, \quad \forall t \in [a, c], \tag{1.1.53}$$

$$\int_c^b \sum_{i=1}^{m_2} q_i G_l(y_i, s) T_{n-2}^{[c,b]}(s, t)\, ds \geq 0, \quad \forall t \in [c, b], \tag{1.1.54}$$

$$\int_a^c \int_a^c \sum_{i=1}^{m_1} p_i G_l(\eta_i, s) T_{n-2}^{[a,c]}(s, t)\, ds dt = \int_c^b \int_c^b \sum_{i=1}^{m_2} q_i G_l(y_i, s) T_{n-2}^{[c,b]}(s, t)\, ds dt, \tag{1.1.55}$$

where $T_n^{[a,c]}$, $T_n^{[c,b]}$, $A_2^{[a,c]}(f)$ *and* $A_2^{[c,b]}(f)$ *are given by* (1.1.33), (1.1.34), (1.1.41) *and* (1.1.42) *respectively. If* $f : [a, b] \to \mathbb{R}$ *is* $(n + 1)$-*convex at point* c, *then*

$$A_2^{[a,c]}(f) \leq A_2^{[c,b]}(f). \tag{1.1.56}$$

If inequalities in (1.1.53) *and* (1.1.54) *are reversed, then* (1.1.56) *is valid with reversed sign of inequality.*

Remark 1.1.2 Similar results can also be stated for integral versions as well by using functionals $A_3^{[a,c]}(\cdot)$, $A_3^{[c,b]}(\cdot)$, $A_4^{[a,c]}(\cdot)$ and $A_4^{[c,b]}(\cdot)$ as defined in (1.1.43), (1.1.44) (1.1.45) and (1.1.46) respectively.

1.1.3 Bounds for Remainders and Functionals

Let $f, g : [a, b] \to \mathbb{R}$ be two Lebesgue integrable functions. We denote Čebyšev functional by $T(f, g)$ which may be define as

$$T(f, g) = \frac{1}{b - a} \int_a^b f(x)g(x)dx$$

$$- \left(\frac{1}{b - a} \int_a^b f(x)dx \right) \left(\frac{1}{b - a} \int_a^b g(x)dx \right). \tag{1.1.57}$$

1.1 Linear Inequalities and the Extension of Montgomery Identity with New...

Here we give several estimations connected with functionals $A_l^{[\cdot,\cdot]}(f)$, $l \in \{1, 2, 3, 4\}$ defined under assumptions of Theorems 1.1.4, 1.1.5, 1.1.8 and 1.1.9 respectively on interval $[a, b]$ as defined in (1.1.39), for example,

$$A_1^{[a,b]}(f) = \frac{1}{(n-3)!} \int_a^b f^{(n)}(t) \left(\int_a^b \sum_{i=1}^m p_i G_l(x_i, s) \tilde{T}_{n-2}^{[a,b]}(s, t) ds \right) dt.$$

In our next results we will use well-known Hölder inequality and a bound for Čebyšev functional $T(f, g)$ which is defined earlier.

Hölder's Integral Inequality Hölder's inequality, named after Otto Hölder, is a basic inequality and an essential tool for the study of L_p spaces. Hölder's inequality was first found by Rogers in 1888, and discovered independently by Hölder in 1889. The integral version of Hölder's Inequality [135] is stated as:

Theorem 1.1.13 *Let $1 \leq p, q \leq \infty$ with $\frac{1}{p} + \frac{1}{q} = 1$. If $f \in L_p$ and $g \in L_q$, then $fg \in L$ and*

$$\int |f(x)g(x)| dx \leq \|f\|_p \|g\|_q.$$

Moreover if $f \in L_1$ and $g \in L_\infty$, then

$$\int |f(x)g(x)| dx \leq \|f\|_1 \|g\|_\infty.$$

Here, the symbol $L_p[a, b]$ ($1 \leq p < \infty$) denotes the space of p-power integrable functions on the interval $[a, b]$ equipped with the norm

$$\|f\|_p = \left(\int_a^b |f(t)|^p dt \right)^{\frac{1}{p}}$$

and $L_\infty[a, b]$ denotes the space of essentially bounded functions on $[a, b]$ with the norm

$$\|f\|_\infty = \operatorname*{ess\,sup}_{t \in [a,b]} |f(t)|.$$

Remark 1.1.3 For special case $p = q = 2$, Hölder's inequality becomes Cauchy-Schwarz's inequality.

For main results of present section we need some notation to work efficiently. Under assumptions of Theorems 1.1.4, 1.1.5, 1.1.8 and 1.1.9 respectively, we introduce following notations

$$\Omega_1(t) = \int_a^b \sum_{i=1}^m p_i G_l(x_i, s) \tilde{T}_{n-2}(s,t) ds, \quad \forall t \in [a,b]; \tag{1.1.58}$$

$$\Omega_2(t) = \int_a^b \sum_{i=1}^m p_i G_l(x_i, s) T_{n-2}(s,t) ds, \quad \forall t \in [a,b]; \tag{1.1.59}$$

$$\Omega_3(t) = \int_a^b \int_\alpha^\beta p(x) G_l(g(x), s) \tilde{T}_{n-2}(s,t) dx\, ds, \quad \forall t \in [a,b]; \tag{1.1.60}$$

$$\Omega_4(t) = \int_a^b \int_\alpha^\beta p(x) G_l(g(x), s) T_{n-2}(s,t) dx\, ds, \quad \forall t \in [a,b]. \tag{1.1.61}$$

Hence by using these notations we may use Čebyšev functional as follows (e.g. using Ω):

$$T(\Omega, \Omega) = \frac{1}{b-a} \int_a^b \Omega^2(s) ds - \left(\frac{1}{b-a} \int_a^b \Omega(s) ds \right)^2.$$

A bound for Čebyšev functional is given in following theorem in which pre-Grüss inequality is given (see [117]).

Theorem 1.1.14 *Let* $f, h : [a,b] \to \mathbb{R}$ *be integrable such that* $fh \in L(a,b)$. *If*

$$\gamma_1 \leq h(x) \leq \gamma_2 \quad \text{for} \quad x \in [a,b],$$

then

$$|T(f,h)| \leq \frac{1}{2}(\gamma_2 - \gamma_1)\sqrt{T(f,f)}.$$

At present stage we denote $A_l^{[a,b]}(\cdot, \cdot, \cdot, \cdot, f)$ by simply $A_l(f)$ for $l \in \{1, 2, 3, 4\}$.

Theorem 1.1.15 *Fix* $l \in \{1, 2, 3, 4\}$. *Let* $f : I \to \mathbb{R}$, $[a,b] \subset I$, *be such that* $f^{(n-1)} \in AC(I)$ *and*

$$\gamma \leq f^{(n)}(x) \leq \Gamma \quad \text{for } x \in [a,b].$$

Then in representation

$$A_l(f) = \frac{[f^{n-1}(b) - f^{n-1}(a)]}{(n-3)!(b-a)} \int_a^b \Omega_l(s) ds + R_1^k(f : a, b, n), \tag{1.1.62}$$

remainder $R_1^l(f; a, b, n)$ *satisfies estimation*

$$|R_1^l(f; a, b, n)| \leq \frac{b-a}{2(n-3)!}(\Gamma - \gamma)\sqrt{T(\Omega_l, \Omega_l)}. \qquad (1.1.63)$$

Proof Fix $l \in \{1, 2, 3, 4\}$. Using definition of A_l and results from previous subsection we have

$$A_l(f) = \frac{1}{(n-3)!} \int_a^b f^{(n)}(s)\Omega_l(s)ds$$

$$= \frac{1}{(n-3)!(b-a)} \int_a^b f^{(n)}(s)ds \int_a^b \Omega_l(s)ds + R_1^l(f; a, b, n)$$

$$= \frac{f^{n-3}(b) - f^{n-1}(a)}{(n-3)!(b-a)} \int_a^b \Omega_l(s)ds + R_1^l(f; a, b, n),$$

where

$$R_1^l(f; a, b, n) = \frac{1}{(n-3)!}$$

$$\times \left(\int_a^b f^{(n)}(s)\Omega_l(s)ds - \frac{1}{b-a} \int_a^b f^{(n)}(s)ds \int_a^b \Omega_l(s)ds \right).$$

Applying Theorem 1.1.14 for $f \to \Omega_l$ and $h \to f^{(n)}$, we obtain

$$|R_1^l(f; a, b, n)| = |T(\Omega_l, f^{(n)})| \leq \frac{b-a}{2(n-3)!}(\Gamma - \gamma)\sqrt{T(\Omega_l, \Omega_l)}.$$

□

Now we state some Ostrowski type inequalities related to our established inequalities. For detailed discussion related to Ostrowski inequality, see Chap. 2.

Theorem 1.1.16 *Let* $l \in \{1, 2, 3, 4\}$. *Let* (q, r) *be such that* $1 \leq q, r \leq \infty$, $\frac{1}{q} + \frac{1}{r} = 1$. *Let* $f^{(n)} \in L_q[a, b]$ *for* $n \geq 3$. *Then*

$$|A_l(f)| \leq \frac{1}{(n-3)!} \|f^{(n)}\|_q \|\Omega_l\|_r. \qquad (1.1.64)$$

The constant on right hand side of (1.1.64) *is sharp for* $1 < q \leq \infty$ *and best possible for* $q = 1$.

Proof Fix $l \in \{1, 2, 3, 4\}$. From definition of A_l and results from second section, applying Hölder inequality we get

$$|A_l(f)| = \left| \frac{1}{(n-3)!} \int_a^b f^{(n)}(s) \Omega_l(s) ds \right| \leq \|f^{(n)}\|_q \|\lambda_l\|_r.$$

Let us denote a quotient $\frac{1}{(n-3)!} \Omega_l$ by λ_l. For proof of sharpness of $\left(\int_a^b |\lambda_l(t)|^r dt \right)^{1/r}$, here we let function f for which equality in (1.1.64) is valid. For $1 < q < \infty$ take f to be such that

$$f^{(n)}(t) = \operatorname{sgn} \lambda_l(t) \cdot |\lambda_l(t)|^{1/(q-1)}.$$

For $q = \infty$, take f such that

$$f^{(n)}(t) = \operatorname{sgn} \lambda_l(t).$$

The fact that (1.1.64) is best possible for $q = 1$ can be proved as in Theorem 5.18 of [101]. □

1.1.4 Mean Value Theorems

Now we state mean value theorems of Lagrange and of Cauchy type for A_l, $l \in \{1, 2, 3, 4\}$. Here $f_0(x) = \dfrac{x^n}{n!}$.

Theorem 1.1.17 *Let $f \in C^{(n)}[a, b]$ and let $A_l : C^{(n)}[a, b] \to \mathbb{R}$ for $l \in \{1, 2, 3, 4\}$ be linear functionals as defined earlier. Then $\exists\, \xi_l \in [a, b]$ for $l \in \{1, 2, 3, 4\}$ such that*

$$A_l(f) = f^{(n)}(\xi_l) A_l(f_0). \tag{1.1.65}$$

Proof Since $f^{(n)}$ is continuous on $[a, b]$, so $m \leq f^{(n)}(x) \leq M$ for $x \in [a, b]$, where $m = \min\limits_{x \in [a,b]} f^{(n)}(x)$ and $M = \max\limits_{x \in [a,b]} f^{(n)}(x)$.

Therefore function

$$F(x) = m \frac{x^n}{n!} - f(x) = M f_0(x) - f(x)$$

produces

$$F^{(n)}(x) = M - f^{(n)}(x) \geq 0,$$

1.1 Linear Inequalities and the Extension of Montgomery Identity with New... 23

i.e., F is n-convex. Hence $A_l(F) \geq 0$ and we conclude that for $l \in \{1, 2, 3, 4\}$

$$A_l(f) \leq M A_l(f_0).$$

Similarly, for $l \in \{1, 2, 3, 4\}$ we have

$$m A_l(f_0) \leq A_l(f).$$

Combining two inequalities we get

$$m A_l(f_0) \leq A_l(f) \leq M A_l(f_0)$$

and we easily arrive at (1.1.65). □

Theorem 1.1.18 *Let $f, h \in C^{(n)}[a, b]$ and let $A_l : C^{(n)}[a, b] \to \mathbb{R}$ for $l \in \{1, 2, 3, 4\}$ be linear functionals as defined earlier. Then $\exists\, \xi_l \in [a, b]$ for $l \in \{1, 2, 3, 4\}$ such that*

$$\frac{A_l(f)}{A_l(h)} = \frac{f^{(n)}(\xi_l)}{h^{(n)}(\xi_l)}$$

assuming non-zero denominators.

Proof Fix $l \in \{1, 2, 3, 4\}$. Let $z \in C^{(n)}[a, b]$ be defined as

$$z = A_l(h) f - A_l(f) h.$$

Using Theorem 1.1.17 $\exists\, \xi_l$ such that

$$0 = A_l(z) = z^{(n)}(\xi_l) A_l(f_0)$$

or

$$[A_l(h) f^{(n)}(\xi_l) - A_l(f) h^{(n)}(\xi_l)] A_k(f_0) = 0$$

which gives us required result. □

Remark 1.1.4 If inverse of $\frac{f^{(n)}}{h^{(n)}}$ exists, then from previous mean value results we may state generalized means as

$$\xi_l = \left(\frac{f^{(n)}}{h^{(n)}}\right)^{-1} \left(\frac{A_l(f)}{A_l(h)}\right), \quad l \in \{1, 2, 3, 4\}. \tag{1.1.66}$$

1.2 Linear Inequalities and the Taylor Formula with New Green Functions

The aim of this section is to present new general linear identities and inequalities for n-convex functions via Taylors formula and new Green functions. We also obtain bounds for the reminders in general linear inequalities by using the Čebyšev type inequalities. We give mean value theorems and n-exponential convexity for functional related to these new general linear inequalities. Most of the contents are part of the article [89].

In the next subsection we will prove inequalities of type (1.0.3) and (1.0.7) for n-convex functions by using the following Taylor's theorem and new Green functions.

Theorem 1.2.1 *Let $n \in \mathbb{N}$, $f : [a, b] \to \mathbb{R}$ be such that $f^{(n-1)}$ is absolutely continuous. Then for all $x \in [a, b]$ the Taylor formula at a point $c \in [a, b]$ is given as*

$$f(x) = \sum_{k=0}^{n-1} \frac{f^{(k)}(c)}{k!}(x-c)^k + \frac{1}{(n-1)!}\int_c^x f^{(n)}(t)(x-t)^{(n-1)}\,dt \qquad (1.2.1)$$

1.2.1 Results Obtained by the Taylor Formula and New Green Functions

In this subsection we obtain some discrete and integral identities and the corresponding linear inequalities using new Green functions and applying the Taylor's formula.

In [102] we proved various results related to general linear inequalities via Taylor's formula using Green's function (see also [101]). Our main objective of present subsection is to further extend results of [102] using new definitions.

Theorem 1.2.2 *Let $n \in \mathbb{N}$, $n \geq 3$, and $f : I \to \mathbb{R}$, $[a, b] \subset I$, be a function such that $f^{(n-1)}$ is absolutely continuous. Furthermore, let $m \in \mathbb{N}$, $x_i \in [a, b]$ and $p_i \in \mathbb{R}$ for $i \in \{1, 2, \ldots, m\}$ be such that*

$$\sum_{i=1}^{m} p_i = 0, \quad \sum_{i=1}^{m} p_i x_i = 0.$$

1.2 Linear Inequalities and the Taylor Formula with New Green Functions

Then for $l \in \{1, 2, 3, 4\}$

$$\sum_{i=1}^{m} p_i f(x_i) = \sum_{k=0}^{n-3} \frac{f^{(k+2)}(a)}{k!} \int_a^b \sum_{i=1}^{m} p_i G_l(x_i, t)(t-a)^k dt$$

$$+ \frac{1}{(n-3)!} \int_a^b f^{(n)}(s) \left(\int_s^b \sum_{i=1}^{m} p_i G_l(x_i, t)(t-s)^{n-3} dt \right) ds$$

(1.2.2)

and

$$\sum_{i=1}^{m} p_i f(x_i) = \sum_{k=0}^{n-3} (-1)^k \frac{f^{(k+2)}(b)}{k!} \int_a^b \sum_{i=1}^{m} p_i G_l(x_i, t)(b-t)^k dt$$

$$- \frac{1}{(n-3)!} \int_a^b f^{(n)}(s) \left(\int_a^s \sum_{i=1}^{m} p_i G_l(x_i, t)(t-s)^{n-3} dt \right) ds.$$

(1.2.3)

Proof First consider four identities (1.1.5), (1.1.10), (1.1.11) and (1.1.12), and putting $x = x_i$ in all these identities, multiplying each with p_i, and then summing over each identity for $i \in \{1, \ldots, m\}$ and using conditions that $\sum_{i=1}^{m} p_i = 0$, $\sum_{i=1}^{m} p_i x_i = 0$ we get by fixing $l \in \{1, 2, 3, 4\}$

$$\sum_{i=1}^{m} p_i f(x_i) = \int_a^b \left(\sum_{i=1}^{m} p_i G_l(x_i, t) \right) f''(t) dt. \qquad (1.2.4)$$

Now differentiating (1.2.1) twice we get

$$f''(x) = \sum_{k=0}^{n-3} \frac{f^{(k+2)}(c)}{k!} (x-c)^k + \frac{1}{(n-3)!} \int_c^x f^{(n)}(s)(x-s)^{n-3} ds. \qquad (1.2.5)$$

Putting in (1.2.5) $c = a$ and $c = b$ respectively and using both in (1.2.4) we get respectively,

$$\sum_{i=1}^{m} p_i f(x_i) = \sum_{k=0}^{n-3} \frac{f^{(k+2)}(a)}{k!} \int_a^b \left(\sum_{i=1}^{m} p_i G_l(x_i, t) \right) (t-a)^k dt$$

$$+ \frac{1}{(n-3)!} \int_a^b \int_a^t f^{(n)}(s)(t-s)^{n-3} \left(\sum_{i=1}^{m} p_i G_l(x_i, t) \right) ds dt$$

and

$$\sum_{i=1}^m p_i f(x_i) = \sum_{k=0}^{n-3} \frac{f^{(k+2)}(b)}{k!} \int_a^b \left(\sum_{i=1}^m p_i G_l(x_i, t)\right)(t-b)^k dt$$
$$+ \frac{1}{(n-3)!} \int_a^b \int_b^t f^{(n)}(s)(t-s)^{n-3} \left(\sum_{i=1}^m p_i G_l(x_i, t)\right) ds dt.$$

Finally, by using Fubini's theorem we obtain identities (1.2.2) and (1.2.3) respectively. □

Now we state inequalities derived from the obtained identities. In the rest of the section we use the following notation:

$$\Omega_5^{[a,b]}(l, m, \mathbf{x}, \mathbf{p}, s) = \int_s^b \sum_{i=1}^m p_i G_l(x_i, t)(t-s)^{n-3} dt, \tag{1.2.6}$$

$$\Omega_6^{[a,b]}(l, m, \mathbf{x}, \mathbf{p}, s) = \int_a^s \sum_{i=1}^m p_i G_l(x_i, t)(t-s)^{n-3} dt, \tag{1.2.7}$$

$$A_5^{[a,b]}(l, m, \mathbf{x}, \mathbf{p}, f) = \sum_{i=1}^m p_i f(x_i) - \sum_{k=0}^{n-3} \frac{f^{(k+2)}(a)}{k!}$$
$$\times \int_a^b \sum_{i=1}^m p_i G_l(x_i, t)(t-a)^k dt, \tag{1.2.8}$$

$$A_6^{[a,b]}(l, m, \mathbf{x}, \mathbf{p}, f) = \sum_{i=1}^m p_i f(x_i) - \sum_{k=0}^{n-3} (-1)^k$$
$$\times \frac{f^{(k+2)}(b)}{k!} \int_a^b \sum_{i=1}^m p_i G_l(x_i, t)(b-t)^k dt. \tag{1.2.9}$$

Theorem 1.2.3 *Fix* $l \in \{1, 2, 3, 4\}$. *Let* $n, m \in \mathbb{N}$, $n \geq 3$, $\mathbf{x} = (x_1, \ldots, x_m) \in [a, b]^m$ *and* $\mathbf{p} = (p_1, \ldots, p_m) \in \mathbb{R}^m$ *be such that*

$$\sum_{i=1}^m p_i = 0, \quad \sum_{i=1}^m p_i x_i = 0. \tag{1.2.10}$$

1.2 Linear Inequalities and the Taylor Formula with New Green Functions

(i) If we take $\Omega_5^{[a,b]}(l, m, \mathbf{x}, \mathbf{p}, s) \geq 0$ for all $s \in [a, b]$, then for every n-convex function $f : I \to \mathbb{R}$ such that $f^{(n-1)}$ is absolutely continuous on $I \subset [a, b]$ the following inequality holds

$$A_5^{[a,b]}(l, m, \mathbf{x}, \mathbf{p}, f) \geq 0, \qquad (1.2.11)$$

and if we take $\Omega_{19}^{[a,b]}(l, m, \mathbf{x}, \mathbf{p}, s) \leq 0$, then inequality (1.2.11) is also reversed.

(ii) If we take $\Omega_6^{[a,b]}(l, m, \mathbf{x}, \mathbf{p}, s) \leq 0$ for all $s \in [a, b]$, then for every n-convex function $f : I \to \mathbb{R}$ such that $f^{(n-1)}$ is absolutely continuous on $I \subset [a, b]$ the following inequality holds

$$A_6^{[a,b]}(l, m, \mathbf{x}, \mathbf{p}, f) \geq 0, \qquad (1.2.12)$$

and if we take $\Omega_6^{[a,b]}(l, m, \mathbf{x}, \mathbf{p}, s) \geq 0$, then inequality (1.2.12) is also reversed.

(iii) If the condition "f is n-convex" is replaced by "f is n-concave", then under the same assumptions about $\Omega_5^{[a,b]}$ and $\Omega_6^{[a,b]}$, inequalities (1.2.11) and (1.2.12) hold in the reversed direction.

Proof If f is n-convex, then $f^{(n)} \geq 0$. Using this fact and identities from Theorem 1.2.2 we get required results. □

If we add new condition on $\mathbf{x}$, then in the previous statements we can remove assumptions about $\Omega_5^{[a,b]}$ and $\Omega_6^{[a,b]}$. More precisely, we have the following result.

Theorem 1.2.4 Fix $l \in \{1, 2, 3, 4\}$. Let $n \in \mathbb{N}$, $n \geq 3$, and $f : I \to \mathbb{R}$, $[a, b] \subset I$, be a function such that $f^{(n-1)}$ is absolutely continuous. Furthermore, let $m \in \mathbb{N}$, $x_i \in [a, b]$ and $p_i \in \mathbb{R}$ for $i \in \{1, 2, \ldots, m\}$ such that (1.0.6) is satisfied.

If f is n-convex, then (1.2.11) holds. If n is even, then (1.2.12) is valid, while if n is odd, then a reversed sign in inequality (1.2.12) holds.

If f is n-concave, then reversed (1.2.11) holds. If n is even, then reversed (1.2.12), while if n is odd, then inequality (1.2.12) holds.

Proof By Remark 1.0.4 m-tuples $\mathbf{x}, \mathbf{p}$ satisfies assumptions of Theorem 1.0.1. Since G_l for $l \in \{1, 2, 3, 4\}$ are convex with respect to the first variable, using that Theorem 1.0.1 we conclude that

$$\sum_{i=1}^{m} p_i G_l(x_i, t) \geq 0 \text{ for } t \in [a, b].$$

Note that $(t - s)^{n-3} \geq 0$ for $t \in [s, b]$ so we get $\Omega_5^{[a,b]}(l, m, \mathbf{x}, \mathbf{p}, s) \geq 0$. By Theorem 1.2.3 (i), we have that $A_5^{[a,b]}(l, m, \mathbf{x}, \mathbf{p}, f) \geq 0$. Other parts are proved in the similar manner. □

The integral versions of the previous three theorems may also be stated. Since the proofs of these results are similar, we omit the details.

Theorem 1.2.5 *Fix $l \in \{1, 2, 3, 4\}$. Let $g : [\alpha, \beta] \to [a, b]$, $p : [\alpha, \beta] \to \mathbb{R}$ be integrable functions such that*

$$\int_\alpha^\beta p(x)dx = 0, \quad \int_\alpha^\beta p(x)g(x)dx = 0. \qquad (1.2.13)$$

Let $n \geq 3$ and $f : I \to \mathbb{R}$, $[a, b] \subset I$, be a function such that $f^{(n-1)}$ is absolutely continuous. Then we get the following identities

$$\int_\alpha^\beta p(x) f(g(x)) dx = \sum_{k=0}^{n-3} \frac{f^{(k+2)}(a)}{k!} \int_a^b \left(\int_\alpha^\beta p(x) G_l(g(x), t) dx \right) (t-a)^k dt$$

$$+ \frac{1}{(n-3)!} \int_a^b f^{(n)}(s) \left(\int_s^b \left(\int_\alpha^\beta p(x) G_l(g(x), t) dx \right) (t-s)^{n-3} dt \right) ds, \qquad (1.2.14)$$

$$\int_\alpha^\beta p(x) f(g(x)) dx = \sum_{k=0}^{n-3} (-1)^k \frac{f^{(k+2)}(b)}{k!}$$

$$\int_a^b \left(\int_\alpha^\beta p(x) G_l(g(x), t) dx \right) (b-t)^k dt$$

$$- \frac{1}{(n-3)!} \int_a^b f^{(n)}(s) \left(\int_a^s \left(\int_\alpha^\beta p(x) G_l(g(x), t) dx \right) (t-s)^{n-3} dt \right) ds. \qquad (1.2.15)$$

Similarly, we introduce here further notations to be used in sequel.

$$\Omega_7^{[a,b]}(l, [\alpha, \beta], g, p, s) = \int_s^b \left(\int_\alpha^\beta p(x) G_l(g(x), t) dx \right) (t-s)^{n-3} dt, \qquad (1.2.16)$$

$$\Omega_8^{[a,b]}(l, [\alpha, \beta], g, p, s) = \int_a^s \left(\int_\alpha^\beta p(x) G_l(g(x), t) dx \right) (t-s)^{n-3} dt, \qquad (1.2.17)$$

$$A_7^{[a,b]}(l, [\alpha, \beta], g, p, f) = \int_\alpha^\beta p(x) f(g(x)) dx$$

$$- \sum_{k=0}^{n-3} \frac{f^{(k+2)}(a)}{k!} \int_a^b \left(\int_\alpha^\beta p(x) G_l(g(x), t) dx \right) (t-a)^k dt, \qquad (1.2.18)$$

1.2 Linear Inequalities and the Taylor Formula with New Green Functions

$$A_8^{[a,b]}(l, [\alpha, \beta], g, p, f)) = \int_\alpha^\beta p(x) f(g(x)) \, dx$$

$$- \sum_{k=0}^{n-3} (-1)^k \frac{f^{(k+2)}(b)}{k!} \int_a^b \left(\int_\alpha^\beta p(x) G_l(g(x), t) dx \right) (b-t)^k dt. \quad (1.2.19)$$

Theorem 1.2.6 Fix $l \in \{1, 2, 3, 4\}$. Let g, p, n satisfy assumptions of Theorem 1.2.5.

(i) If we take $\Omega_7^{[a,b]}(l, [\alpha, \beta], g, p, s) \geq 0$ for all $s \in [a, b]$, then for every n-convex function $f : I \to \mathbb{R}$, $[a, b] \subset I$, such that $f^{(n-1)}$ is absolutely continuous, the following inequality holds

$$A_7^{[a,b]}(l, [\alpha, \beta], g, p, f) \geq 0, \quad (1.2.20)$$

and if we take $\Omega_7^{[a,b]}(l, [\alpha, \beta], g, p, s) \leq 0$, then inequality (1.2.20) is also reversed.

(ii) If we take $\Omega_8^{[a,b]}(l, [\alpha, \beta], g, p, s) \leq 0$ for all $s \in [a, b]$, then for every n-convex function $f : I \to \mathbb{R}$, $[a, b] \subset I$, such that $f^{(n-1)}$ is absolutely continuous, the following inequality holds

$$A_8^{[a,b]}(l, [\alpha, \beta], g, p, f)) \geq 0, \quad (1.2.21)$$

and if we take $\Omega_8^{[a,b]}(l, [\alpha, \beta], g, p, s) \geq 0$, then inequality (1.2.21) is also reversed.

(iii) If the condition "f is n-convex" is replaced by "f is n-concave", then under the same assumptions about $\Omega_7^{[a,b]}$ and $\Omega_8^{[a,b]}$, inequalities (1.2.20) and (1.2.21) hold in the reversed direction.

Theorem 1.2.7 Let all the assumptions of Theorem 1.2.5 hold. Additionally, let

$$\int_\alpha^\beta p(x)(g(x) - t)_+^{n-1} dx \geq 0 \text{ for all } t \in [a, b].$$

If f is n-convex, then (1.2.20) holds. If n is even, then (1.2.21), while if n is odd, then a reversed sign in inequality (1.2.21) holds.

If f is n-concave, then reversed (1.2.20) holds. If n is even, then reversed (1.2.21), while if n is odd, then inequality (1.2.21) holds.

1.2.2 Inequalities for n-Convex Functions at a Point

In this subsection we give related results for the class of n-convex functions at a point which was introduced in [160]. Recall the Definition 1.1.1 as given in the first section of this chapter.

It is known that a function is n-convex on an interval iff it is n-convex at every point of the interval. Necessary and sufficient conditions on two linear functionals $A : C[a, c] \to \mathbb{R}$ and $B : C[c, b] \to \mathbb{R}$ so that the inequality $A(f) \leq B(f)$ holds for every function f that is n-convex at c are studied in [160]. In this subsection we give inequalities of this type for particular linear functionals related to inequalities obtained in the previous subsection.

First we state our main theorem of this subsection for the discrete case.

Theorem 1.2.8 *Let* $c \in \langle a, b \rangle$, $\mathbf{x} \in [a, c]^{m_1}$, $\mathbf{y} \in [c, b]^{m_2}$, $\mathbf{p} \in \mathbb{R}^{m_1}$, $\mathbf{q} \in \mathbb{R}^{m_2}$ *and* $f : [a, b] \to \mathbb{R}$ *be a function such that* $f^{(n-1)}$ *is absolutely continuous.*

(i) *Let* $\Omega_5^{[\cdot,\cdot]}(\cdot, \cdot, \cdot, \cdot, s)$ *and* $A_5^{[\cdot,\cdot]}(\cdot, \cdot, \cdot, \cdot, f)$ *be defined as in* (1.2.6) *and* (1.2.8) *respectively and satisfy the following conditions:*

$$\Omega_5^{[a,c]}(l, m_1, \mathbf{x}, \mathbf{p}, s) \geq 0, \quad \text{for every } s \in [a, c], \tag{1.2.22}$$

$$\Omega_5^{[c,b]}(l, m_2, \mathbf{y}, \mathbf{q}, s) \geq 0, \quad \text{for every } s \in [c, b], \tag{1.2.23}$$

and

$$A_5^{[a,c]}(l, m_1, \mathbf{x}, \mathbf{p}, e_n) = A_5^{[c,b]}(l, m_2, \mathbf{y}, \mathbf{q}, e_n). \tag{1.2.24}$$

If f *is* $(n+1)$-*convex at point* c, *then*

$$A_5^{[a,c]}(l, m_1, \mathbf{x}, \mathbf{p}, f) \leq A_5^{[c,b]}(l, m_2, \mathbf{y}, \mathbf{q}, f). \tag{1.2.25}$$

If inequalities in (1.2.22) *and* (1.2.23) *are reversed, then* (1.2.25) *holds with the reversed sign of inequality.*

(ii) *Let* $\Omega_6^{[\cdot,\cdot]}(\cdot, \cdot, \cdot, s)$ *and* $A_6^{[\cdot,\cdot]}(\cdot, \cdot, \cdot, f)$ *be defined as in* (1.2.7) *and* (1.2.9) *respectively and satisfy the following conditions:*

$$\Omega_6^{[a,c]}(l, m_1, \mathbf{x}, \mathbf{p}, s) \leq 0, \quad \text{for every } s \in [a, c], \tag{1.2.26}$$

$$\Omega_6^{[c,b]}(l, m_2, \mathbf{y}, \mathbf{q}, s) \leq 0, \quad \text{for every } s \in [c, b], \tag{1.2.27}$$

and

$$A_6^{[a,c]}(l, m_1, \mathbf{x}, \mathbf{p}, e_n) = A_6^{[c,b]}(l, m_2, \mathbf{y}, \mathbf{q}, e_n). \tag{1.2.28}$$

1.2 Linear Inequalities and the Taylor Formula with New Green Functions

If f is $(n+1)$-convex at point c, then

$$A_6^{[a,c]}(l, m_1, \mathbf{x}, \mathbf{p}, f) \leq A_6^{[c,b]}(l, m_2, \mathbf{y}, \mathbf{q}, f). \tag{1.2.29}$$

If inequalities in (1.2.26) and (1.2.27) are reversed, then (1.2.29) holds with the reversed sign of inequality.

Remark 1.2.1 Since the proof of the Theorem 1.2.8 is similar to proof of Theorem 1.1.11 so we omit the details. Further, a remark similar to Remark 1.1.1 is also valid for Theorem 1.2.8.

Integral analogous of previous theorem may be stated as:

Theorem 1.2.9 *Let $\alpha \leq \beta, \gamma \leq \delta, a < c < b$, $g : [\alpha, \beta] \to [a,c]$, $p : [\alpha, \beta] \to \mathbb{R}$, $h : [\gamma, \delta] \to [c,b]$, $q : [\gamma, \delta] \to \mathbb{R}$ be integrable. Let $f : I \to \mathbb{R}$, $[a,b] \subset I$, be a function such that $f^{(n-1)}$ is absolutely continuous.*

(i) *Let $\Omega_7^{[\cdot, \cdot]}(\cdot, \cdot, \cdot, \cdot, s)$ and $A_7^{[\cdot, \cdot]}(\cdot, \cdot, \cdot, \cdot, f)$ be defined as in (1.2.16) and (1.2.18) respectively and satisfy the following conditions:*

$$\Omega_7^{[a,c]}(l, [\alpha, \beta], g, p, s) \geq 0, \quad \text{for every } s \in [a,c], \tag{1.2.30}$$

$$\Omega_7^{[c,b]}(l, [\gamma, \delta], h, q, s) \geq 0, \quad \text{for every } s \in [c,b], \tag{1.2.31}$$

$$A_7^{[a,c]}(l, [\alpha, \beta], g, p, e_n) = A_7^{[c,b]}(l, [\gamma, \delta], h, q, e_n). \tag{1.2.32}$$

If f is $(n+1)$-convex at point c, then

$$A_7^{[a,c]}(l, [\alpha, \beta], g, p, f) \leq A_7^{[c,b]}(l, [\gamma, \delta], h, q, f). \tag{1.2.33}$$

If the inequalities in (1.2.30) and (1.2.31) are reversed, then the reversed sign in (1.2.33) holds.

(ii) *Let $\Omega_8^{[\cdot, \cdot]}(\cdot, \cdot, \cdot, \cdot, s)$ and $A_8^{[\cdot, \cdot]}(\cdot, \cdot, \cdot, \cdot, f)$ be defined as in (1.2.17) and (1.2.19) respectively and satisfy the following conditions:*

$$\Omega_8^{[a,c]}(l, [\alpha, \beta], g, p, s) \leq 0, \quad \text{for every } s \in [a,c], \tag{1.2.34}$$

$$\Omega_8^{[c,b]}(l, [\gamma, \delta], h, q, s) \leq 0, \quad \text{for every } s \in [c,b], \tag{1.2.35}$$

$$A_8^{[a,c]}(l, [\alpha, \beta], g, p, e_n) = A_8^{[c,b]}(l, [\gamma, \delta], h, q, e_n). \tag{1.2.36}$$

If f is $(n+1)$-convex at point c, then

$$A_8^{[a,c]}(l, [\alpha, \beta], g, p, f) \leq A_8^{[c,b]}(l, [\gamma, \delta], h, q, f). \tag{1.2.37}$$

If the inequalities in (1.2.34) and (1.2.35) are reversed, then the reversed sign in (1.2.37) holds.

1.2.3 Bounds for Remainders and Functionals

For the sake of brevity, in present and next subsection at some places we will use the notations $A_l(f) = A_l^{[\cdot,\cdot]}(\cdot,\cdot,\cdot,\cdot,f)$ and $\Omega_l(s) = \Omega_l^{[\cdot,\cdot]}(\cdot,\cdot,\cdot,\cdot,s)$ for $l \in \{5, 6, 7, 8\}$.

Here we state some results involving Čebyšev functional as given in (1.1.57). The following results can be found in [35]:

Theorem 1.2.10 *Let $f : [a, b] \to \mathbb{R}$ be a Lebesgue integrable function and let $h : [a, b] \to \mathbb{R}$ be an absolutely continuous function with $(\cdot - a)(b - \cdot)[h']^2 \in L[a, b]$. Then we have inequality*

$$|T(f, h)| \leq \frac{1}{\sqrt{2}} \left(\frac{1}{b-a} |T(f, f)| \int_a^b (x-a)(b-x)[h'(x)]^2 dx \right)^{1/2}. \qquad (1.2.38)$$

The constant $\frac{1}{\sqrt{2}}$ in (1.2.38) is best possible.

Theorem 1.2.11 *Let $h : [a, b] \to \mathbb{R}$ be a monotonic nondecreasing function and let $f : [a, b] \to \mathbb{R}$ be an absolutely continuous function such that $f' \in L_\infty[a, b]$. Then we have inequality*

$$|T(f, h)| \leq \frac{1}{2(b-a)} \|f'\|_\infty \int_a^b (x-a)(b-x) dh(x). \qquad (1.2.39)$$

The constant $\frac{1}{2}$ in (1.2.39) is best possible.

Now, we are ready to state main results of this subsection:

Theorem 1.2.12 *Let $f : [a, b] \to \mathbb{R}$ be such that $f^{(n)}$ is an absolutely continuous function for $n \in \mathbb{N}$ with $(\cdot - a)(b - \cdot)[f^{(n+1)}]^2 \in L[a, b]$ and let $A_l(f) = A_l^{[\cdot,\cdot]}(\cdot,\cdot,\cdot,\cdot,f)$ and $\Omega_l(t) = \Omega_l^{[\cdot,\cdot]}(\cdot,\cdot,\cdot,\cdot,t)$ ($l \in \{5, 6, 7, 8\}$) be as defined in Theorems 1.2.3 and 1.2.6. Then it holds for $l \in \{5, 6, 7, 8\}$*

$$A_l(f) = \frac{[f^{(n-1)}(b) - f^{(n-1)}(a)]}{(n-3)!(b-a)} \int_a^b \Omega_l(s) ds + R_2^l(f; a, b, n),$$

where the remainder $R_2^l(f; a, b, n)$ satisfies the estimation

$$|R_2^l(f; a, b, n)| \leq \frac{1}{(n-3)!}$$
$$\left(\frac{(b-a)}{2} \left| T(\Omega_l, \Omega_l) \int_a^b (s-a)(b-s)[f^{(n+1)}(s)]^2 ds \right| \right)^{1/2}. \qquad (1.2.40)$$

1.2 Linear Inequalities and the Taylor Formula with New Green Functions

Proof Fix $l \in \{5, 6, 7, 8\}$. If we apply Theorem 1.2.10 for $f \to \Omega_l$ and $h \to f^{(n)}$, then we obtain

$$\left| \frac{1}{b-a} \int_a^b \Omega_l(t) f^{(n)}(t) dt - \left(\frac{1}{b-a} \int_a^b \Omega_l(t) dt \right) \left(\frac{1}{b-a} \int_a^b f^{(n)}(t) dt \right) \right|$$

$$\leq \frac{1}{\sqrt{2}} \left(\frac{1}{b-a} |T(\Omega_l, \Omega_l)| \int_a^b (t-a)(b-t) [f^{(n+1)}(t)]^2 dt \right)^{1/2}.$$

Therefore we have

$$\frac{1}{(n-3)!} \int_a^b \Omega_l(t) f^{(n)}(t) dt = \frac{[f^{(n-1)}(b) - f^{(n-1)}(a)]}{(n-3)!(b-a)}$$

$$\times \int_a^b \Omega_l(t) dt + R_2^l(f; a, b, n).$$

where $R_2^l(f; a, b, n)$ satisfies inequality (1.2.40). Now from identities (1.2.2), (1.2.3), (1.2.14) and (1.2.15) for $l = 5, 6, 7$ and 8 respectively, we obtain (1.2.40). □

By using Theorem 1.2.11 we obtain the following Grüss type inequality.

Theorem 1.2.13 *Let $f : [a, b] \to \mathbb{R}$ be such that $f^{(n)}$ is an absolutely continuous function for $n \in \mathbb{N}$ with $(\cdot - a)(b - \cdot)[f^{(n+1)}]^2 \in L[a, b]$ with $f^{(n+1)} \geq 0$ on $[a, b]$ and let $A_l(f) = A_l^{[\cdot,\cdot]}(\cdot, \cdot, \cdot, \cdot, f)$ and $\Omega_l(t) = \Omega_l^{[\cdot,\cdot]}(\cdot, \cdot, \cdot, \cdot, t)$ ($l \in \{5, 6, 7, 8\}$) be as defined in Theorems 1.2.3 and 1.2.6. Then we have the representation (1.2.40) and the remainder $R_3^l(f; a, b, n)$ satisfies the following condition for $l \in \{5, 6, 7, 8\}$*

$$|R_3^l(f; a, b, n)| \leq \frac{1}{(n-3)!} \|\Omega_l'\|_\infty$$

$$\times \left\{ \frac{b-a}{2} \left[f^{(n-1)}(b) + f^{(n-1)}(a) \right] - \left[f^{(n-2)}(b) - f^{(n-2)}(a) \right] \right\}. \quad (1.2.41)$$

Proof Fix $l \in \{5, 6, 7, 8\}$. If we apply Theorem 1.2.11 for $f \to \Omega_l$ and $h \to f^{(n)}$, then we obtain

$$\left| \frac{1}{b-a} \int_a^b \Omega_l(t) f^{(n)}(t) dt - \left(\frac{1}{b-a} \int_a^b \Omega_l(t) dt \right) \left(\frac{1}{b-a} \int_a^b f^{(n)}(t) dt \right) \right|$$

$$\leq \frac{1}{2(b-a)} \|\Omega_l'\|_\infty \int_a^b (t-a)(b-t) f^{(n+1)}(t) dt.$$

Since

$$\int_a^b (t-a)(b-t) f^{(n+1)}(t) dt = \int_a^b (2t - a - b) f^{(n)}(t) dt$$
$$= (b-a)\left[f^{(n-1)}(b) + f^{(n-1)}(a)\right] - 2\left[f^{(n-2)}(b) - f^{(n-2)}(a)\right]. \quad (1.2.42)$$

Therefore, by using the identities (1.2.2), (1.2.3), (1.2.14) and (1.2.15) for $l = 5, 6, 7$ and 8 respectively and (1.2.42) we deduce (1.2.41). □

Now we state some Ostrowski-type inequalities related to the generalized linear inequalities.

Theorem 1.2.14 *Let $A_l(f) = A_l^{[\cdot,\cdot]}(\cdot, \cdot, \cdot, \cdot, f)$ and $\Omega_l(t) = \Omega_l^{[\cdot,\cdot]}(\cdot, \cdot, \cdot, \cdot, t)$ ($l \in \{5, 6, 7, 8\}$) be as defined in Theorems 1.2.3 and 1.2.6. Furthermore, let (q, r) be a pair of conjugate exponents, i.e., $1 \leq q, r \leq \infty$, $\frac{1}{q} + \frac{1}{r} = 1$. Let $f^{(n)} \in L_q[a, b]$ for some $n \in \mathbb{N}$. Then we have for $l \in \{5, 6, 7, 8\}$*

$$|A_l(f)| \leq \frac{1}{(n-3)!} \|f^{(n)}\|_q \|\Omega_l\|_r. \quad (1.2.43)$$

The constant on the right hand side of (1.2.43) is sharp for $1 < q \leq \infty$ and the best possible for $q = 1$.

Proof Fix $l \in \{5, 6, 7, 8\}$. Let us denote

$$\lambda_l(t) = \frac{1}{(n-3)!} \Omega_l(t).$$

Now, by using the identities (1.2.2), (1.2.3), (1.2.14) and (1.2.15) for $l = 1, 2, 3$ and 4 respectively, and applying Hölder's inequality we obtain

$$|A_l(f)| \leq \|f^{(n)}\|_q \|\lambda_l\|_r. \quad (1.2.44)$$

For the proof of the sharpness of the constant $\left(\int_a^b |\lambda_l(t)|^r dt\right)^{1/r}$, let us find a function f for which the equality in (1.2.44) is obtained.

For $1 < q < \infty$ take f to be such that

$$f^{(n)}(t) = sgn\lambda_l(t) \cdot |\lambda_l(t)|^{1/(q-1)}.$$

For $q = \infty$, take f such that

$$f^{(n)}(t) = sgn\lambda_l(t).$$

1.2 Linear Inequalities and the Taylor Formula with New Green Functions

Finally, for $q = 1$, we prove that

$$\left|\int_a^b \lambda_l(t) f^{(n)}(t) dt\right| \leq \max_{t \in [a,b]} |\lambda_l(t)| \int_a^b f^{(n)}(t) dt \quad (1.2.45)$$

is the best possible inequality.

Suppose that $|\lambda(t)|$ attains its maximum at $t_0 \in [a, b]$. First we consider the case $\lambda_l(t_0) > 0$. For ϵ small enough we define $f_\epsilon(t)$ by

$$f_\epsilon(t) = \begin{cases} 0, & a \leq t \leq t_0; \\ \frac{1}{\epsilon n!}(t - t_0)^n, & t_0 \leq t \leq t_0 + \epsilon; \\ \frac{1}{(n-1)!}(t - t_0)^{n-1}, & t_0 + \epsilon \leq t \leq b. \end{cases}$$

So, we have

$$\left|\int_a^b \lambda_l(t) f_\epsilon^{(n)}(t) dt\right| = \left|\int_{t_0}^{t_0+\epsilon} \lambda_l(t) \frac{1}{\epsilon} dt\right| = \frac{1}{\epsilon} \int_{t_0}^{t_0+\epsilon} \lambda_l(t) dt$$

Now from inequality (1.2.45) we have

$$\frac{1}{\epsilon} \int_{t_0}^{t_0+\epsilon} \lambda_l(t) dt \leq \lambda_l(t_0) \frac{1}{\epsilon} \int_{t_0}^{t_0+\epsilon} dt = \lambda_l(t_0)$$

Since

$$\lim_{\epsilon \to 0} \frac{1}{\epsilon} \int_{t_0}^{t_0+\epsilon} \lambda_l(t) dt = \lambda_l(t_0)$$

the statement follows.

In the case $\lambda_l(t_0) < 0$, we define $f_\epsilon(t)$ by

$$f_\epsilon(t) = \begin{cases} \frac{1}{(n-1)!}(t - t_0 - \epsilon)^{n-1}, & a \leq t \leq t_0; \\ -\frac{1}{\epsilon n!}(t - t_0 - \epsilon)^n, & t_0 \leq t \leq t_0 + \epsilon; \\ 0, & t_0 + \epsilon \leq t \leq b; \end{cases}$$

and the rest of the proof is the same as above. □

1.2.4 Mean Value Theorems and Exponential Convexity

Mean Value Theorems

Now we give mean value theorems for $A_l(f) = A_l^{[\cdot,\cdot]}(\cdot,\cdot,\cdot,\cdot,f)$, $l \in \{5,6,7,8\}$. Here $f_0(x) = \frac{x^n}{n!}$. Since proving techniques are same as done in previous sections so we omit the details in following theorems (see [5]).

Theorem 1.2.15 *Let $f \in C^{(n)}[a,b]$ and let $A_l : C^{(n)}[a,b] \to \mathbb{R}$ for $l \in \{5,6,7,8\}$ be linear functionals as defined in Theorems 1.2.3 and 1.2.6 respectively. Then there exists $\xi_l \in [a,b]$ for $l \in \{5,6,7,8\}$ such that*

$$A_l(f) = f^{(n)}(\xi_l) A_l(f_0), \quad l \in \{5,6,7,8\}. \tag{1.2.46}$$

Theorem 1.2.16 *Let $f, h \in C^{(n)}[a,b]$ and let $A_l : C^{(n)}[a,b] \to \mathbb{R}$ for $l \in \{5,6,7,8\}$ be linear functionals as defined in Theorems 1.2.3 and 1.2.6 respectively. Then there exists $\xi_l \in [a,b]$ for $l \in \{5,6,7,8\}$ such that*

$$\frac{A_l(f)}{A_l(h)} = \frac{f^{(n)}(\xi_l)}{h^{(n)}(\xi_k)}$$

assuming that both the denominators are non-zero.

Remark 1.2.2 If the inverse of $\frac{f^{(n)}}{h^{(n)}}$ exists, then from the above mean value theorems we can give generalized means

$$\xi_l = \left(\frac{f^{(n)}}{h^{(n)}}\right)^{-1} \left(\frac{A_l(f)}{A_l(h)}\right), \quad l \in \{5,6,7,8\}. \tag{1.2.47}$$

Logarithmically Convex Functions

A number of important inequalities arise from the logarithmic convexity of some functions as one can see in [114].

Now, we recall some definitions. Here I is an interval in $\mathbb{R}$.

Definition 1.2.1 A function $f : I \to (0, \infty)$ is called log-convex in J-sense if the following inequality holds for each $x_1, x_2 \in I$,

$$f^2\left(\frac{x_1 + x_2}{2}\right) \le f(x_1) f(x_2).$$

Definition 1.2.2 ([158, p. 7]) A function $f : I \to (0, \infty)$ is called log-*convex* if the following inequality holds for each $x_1, x_2 \in I$ and $\lambda \in [0,1]$,

$$f(\lambda x_1 + (1-\lambda)x_2) \le [f(x_1)]^{\lambda}[f(x_2)]^{(1-\lambda)}.$$

1.2 Linear Inequalities and the Taylor Formula with New Green Functions

Remark 1.2.3 A function log-convex in the J-sense is log-convex if it is continuous as well.

n-Exponentially Convex Functions

Widder [186] and Bernstein [22] independently introduced an important class of functions named "exponentially convex functions" which is a sub-class of convex functions. These exponential convex functions possess a lot of important characteristics, for example the functions are analytic in their domain. Positive-semidefinite matrices are also provided by these functions. Further, these act as an effective part in studying the characteristics of Cauchy and Stolarsky means, such as monotonicity of their means etc. For more detailed information see [3, 81] and [135].

In [153], Perić and Pečarić introduced the notion of n-exponentially convex functions, which is the concept to generalize the exponentially convex functions. In recent given sub-section of this chapter, we would like to describe the same notion of n-exponential convexity using related definitions and some significant consequences with few special remarks from [153]. Here I is an interval in $\mathbb{R}$.

Definition 1.2.3 A function $f : I \to \mathbb{R}$ is *n-exponentially convex* in the J−sense if the following inequality holds for each $t_i \in I$ and $u_i \in \mathbb{R}$, $i \in \{1, \ldots, n\}$,

$$\sum_{i,j=1}^{n} u_i u_j f\left(\frac{t_i + t_j}{2}\right) \geq 0.$$

Definition 1.2.4 A function $f : I \to \mathbb{R}$ is *n-exponentially convex* if it is n-exponentially convex in the J-sense and continuous on I.

Remark 1.2.4 We can see from the definition that 1-exponentially convex functions in the J-sense are in fact nonnegative functions. Also, n-exponentially convex functions in the J-sense are k-exponentially convex in the J-sense for every $k \in \mathbb{N}$ such that $k \leq n$. It follows that a positive function is log-convex in the J-sense iff it is 2-exponentially convex in the J-sense. Also, using basic theory of convex functions, it follows that a positive function is log-convex iff it is 2-exponentially convex.

Definition 1.2.5 A function $f : I \to \mathbb{R}$ is *exponentially convex in the J-sense*, if it is n- exponentially convex in the J-sense for each $n \in \mathbb{N}$.

Remark 1.2.5 A function $f : I \to \mathbb{R}$ is *exponentially convex* if it is n-exponentially convex in the J-sense and continuous on I.

Example 1.2.1 A function $x \mapsto ce^{kx}$ is an example of exponentially convex function for $k \in \mathbb{R}$ and constant $c \geq 0$.

Theorem 1.2.17 *If function $f : I \to \mathbb{R}$ is n-exponentially convex in the J-sense, then the matrix*

$$\left[f\left(\frac{t_i + t_j}{2} \right) \right]_{i,j=1}^{m}$$

is positive-semidefinite. Particularly

$$\det \left[f\left(\frac{t_i + t_j}{2} \right) \right]_{i,j=1}^{m} \geq 0$$

for each $m \in \mathbb{N}$, $m \leq n$ and $t_i \in I$ for $i \in \{1, 2, \ldots, m\}$.

Corollary 1.2.18 *If function $f : I \to \mathbb{R}$ is exponentially convex, then the matrix*

$$\left[f\left(\frac{t_i + t_j}{2} \right) \right]_{i,j=1}^{m}$$

is positive-semidefinite. Particularly

$$\det \left[f\left(\frac{t_i + t_j}{2} \right) \right]_{i,j=1}^{m} \geq 0$$

for each $m \in \mathbb{N}$ and $t_i \in I$ for $i \in \{1, 2, \ldots, m\}$.

Corollary 1.2.19 *If function $f : I \to (0, \infty)$ is exponentially convex, then f is log-convex, where $\forall\, t_1, t_2 \in I$; $\forall\, \lambda \in [0, 1]$, we have*

$$f(\lambda t_1 + (1 - \lambda)t_2) \leq f^{\lambda}(t_1) f^{1-\lambda}(t_2).$$

Remark 1.2.6 A function $f : I \to (0, \infty)$ is log-convex in J-sense iff the inequality

$$u_1^2 f(t_1) + 2u_1 u_2 f\left(\frac{t_1 + t_2}{2} \right) + u_2^2 f(t_2) \geq 0$$

holds for each $t_1, t_2 \in I$ and $u_1, u_2 \in \mathbb{R}$. It follows that a positive function is log-convex in the J-sense iff it is 2-exponentially convex in the J-sense. Also, using basic convexity theory it follows that a positive function is log-convex iff it is 2-exponentially convex.

Here, we get our results concerning the n-exponential convexity and exponential convexity for our functionals A_l, $l \in \{5, 6, 7, 8\}$ as defined in Theorems 1.2.3 and 1.2.6. Throughout the section I is an interval in $\mathbb{R}$.

Theorem 1.2.20 *Let $D_1 = \{f_t : t \in I\}$ be a class of functions such that the function $t \mapsto [z_0, z_1, \ldots, z_n; f_t]$ is n-exponentially convex in the J-sense on I for*

1.2 Linear Inequalities and the Taylor Formula with New Green Functions

any $n + 1$ mutually distinct points $z_0, z_1, \ldots, z_n \in [a, b]$. Let A_l be the linear functionals for $l \in \{5, 6, 7, 8\}$ as defined in Theorems 1.2.3 and 1.2.6. Then the following statements are valid:

(a) The function $t \mapsto A_l(f_t)$ is n-exponentially convex function in the J-sense on I.
(b) If the function $t \mapsto A_l(f_t)$ is continuous on I, then the function $t \mapsto A_l(f_t)$ is n-exponentially convex on I.

Proof

(a) Fix $l \in \{5, 6, 7, 8\}$. Let us define the function p for $t_i \in I$, $u_i \in \mathbb{R}$, $i \in I_n$ as follows

$$p = \sum_{i,j=1}^{n} u_i u_j f_{\frac{t_i+t_j}{2}}.$$

Since the function $t \mapsto [z_0, z_1, \ldots, z_n; f_t]$ is n-exponentially convex in the J-sense, therefore

$$[z_0, z_1, \ldots, z_n; p] = \sum_{i,j=1}^{n} u_i u_j [z_0, z_1, \ldots, z_n; f_{\frac{t_i+t_j}{2}}] \geq 0$$

which implies that p is n-convex function on I and therefore $A_l(p) \geq 0$. Hence

$$\sum_{i,j=1}^{n} u_i u_j A_l(f_{\frac{t_i+t_j}{2}}) \geq 0.$$

We conclude that the function $t \mapsto A_l(f_t)$ is an n-exponentially convex function on I in J-sense.

(b) This part easily follows from definition of n-exponentially convex function. □

As a consequence of the above theorem we give the following corollaries:

Corollary 1.2.21 Let $D_2 = \{f_t : t \in I\}$ be a class of functions such that the function $t \mapsto [z_0, z_1, \ldots, z_n; f_t]$ is an exponentially convex in the J-sense on I for any $n+1$ mutually distinct points $z_0, z_1, \ldots, z_n \in [a, b]$. Let A_l be the linear functionals for $l \in \{5, 6, 7, 8\}$ as defined in Theorems 1.2.3 and 1.2.6. Then the following statements are valid:

(a) The function $t \mapsto A_l(f_t)$ is exponentially convex in the J-sense on I.
(b) If the function $t \mapsto A_l(f_t)$ is continuous on I, then the function $t \mapsto A_l(f_t)$ is exponentially convex on I.

(c) The matrix $\left[A_l \left(f_{\frac{t_i+t_j}{2}} \right) \right]_{i,j=1}^m$ is positive-semidefinite. Particularly,

$$\det \left[A_l \left(f_{\frac{t_i+t_j}{2}} \right) \right]_{i,j=1}^m \geq 0$$

for each $m \in \mathbb{N}$ and $t_i \in I$ where $i \in \{1, 2, \ldots, m\}$.

Proof Proof follows directly from Theorem 1.2.20 by using definition of exponential convexity and Corollary 1.2.18. □

Corollary 1.2.22 *Let $D_3 = \{f_t : t \in I\}$ be a class of functions such that the function $t \mapsto [z_0, z_1, \ldots, z_n; f_t]$ is 2-exponentially convex in the J-sense on I for any $n + 1$ mutually distinct points $z_0, z_1, \ldots, z_n \in [a, b]$. Let A_l be the linear functionals for $l \in \{5, 6, 7, 8\}$ as defined in Theorems 1.2.3 and 1.2.6. Then the following statements are valid:*

(a) *If the function $t \mapsto A_l(f_t)$ is continuous on I, then it is 2-exponentially convex on I. If the function $t \mapsto A_l(f_t)$ is additionally positive, then it is also log-convex on I. Moreover, the following Lyapunov's inequality holds for $r < s < t$; $r, s, t \in I$*

$$[A_l(f_s)]^{t-r} \leq [A_l(f_r)]^{t-s} [A_l(f_t)]^{s-r}. \tag{1.2.48}$$

(b) *If the function $t \mapsto A_l(f_t)$ is positive and differentiable on I, then for every $s, t, u, v \in I$ such that $s \leq u$ and $t \leq v$, we have*

$$\mu_{s,t}(A_l, D_3) \leq \mu_{u,v}(A_l, D_3), \tag{1.2.49}$$

where $\mu_{s,t}$ is defined as

$$\mu_{s,t}(A_l, D_3) = \begin{cases} \left(\frac{A_l(f_s)}{A_l(f_t)} \right)^{\frac{1}{s-t}} &, s \neq t; \\ \exp\left(\frac{\frac{d}{ds} A_l(f_s)}{A_l(f_s)} \right) &, s = t, \end{cases} \tag{1.2.50}$$

for $f_s, f_t \in D_3$.

Proof

(a) It follows directly form Theorem 1.2.20 and Corollary 1.2.19. As the function $t \mapsto A_l(f_t)$ is log-convex, i.e., $\ln A_l(f_t)$ is convex, so we have

$$\ln[A_l(f_s)]^{t-r} \leq \ln[A_l(f_r)]^{t-s} + \ln[A_l(f_t)]^{s-r}, \quad l \in \{5, 6, 7, 8\}$$

which gives us (1.2.48).

1.2 Linear Inequalities and the Taylor Formula with New Green Functions

(b) For convex function f, the inequality

$$\frac{f(s) - f(t)}{s - t} \leq \frac{f(u) - f(v)}{u - v} \tag{1.2.51}$$

holds $\forall\, s, t, u, v \in I \subset \mathbb{R}$ such that $s \leq u,\ t \leq v,\ s \neq t,\ u \neq v$.
Since by (a), $\Gamma_k(f_t)$ are log-convex for $l \in \{5, 6, 7, 8\}$, so set $f(t) = \ln \Gamma_k(f_t)$ in (1.2.51) we have

$$\frac{\ln A_l(f_s) - \ln A_l(f_t)}{s - t} \leq \frac{\ln A_l(f_u) - \ln A_l(f_v)}{u - v}, \quad l \in \{5, 6, 7, 8\} \tag{1.2.52}$$

for $s \leq u,\ t \leq v,\ s \neq t,\ u \neq v$, which is equivalent to (1.2.49). The cases for $s = t$ and/or $u = v$ are easily followed from (1.2.52) by taking respective limits.

□

Remark 1.2.7 The results from Theorem 1.2.20 and Corollaries 1.2.21 and 1.2.22 still hold when any two (all) points $z_0, z_1, \ldots, z_n \in [a, b]$ coincide for a family of differentiable (n-times differentiable) functions f_t such that the function $t \mapsto [z_0, z_1, \ldots, z_n; f_t]$ is n-exponentially convex, exponentially convex and 2-expoenetially convex in the J-sense respectively.

Now, we give two important remarks and one useful corollary from [81], which we will use in some examples in next section.

Remark 1.2.8 For $\mu_{s,t}(A_l, F)$ defined with (1.2.49) we will refer as mean if

$$a \leq \mu_{s,t}(A_l, F) \leq b$$

for $s, t \in I$ and $l \in \{5, 6, 7, 8\}$ where $F = \{f_t : t \in I\}$ be a family of functions and $[a, b] \subset \text{Dom}(f_t)$.

□

Theorem 1.2.20 gives us the following corollary.

Corollary 1.2.23 *Let $a, b \in \mathbb{R}$ and A_l be linear functionals for $l \in \{5, 6, 7, 8\}$. Let $F = \{f_t : t \in I\}$ be a family of functions in $C^{(2)}[a, b]$. If*

$$a \leq \left(\frac{\frac{d^2 f_s}{dx^2}}{\frac{d^2 f_t}{dx^2}} \right)^{\frac{1}{s-t}} (\xi) \leq b,$$

for $\xi \in [a, b]$, $s, t \in I$, then $\mu_{s,t}(A_l, F)$ is a mean for $l \in \{5, 6, 7, 8\}$.

Remark 1.2.9 In some examples, we will get means of this type:

$$\left(\frac{\frac{d^2 f_s}{dx^2}}{\frac{d^2 f_t}{dx^2}}\right)^{\frac{1}{s-t}}(\xi) = \xi, \quad \xi \in [a,b], \quad s \neq t.$$

1.2.5 Examples with Applications

In this section, we use various classes of functions $F = \{f_t : t \in I\}$ for any open interval $I \subset \mathbb{R}$ to construct different examples of exponentially convex functions and applications to Stolarsky-type means. Let us consider some examples:

Example 1.2.2 Let $F_1 = \{\psi_t : [a,b] \subset \mathbb{R} \to [0,\infty) : t \in \mathbb{R}\}$ be a family of functions defined by

$$\psi_t(x) = \begin{cases} \frac{e^{tx}}{t^n}, & t \neq 0; \\ \frac{x^n}{n!}, & t = 0. \end{cases}$$

Since $\frac{d^n}{dx^n}\psi_t(x) = e^{tx} > 0$ for $x \in [a,b] \subset \mathbb{R}$, the function $\psi_t(x)$ is a n-convex on $\mathbb{R}$ for every $t \in \mathbb{R}$ and $t \to \frac{d^n}{dx^n}\psi_t(x)$ is exponentially convex by definition. Using analogous arguing as in the proof of Theorems 1.2.20, we have that $t \mapsto [z_0, z_1, \ldots, z_n; \psi_t]$ is exponentially convex (and so exponentially convex in the J-sense). Using Corollary 1.2.21 we conclude that $t \mapsto A_l(\psi_t)$, $l \in \{5,6,7,8\}$ are exponentially convex in the J-sense. It is easy to see that these mappings are continuous, so they are exponentially convex.

Assume that $t \mapsto A_l(\psi_t) > 0$ for $l \in \{5,6,7,8\}$. By introducing convex functions ψ_t in (1.2.47), we obtain the following means: for $l \in \{5,6,7,8\}$

$$\mathfrak{M}_{s,t}(A_l, F_1) = \begin{cases} \frac{1}{s-t}\ln\left(\frac{A_l(\psi_s)}{A_l(\psi_t)}\right), & s \neq t; \\ \frac{A_l(id.\psi_s)}{A_l(\psi_s)} - \frac{n}{s}, & s = t \neq 0; \\ \frac{A_l(id.\psi_0)}{(n+1)A_l(\psi_0)}, & s = t = 0; \end{cases}$$

where id stands for identity function on $[a,b] \subset \mathbb{R}$. Here $\mathfrak{M}_{s,t}(A_l, F_1) = \ln(\mu_{s,t}(A_l, F_1))$, $l \in \{5,6,7,8\}$ are in fact means.

Remark 1.2.10 We observe here that $\left(\frac{\frac{d^n \psi_s}{dx^n}}{\frac{d^n \psi_t}{dx^n}}\right)^{\frac{1}{s-t}}(\ln \xi) = \xi$ is a mean for $\xi \in [a,b]$ where $a,b \in (0,\infty)$.

1.2 Linear Inequalities and the Taylor Formula with New Green Functions

Example 1.2.3 Let $F_2 = \{\varphi_t : [a,b] \subset (0,\infty) \to \mathbb{R} : t \in \mathbb{R}\}$ be a family of functions defined as

$$\varphi_t(x) = \begin{cases} \frac{x^t}{t(t-1)\cdots(t-n+1)}, & t \notin \{0,\ldots,n-1\}; \\ \frac{(x)^j \ln(x)}{(-1)^{n-1-j} j!(n-1-j)!}, & t = j \in \{0,\ldots,n-1\}. \end{cases}$$

Since $\varphi_t(x)$ is n-convex function for $x \in [a,b] \subset (0,\infty)$ and $t \mapsto \frac{d^2}{dx^2}\varphi_t(x)$ is exponentially convex, so by the same arguments given in previous example we conclude that $A_l(\varphi_t)$, $l \in \{5,6,7,8\}$ are exponentially convex.

We assume that $A_l(\varphi_t) > 0$ for $l \in \{5,6,7,8\}$. For this family of convex functions we obtain the following means: for $l \in \{5,6,7,8\}$, $J = \{0,1,\ldots,n-1\}$

$$\mathfrak{M}_{s,t}(A_l, F_2) = \begin{cases} \left(\frac{A_l(\varphi_s)}{A_l(\varphi_t)}\right)^{\frac{1}{s-t}}, & s \neq t, \\ \exp\left((-1)^{n-1}(n-1)! \frac{A_l(\varphi_0 \varphi_s)}{A_l(\varphi_s)} + \sum_{k=0}^{n-3} \frac{1}{k-t}\right), & s = t \notin J, \\ \exp\left((-1)^{n-1}(n-1)! \frac{A_l(\varphi_0 \varphi_s)}{2 A_l(\varphi_s)} + \sum_{k=0, k \neq t}^{n-1} \frac{1}{k-t}\right), & s = t \in J. \end{cases}$$

Here $\mathfrak{M}_{s,t}(A_l, F_2) = \mu_{s,t}(A_l, F_2)$, $l \in \{5,6,7,8\}$ are in fact means.

Remark 1.2.11 Further, in this choice of family F_2, we have

$$\left(\frac{\frac{d^n \varphi_s}{dx^n}}{\frac{d^n \varphi_t}{dx^n}}\right)^{\frac{1}{s-t}}(\xi) = \xi, \quad \xi \in [a,b], \ s \neq t, \text{ where } a,b \in (0,\infty).$$

So, using Remark 1.2.9 we have an important conclusion that $\mu_{s,t}(A_l, F_2)$ is in fact mean for $l \in \{1,2,3,4\}$.

Example 1.2.4 Let $F_3 = \{\theta_t : [a,b] \subset (0,\infty) \to (0,\infty) : t \in (0,\infty)\}$ be a family of functions defined by

$$\theta_t(x) = \frac{e^{-x\sqrt{t}}}{t^{n/2}}.$$

Since $t \mapsto \frac{d^n}{dx^n}\theta_t(x) = e^{-x\sqrt{t}}$ is exponentially convex for $x \in [a,b] \subset (0,\infty)$, being the Laplace transform of a nonnegative function [81]. So, by same argument given in Example 1.2.2 we conclude that $A_l(\theta_t)$, $l \in \{5,6,7,8\}$ are exponentially convex.

We assume that $A_l(\theta_t) > 0$ for $l \in \{5,6,7,8\}$. For this family of functions we have the following possible cases of $\mu_{s,t}(\Lambda_k, F_3)$: for $l \in \{5,6,7,8\}$

$$\mathfrak{M}_{s,t}(A_l, F_3) = \begin{cases} \left(\frac{A_l(\theta_s)}{A_l(\theta_t)}\right)^{\frac{1}{s-t}}, & s \neq t, \\ \exp\left(-\frac{A_l(id.\theta_s)}{2\sqrt{s} A_l(\theta_s)} - \frac{n}{2s}\right), & s = t. \end{cases}$$

By (1.2.47), $\mathfrak{M}_{s,t}(A_l, F_3) = -(\sqrt{s} + \sqrt{t}) \ln \mu_{s,t}(A_l, F_3)$, $l \in \{5, 6, 7, 8\}$ defines a class of means.

Example 1.2.5 Let $F_4 = \{\phi_t : [a, b] \subset (0, \infty) \to (0, \infty) : t \in (0, \infty)\}$ be a family of functions defined by

$$\phi_t(x) = \begin{cases} \frac{t^{-x}}{(\ln t)^n}, & t \neq 1; \\ \frac{x^n}{n}, & t = 1. \end{cases}$$

Since $\frac{d^n}{dx^n} \phi_t(x) = t^{-x} = e^{-x \ln t} > 0$ for $x \in [a, b] \subset (0, \infty)$, so by same argument given in Example 1.2.2 we conclude that $t \mapsto A_l(\phi_t)$, $l \in \{5, 6, 7, 8\}$ are exponentially convex.

We assume that $A_l(\phi_t) > 0$ for $l \in \{5, 6, 7, 8\}$. For this family of functions we have the following possible cases of $\mu_{s,t}(\Lambda_k, F_4)$: for $l \in \{5, 6, 7, 8\}$

$$\mathfrak{M}_{s,t}(A_l, F_4) = \begin{cases} \left(\frac{A_l(\phi_s)}{A_l(\phi_t)}\right)^{\frac{1}{s-t}}, & s \neq t; \\ \exp\left(-\frac{A_l(id.\phi_s)}{s A_l(\phi_s)} - \frac{n}{s \ln s}\right), & s = t \neq 1; \\ \exp\left(-\frac{1}{(n+1)} \frac{A_l(id.\phi_1)}{A_l(\phi_1)}\right), & s = t = 1. \end{cases}$$

By (1.2.47), $\mathfrak{M}_{s,t}(A_l, F_4) = -L(s, t) \ln \mu_{s,t}, (A_l, F_4)$, $l \in \{5, 6, 7, 8\}$ defines a class of means, where $L(s, t)$ is Logarithmic mean defined as:

$$L(s, t) = \begin{cases} \frac{s-t}{\ln s - \ln t}, & s \neq t; \\ s, & s = t. \end{cases}$$

Remark 1.2.12 Monotonicity of $\mu_{s,t}(A_l, F_j)$ follow form (1.2.49) for $l \in \{5, 6, 7, 8\}$, $j \in \{1, 2, 3, 4\}$.

1.3 Linear Inequalities and Hermite Interpolation with New Green Functions

In this section, we state new general linear identities and inequalities involving n-convex functions using Hermite interpolation polynomials and Green functions. We will proceed in the same manner as done in the last two sections of this chapter. It is worth mentioning that this section is mainly based on results collected from [90].

Here we recall some basic definitions, facts and results from [101]. Let $-\infty < a \leq a_1 < a_2 < \cdots < a_r \leq b < \infty$, $r \geq 2$. The Hermite interpolation of a function $H \in C^{(n)}[a, b]$ is of the form

$$H(x) = P_H(x) + e_H(x)$$

1.3 Linear Inequalities and Hermite Interpolation with New Green Functions

where P_H is the unique polynomial of degree $n-1$, called the Hermite interpolating polynomial of H, satisfying

$$P_H^{(i)}(a_j) = H^{(i)}(a_j), \quad 0 \le i \le k_j,\ 1 \le j \le r,\ \sum_{j=1}^{r} k_j + r = n.$$

The associated error $e_H(x)$ can be represented in terms of the Green function $G_{H,n}(x,s)$ for the multipoint boundary value problem

$$z^{(n)}(x) = 0, \quad z^{(i)}(a_j) = 0, \quad 0 \le i \le k_j, \quad 1 \le j \le r,$$

that is, the following result holds (see also [1]):

Theorem 1.3.1 *Let $H \in C^{(n)}[a,b]$ and let P_H be its Hermite interpolating polynomial. Then*

$$H(x) = P_H(x) + e_H(x)$$

$$= \sum_{j=1}^{r} \sum_{i=0}^{k_j} H_{ij}(x) H^{(i)}(a_j) + \int_a^b G_{H,n}(x,s) H^{(n)}(s)\, ds, \quad (1.3.1)$$

where H_{ij} are the fundamental polynomials of the Hermite basis defined by

$$H_{ij}(x) = \frac{1}{i!} \frac{w(x)}{(x-a_j)^{k_j+1-i}} \sum_{k=0}^{k_j-i} \frac{1}{k!} \frac{d^k}{dx^k}\left(\frac{(x-a_j)^{k_j+1}}{w(x)}\right)\bigg|_{x=a_j} (x-a_j)^k, \quad (1.3.2)$$

where

$$w(x) = \prod_{j=1}^{r} (x-a_j)^{k_j+1} \quad (1.3.3)$$

and $G_{H,n}$ is the Green function defined by

$$G_{H,n}(x,s) = \begin{cases} \displaystyle\sum_{j=1}^{j_1} \sum_{i=0}^{k_j} \frac{(a_j-s)^{n-i-1}}{(n-i-1)!} H_{ij}(x), & s \le x; \\ \displaystyle -\sum_{j=j_1+1}^{r} \sum_{i=0}^{k_j} \frac{(a_j-s)^{n-i-1}}{(n-i-1)!} H_{ij}(x), & s \ge x; \end{cases} \quad (1.3.4)$$

for all $a_{j_1} \le s \le a_{j_1+1}$, $j_1 \in \{0, 1, \ldots, r\}$ $(a_{r+1} = b)$.

The following are some special cases of the Hermite interpolation of functions:

(i) $(m, n-m)$ conditions: $r = 2$, $a_1 = a$, $a_2 = b$, $1 \leq m \leq n-1$, $k_1 = m-1$ and $k_2 = n-m-1$. In this case

$$H(x) = \sum_{i=0}^{m-1} \tau_i(x) H^{(i)}(a) + \sum_{i=0}^{n-m-1} \eta_i(x) H^{(i)}(b) + \int_a^b G_{m,n}(x,s) H^{(n)}(s) ds,$$

where

$$\tau_i(x) = \frac{1}{i!}(x-a)^i \left(\frac{x-b}{a-b}\right)^{n-m} \sum_{k=0}^{m-1-i} \binom{n-m+k-1}{k} \left(\frac{x-a}{b-a}\right)^k, \quad (1.3.5)$$

$$\eta_i(x) = \frac{1}{i!}(x-b)^i \left(\frac{x-a}{b-a}\right)^m \sum_{k=0}^{n-m-1-i} \binom{m+k-1}{k} \left(\frac{x-b}{a-b}\right)^k, \quad (1.3.6)$$

and the Green function $G_{m,n}$ is of the form

$$G_{m,n}(x,s) = \begin{cases} \sum_{j=0}^{m-1} \left[\sum_{p=0}^{m-1-j} \binom{n-m+p-1}{p} \left(\frac{x-a}{b-a}\right)^p \right] \\ \quad \times \frac{(x-a)^j (a-s)^{n-j-1}}{j!(n-j-1)!} \left(\frac{b-x}{b-a}\right)^{n-m}, \quad s \leq x; \\ -\sum_{i=0}^{n-m-1} \left[\sum_{q=0}^{n-m-1-i} \binom{m+q-1}{q} \left(\frac{b-x}{b-a}\right)^q \right] \\ \quad \times \frac{(x-b)^i (b-s)^{n-i-1}}{i!(n-i-1)!} \left(\frac{x-a}{b-a}\right)^m, \quad s \geq x. \end{cases}$$

(ii) Taylor's two-point condition: $m \in \mathbb{N}$, $n = 2m$, $r = 2$, $a_1 = a$, $a_2 = b$ and $k_1 = k_2 = m-1$. In this case

$$H(x) = \sum_{i=0}^{m-1} \sum_{k=0}^{m-i-1} \binom{m+k-1}{k} \left[\frac{(x-a)^i}{i!} \left(\frac{x-b}{a-b}\right)^m \left(\frac{x-a}{b-a}\right)^k H^{(i)}(a) \right.$$

$$\left. + \frac{(x-b)^i}{i!} \left(\frac{x-a}{b-a}\right)^m \left(\frac{x-b}{a-b}\right)^k H^{(i)}(b) \right] + \int_a^b G_{2T,m}(x,s) H^{(2m)}(s) ds,$$

1.3 Linear Inequalities and Hermite Interpolation with New Green Functions

where the Green function $G_{2T,m}$ is of the form

$$G_{2T,m}(x,s) = \frac{(-1)^m}{(2m-1)!}$$

$$\begin{cases} p^m(x,s) \sum_{k=0}^{m-1} \binom{m+k-1}{k}(x-s)^{m-1-k}q^k(x,s), & s \le x; \\ q^m(x,s) \sum_{k=0}^{m-1} \binom{m+k-1}{k}(s-x)^{m-1-k}p^k(x,s), & x \le s; \end{cases}$$

where $p(x,s) = \dfrac{(s-a)(b-x)}{(b-a)}$ and $q(x,s) = p(s,x)$.

In this section, we would state and prove new results involving new Green functions and inequalities of type (1.0.3) and (1.0.7) for n-convex functions by making use of the Hermite interpolation. Similar results can be found in [103].

The following lemma yields the sign of the Green function (1.3.4) on certain intervals (see Lemma 2.3.3, page 75, in [1]).

Lemma 1.3.1 *The Green function $G_{H,n}$ given by (1.3.4) and w given by (1.3.3) satisfy*

$$\frac{G_{H,n}(x,s)}{w(x)} > 0, \quad \text{for } a_1 \le x \le a_r, \, a_1 < s < a_r.$$

1.3.1 Results Obtained by the Hermite Interpolation Polynomial and Green Functions

We start this subsection with our first main discrete identity.

Theorem 1.3.2 *Let all the assumptions of Theorem 1.3.1 be valid with additional conditions $\mathbf{x} \in [a,b]^m$ and $\mathbf{p} \in \mathbb{R}^m$. Then*

$$\sum_{k=1}^{m} p_k H(x_k) = (H(a) - aH'(b))\sum_{k=1}^{m} p_k + H'(b)\sum_{k=1}^{m} p_k x_k$$

$$+ \sum_{j=1}^{r}\sum_{i=0}^{k_j} H^{(i+2)}(a_j) \int_a^b \sum_{k=1}^{m} p_k G_1(x_k,s) H_{ij}(s)\,ds$$

$$+ \int_a^b \int_a^b \sum_{k=1}^{m} p_k G_1(x_k,s) G_{H,n-2}(s,t) H^{(n)}(t)\,dt\,ds, \quad (1.3.7)$$

$$\sum_{k=1}^{m} p_k H(x_k) = (H(b) - bH'(a)) \sum_{k=1}^{m} p_k - H'(a) \sum_{k=1}^{m} p_k x_k$$

$$+ \sum_{j=1}^{r} \sum_{i=0}^{k_j} H^{(i+2)}(a_j) \int_a^b \sum_{k=1}^{m} p_k G_2(x_k, s) H_{ij}(s)\, ds$$

$$+ \int_a^b \int_a^b \sum_{k=1}^{m} p_k G_2(x_k, s) G_{H,n-2}(s, t) H^{(n)}(t)\, dt\, ds, \quad (1.3.8)$$

$$\sum_{k=1}^{m} p_k H(x_k) = (H(b) - bH'(b) + (H'(b) - H'(a))a) \sum_{k=1}^{m} p_k + H'(a) \sum_{k=1}^{m} p_k x_k$$

$$+ \sum_{j=1}^{r} \sum_{i=0}^{k_j} H^{(i+2)}(a_j) \int_a^b \sum_{k=1}^{m} p_k G_3(x_k, s) H_{ij}(s)\, ds$$

$$+ \int_a^b \int_a^b \sum_{k=1}^{m} p_k G_3(x_k, s) G_{H,n-2}(s, t) H^{(n)}(t)\, dt\, ds, \quad (1.3.9)$$

$$\sum_{k=1}^{m} p_k H(x_k) = (H(a) - aH'(a) - (H'(b) - H'(a))b) \sum_{k=1}^{m} p_k + H'(b) \sum_{k=1}^{m} p_k x_k$$

$$+ \sum_{j=1}^{r} \sum_{i=0}^{k_j} H^{(i+2)}(a_j) \int_a^b \sum_{k=1}^{m} p_k G_4(x_k, s) H_{ij}(s)\, ds$$

$$+ \int_a^b \int_a^b \sum_{k=1}^{m} p_k G_4(x_k, s) G_{H,n-2}(s, t) H^{(n)}(t)\, dt\, ds. \quad (1.3.10)$$

Proof Applying identities (1.1.5), (1.1.10), (1.1.11) and (1.1.12) for $f = H$ and $x = x_k$, multiplying it by p_k and summing up over k from 1 to m, we obtain respectively

$$\sum_{k=1}^{m} p_k H(x_k) = (H(a) - aH'(b)) \sum_{k=1}^{m} p_k + H'(b) \sum_{k=1}^{m} p_k x_k$$

$$+ \int_a^b \sum_{k=1}^{m} p_k G_1(x_k, s) H''(s)\, ds, \quad (1.3.11)$$

1.3 Linear Inequalities and Hermite Interpolation with New Green Functions

$$\sum_{k=1}^{m} p_k H(x_k) = (H(b) - bH'(a)) \sum_{k=1}^{m} p_k - H'(a) \sum_{k=1}^{m} p_k x_k$$

$$+ \int_a^b \sum_{k=1}^{m} p_k G_2(x_k, s) H''(s) \, ds, \quad (1.3.12)$$

$$\sum_{k=1}^{m} p_k H(x_k) = (H(b) - bH'(b) + (H'(b) - H'(a))a) \sum_{k=1}^{m} p_k + H'(a) \sum_{k=1}^{m} p_k x_k$$

$$+ \int_a^b \sum_{k=1}^{m} p_k G_3(x_k, s) H''(s) \, ds, \quad (1.3.13)$$

$$\sum_{k=1}^{m} p_k H(x_k) = (H(a) - aH'(a) - (H'(b) - H'(a))b) \sum_{k=1}^{m} p_k + H'(b) \sum_{k=1}^{m} p_k x_k$$

$$+ \int_a^b \sum_{k=1}^{m} p_k G_4(x_k, s) H''(s) \, ds. \quad (1.3.14)$$

By Theorem 1.3.1, $H''(s)$ can be expressed as

$$H''(s) = \sum_{j=1}^{r} \sum_{i=0}^{k_j} H_{ij}(s) H^{(i+2)}(a_j) + \int_a^b G_{H,n-2}(s,t) H^{(n)}(t) \, dt. \quad (1.3.15)$$

Inserting one by one expression from (1.3.15) in (1.3.11)−(1.3.14) we get (1.3.7)−(1.3.10) respectively. □

Under the assumptions of Theorem 1.3.2 here we define some linear functional as follows:

$$A_9^{[a,b]}(m, \mathbf{x}, \mathbf{p}, H) = \sum_{k=1}^{m} p_k H(x_k) - (H(a) - aH'(b)) \sum_{k=1}^{m} p_k + H'(b) \sum_{k=1}^{m} p_k x_k$$

$$- \sum_{j=1}^{r} \sum_{i=0}^{k_j} H^{(i+2)}(a_j) \int_a^b \sum_{k=1}^{m} p_k G_1(x_k, s) H_{ij}(s) \, ds$$

$$= \int_a^b \int_a^b \sum_{k=1}^{m} p_k G_1(x_k, s) G_{H,n-2}(s,t) H^{(n)}(t) \, dt \, ds, \quad (1.3.16)$$

$$A_{10}^{[a,b]}(m, \mathbf{x}, \mathbf{p}, H) = \sum_{k=1}^{m} p_k H(x_k) - (H(b) - bH'(a)) \sum_{k=1}^{m} p_k - H'(a) \sum_{k=1}^{m} p_k x_k$$

$$- \sum_{j=1}^{r} \sum_{i=0}^{k_j} H^{(i+2)}(a_j) \int_a^b \sum_{k=1}^{m} p_k G_2(x_k, s) H_{ij}(s)\, ds$$

$$= \int_a^b \int_a^b \sum_{k=1}^{m} p_k G_2(x_k, s) G_{H,n-2}(s, t) H^{(n)}(t)\, dt\, ds, \tag{1.3.17}$$

$$A_{11}^{[a,b]}(m, \mathbf{x}, \mathbf{p}, H) =$$

$$\sum_{k=1}^{m} p_k H(x_k) - (H(b) - bH'(b) + (H'(b) - H'(a))a) \sum_{k=1}^{m} p_k + H'(a) \sum_{k=1}^{m} p_k x_k$$

$$- \sum_{j=1}^{r} \sum_{i=0}^{k_j} H^{(i+2)}(a_j) \int_a^b \sum_{k=1}^{m} p_k G_3(x_k, s) H_{ij}(s)\, ds$$

$$= \int_a^b \int_a^b \sum_{k=1}^{m} p_k G_3(x_k, s) G_{H,n-2}(s, t) H^{(n)}(t)\, dt\, ds, \tag{1.3.18}$$

$$A_{12}^{[a,b]}(m, \mathbf{x}, \mathbf{p}, H) =$$

$$\sum_{k=1}^{m} p_k H(x_k) - (H(a) - aH'(a) - (H'(b) - H'(a))b) \sum_{k=1}^{m} p_k + H'(b) \sum_{k=1}^{m} p_k x_k$$

$$- \sum_{j=1}^{r} \sum_{i=0}^{k_j} H^{(i+2)}(a_j) \int_a^b \sum_{k=1}^{m} p_k G_4(x_k, s) H_{ij}(s)\, ds$$

$$= \int_a^b \int_a^b \sum_{k=1}^{m} p_k G_4(x_k, s) G_{H,n-2}(s, t) H^{(n)}(t)\, dt\, ds. \tag{1.3.19}$$

Now we would use the identities stated above to prove our next results. Here we also need a remark as well as follows:

Remark 1.3.1 It is known that the Bernstein polynomials B_n defined as in [158]

$$P_n(H, x) = \sum_{i=0}^{n} \binom{n}{i} H(a + ih)(x - a)^i (b - x)^{n-i},$$

1.3 Linear Inequalities and Hermite Interpolation with New Green Functions

(where $h = \frac{b-a}{n}$) converges uniformly to H on $[a, b]$ as $n \to \infty$ provided that H is continuous. Further, if H is m-convex function these polynomials have nonnegative derivatives of order m, i.e., $\frac{d^m}{dx^x} P_n \geq 0$ which can be prove by induction by using the following formula:

$$\frac{d^m}{d^m x} P_n(H, x) = m! \binom{n}{m} \sum_{i=0}^{n-m} \binom{n-m}{i} \Delta_h^m H(a+ih)(x-a)^i (b-x)^{n-m-i},$$

where the finite difference of the function H defined on $[a, b]$, of order m for $m \in \{0, 1, \ldots\}$, is defined by

$$\Delta_h^0 H(x) = H(x), \quad \Delta_h^m H(x) = \Delta_h^{m-1} H(x+h) - \Delta_h^{m-1} H(x),$$

where h is a non-zero real number. It is easy to see that

$$\Delta_h^m H(x) = \sum_{i=0}^{m} (-1)^{m-i} \binom{m}{i} H(x+ih)$$

for $x \in [a, b]$ and $h \in \mathbb{R}$ provided $x + ih \in [a, b]$ for $i \in \{0, \ldots, m\}$.

Theorem 1.3.3 *Let all the assumptions of Theorem 1.3.2 be valid. Further let $H : [a, b] \to \mathbb{R}$ be n-convex and for $l \in \{1, 2, 3, 4\}$*

$$\sum_{k=1}^{m} p_k G_l(x_k, s) \geq 0 \quad \text{for all } s \in [a, b], \tag{1.3.20}$$

and consider the inequality

$$A_{l+8}^{[a,b]}(m, \mathbf{x}, \mathbf{p}, H) \geq 0, \tag{1.3.21}$$

where functionals A_l for $l \in \{1, 2, 3, 4\}$ are defined in (1.3.16)–(1.3.19) respectively.

(i) *If k_j for $j \in \{2, \ldots, r\}$ are odd, then inequality (1.3.21) holds for each $l \in \{1, 2, 3, 4\}$.*
(ii) *If k_j for $j \in \{2, \ldots, r-1\}$ are odd and k_r is even, then the reverse inequality hold in (1.3.21) for each $l \in \{1, 2, 3, 4\}$.*

Proof Fix $l \in \{1, 2, 3, 4\}$. (i) First we consider the case if $H \in C^{(n)}[a, b]$, then by given assumptions we have w satisfies $w(x) \geq 0$ for all x where w is defined in (1.3.3) and hence, by Lemma 1.3.1, $G_{H,n-2}(s, t) \geq 0$ for all $s, t \in [a, b]$. Therefore, the last terms on the right hand side of (1.3.7)−(1.3.10) are nonnegative, so inequality (1.3.21) holds. By Remark 1.3.1 the inequality for general H follows since every n-convex function can be obtained, by making use of the Bernstein polynomials, as a uniform limit of n-convex functions with a continuous n-th derivative.

(ii) Under these assumptions $w(x) \leq 0$, so $G_{H,n-2}(s,t) \leq 0$. The rest of the proof is same as in (i). □

In the case of the $(m, n-m)$ conditions we have the following corollary.

Corollary 1.3.4 *Let τ_i and η_i be given by* (1.3.5) *and* (1.3.6) *and let* $\mathbf{x} \in [a,b]^m$ *and* $\mathbf{p} \in \mathbb{R}^m$ *be such that* (1.3.20) *holds. Let* $H : [a,b] \to \mathbb{R}$ *be n-convex and consider the inequalities*

$$\sum_{k=1}^{m} p_k H(x_k) \geq (H(a) - aH'(b)) \sum_{k=1}^{m} p_k + H'(b) \sum_{k=1}^{m} p_k x_k$$

$$+ \int_a^b \left(\sum_{k=1}^{m} p_k G_1(x_k, s) \right) \left(\sum_{i=0}^{j-1} \tau_i(s) H^{(i+2)}(a) \right.$$

$$\left. + \sum_{i=0}^{n-j-1} \eta_i(s) H^{(i+2)}(b) \right) ds, \qquad (1.3.22)$$

$$\sum_{k=1}^{m} p_k H(x_k) \geq (H(b) - bH'(a)) \sum_{k=1}^{m} p_k - H'(a) \sum_{k=1}^{m} p_k x_k$$

$$+ \int_a^b \left(\sum_{k=1}^{m} p_k G_2(x_k, s) \right) \left(\sum_{i=0}^{j-1} \tau_i(s) H^{(i+2)}(a) \right.$$

$$\left. + \sum_{i=0}^{n-j-1} \eta_i(s) H^{(i+2)}(b) \right) ds, \qquad (1.3.23)$$

$$\sum_{k=1}^{m} p_k H(x_k) \geq (H(b) - bH'(b) + (H'(b) - H'(a))a) \sum_{k=1}^{m} p_k + H'(a) \sum_{k=1}^{m} p_k x_k$$

$$+ \int_a^b \left(\sum_{k=1}^{m} p_k G_3(x_k, s) \right) \left(\sum_{i=0}^{j-1} \tau_i(s) H^{(i+2)}(a) \right.$$

$$\left. + \sum_{i=0}^{n-j-1} \eta_i(s) H^{(i+2)}(b) \right) ds, \qquad (1.3.24)$$

1.3 Linear Inequalities and Hermite Interpolation with New Green Functions

$$\sum_{k=1}^{m} p_k H(x_k) \geq (H(a) - aH'(a) - (H'(b) - H'(a))b) \sum_{k=1}^{m} p_k + H'(b) \sum_{k=1}^{m} p_k x_k$$

$$+ \int_a^b \left(\sum_{k=1}^{m} p_k G_4(x_k, s) \right) \left(\sum_{i=0}^{j-1} \tau_i(s) H^{(i+2)}(a) \right.$$

$$\left. + \sum_{i=0}^{n-j-1} \eta_i(s) H^{(i+2)}(b) \right) ds. \qquad (1.3.25)$$

(i) If $n - j$ is even, then (1.3.22)–(1.3.25) hold.
(ii) If $n - j$ is odd, then the reverse of (1.3.22)–(1.3.25) hold.

In the case of Taylor's two point conditions we have the following corollary.

Corollary 1.3.5 *Let* $\mathbf{x} \in [a, b]^m$ *and* $\mathbf{p} \in \mathbb{R}^m$ *be such that (1.3.20) holds. Let* $H : [a, b] \to \mathbb{R}$ *be n-convex and consider the inequalities*

$$\sum_{k=1}^{m} p_k H(x_k) \geq (H(a) - aH'(b)) \sum_{k=1}^{m} p_k + H'(b) \sum_{k=1}^{m} p_k x_k$$

$$+ \int_a^b \left(\sum_{k=1}^{m} p_k G_{1s}(x_k, s) \right) \left(\sum_{i=0}^{j-1} \sum_{k=0}^{j-i-1} \binom{j+k-1}{k} \right.$$

$$\times \left[\frac{(s-a)^i}{i!} \left(\frac{s-b}{a-b} \right)^j \left(\frac{s-a}{b-a} \right)^k H^{(i+2)}(a) \right.$$

$$\left. \left. + \frac{(s-b)^i}{i!} \left(\frac{s-a}{b-a} \right)^j \left(\frac{s-b}{a-b} \right)^k H^{(i+2)}(b) \right] \right) ds, \quad (1.3.26)$$

$$\sum_{k=1}^{m} p_k H(x_k) \geq (H(b) - bH'(a)) \sum_{k=1}^{m} p_k - H'(a) \sum_{k=1}^{m} p_k x_k$$

$$+ \int_a^b \left(\sum_{k=1}^{m} p_k G_2(x_k, s) \right) \left(\sum_{i=0}^{j-1} \sum_{k=0}^{j-i-1} \binom{j+k-1}{k} \right.$$

$$\times \left[\frac{(s-a)^i}{i!} \left(\frac{s-b}{a-b} \right)^j \left(\frac{s-a}{b-a} \right)^k H^{(i+2)}(a) \right.$$

$$\left. \left. + \frac{(s-b)^i}{i!} \left(\frac{s-a}{b-a} \right)^j \left(\frac{s-b}{a-b} \right)^k H^{(i+2)}(b) \right] \right) ds, \quad (1.3.27)$$

$$\sum_{k=1}^{m} p_k H(x_k) \geq (H(b) - bH'(b) + (H'(b) - H'(a))a) \sum_{k=1}^{m} p_k + H'(a) \sum_{k=1}^{m} p_k x_k$$

$$+ \int_a^b \left(\sum_{k=1}^{m} p_k G_3(x_k, s) \right) \left(\sum_{i=0}^{j-1} \sum_{k=0}^{j-i-1} \binom{j+k-1}{k} \right.$$

$$\times \left[\frac{(s-a)^i}{i!} \left(\frac{s-b}{a-b} \right)^j \left(\frac{s-a}{b-a} \right)^k H^{(i+2)}(a) \right.$$

$$\left. \left. + \frac{(s-b)^i}{i!} \left(\frac{s-a}{b-a} \right)^j \left(\frac{s-b}{a-b} \right)^k H^{(i+2)}(b) \right] \right) ds, \qquad (1.3.28)$$

$$\sum_{k=1}^{m} p_k H(x_k) \geq (H(a) - aH'(a) - (H'(b) - H'(a))b) \sum_{k=1}^{m} p_k + H'(b) \sum_{k=1}^{m} p_k x_k$$

$$+ \int_a^b \left(\sum_{k=1}^{m} p_k G_4(x_k, s) \right) \left(\sum_{i=0}^{j-1} \sum_{k=0}^{j-i-1} \binom{j+k-1}{k} \right.$$

$$\times \left[\frac{(s-a)^i}{i!} \left(\frac{s-b}{a-b} \right)^j \left(\frac{s-a}{b-a} \right)^k H^{(i+2)}(a) \right.$$

$$\left. \left. + \frac{(s-b)^i}{i!} \left(\frac{s-a}{b-a} \right)^j \left(\frac{s-b}{a-b} \right)^k H^{(i+2)}(b) \right] \right) ds. \qquad (1.3.29)$$

(i) If j is even, then (1.3.26)–(1.3.29) hold.
(ii) If j is odd, then the reverse of (1.3.26)–(1.3.29) hold.

Theorem 1.3.6 *Let all the assumptions of Theorem 1.3.1 be valid and let* $\mathbf{x} \in [a,b]^m$ *and* $\mathbf{p} \in \mathbb{R}^m$ *satisfy* (1.0.6). *Further let* $H : [a, b] \to \mathbb{R}$ *be n-convex and consider the inequality for* $l \in \{1, 2, 3, 4\}$

$$\sum_{k=1}^{m} p_k H(x_k) \geq \sum_{j=1}^{r} \sum_{i=0}^{k_j} H^{(i+2)}(a_j) \int_a^b \sum_{k=1}^{m} p_k G_l(x_k, s) H_{ij}(s)\, ds \qquad (1.3.30)$$

and the function

$$F(x) = \sum_{j=1}^{r} \sum_{i=0}^{k_j} H^{(i+2)}(a_j) \int_a^b G_l(x, s) H_{ij}(s)\, ds. \qquad (1.3.31)$$

1.3 Linear Inequalities and Hermite Interpolation with New Green Functions

(i) *If k_j for $j \in \{2, \ldots, r\}$ are odd, then (1.3.30) holds. Furthermore, if the function F is convex, then inequality $\sum_{k=1}^{m} p_k H(x_k) \geq 0$ holds.*

(ii) *If k_j for $j \in \{2, \ldots, r-1\}$ are odd and k_r is even, then the reverse of (1.3.30) holds. Furthermore, if the function F is concave, then inequality $\sum_{k=1}^{m} p_k H(x_k) \leq 0$ holds.*

Proof Fix $l \in \{1, 2, 3, 4\}$. The functions $G_l(x, s)$ are convex in the first variable, so assumption (1.3.20) is satisfied by Remark 1.0.4. Now, the claims of the theorem follow from Theorem 1.3.3. □

Here we state the integral version of all the results stated in the start of this subsection. Since proving techniques are quit similar so we omit the details.

Theorem 1.3.7 *Let all the assumptions of Theorem 1.3.1 be valid with additional conditions $g : [\alpha, \beta] \to [a, b]$, $p : [\alpha, \beta] \to \mathbb{R}$. Then for $l \in \{1, 2, 3, 4\}$*

$$\int_\alpha^\beta p(x) H(g(x))\, dx = (H(a) - aH'(b)) \int_\alpha^\beta p(x)\, dx + H'(b) \int_\alpha^\beta p(x) g(x)\, dx$$

$$+ \sum_{j=1}^{r} \sum_{i=0}^{k_j} H^{(i+2)}(a_j) \int_a^b \left(\int_\alpha^\beta p(x) G_1(g(x), s)\, dx \right) H_{ij}(s)\, ds$$

$$+ \int_a^b \int_a^b \left(\int_\alpha^\beta p(x) G_1(g(x), s)\, dx \right) G_{H, n-2}(s, t) H^{(n)}(t)\, dt\, ds,$$

$$\int_\alpha^\beta p(x) H(g(x))\, dx =$$

$$(H(b) - bH'(a)) \int_\alpha^\beta p(x)\, dx - H'(a) \int_\alpha^\beta p(x) g(x)\, dx$$

$$+ \sum_{j=1}^{r} \sum_{i=0}^{k_j} H^{(i+2)}(a_j) \int_a^b \left(\int_\alpha^\beta p(x) G_2(g(x), s)\, dx \right) H_{ij}(s)\, ds$$

$$+ \int_a^b \int_a^b \left(\int_\alpha^\beta p(x) G_2(g(x), s)\, dx \right) G_{H, n-2}(s, t) H^{(n)}(t)\, dt\, ds,$$

$$\int_\alpha^\beta p(x) H(g(x))\, dx =$$

$$(H(b) - bH'(b) + (H'(b) - H'(a))a) \int_\alpha^\beta p(x)\, dx + H'(a) \int_\alpha^\beta p(x) g(x)\, dx$$

$$+ \sum_{j=1}^{r} \sum_{i=0}^{k_j} H^{(i+2)}(a_j) \int_a^b \left(\int_\alpha^\beta p(x) G_3(g(x), s) \, dx \right) H_{ij}(s) \, ds$$

$$+ \int_a^b \int_a^b \left(\int_\alpha^\beta p(x) G_3(g(x), s) \, dx \right) G_{H,n-2}(s, t) H^{(n)}(t) \, dt \, ds,$$

$$\int_\alpha^\beta p(x) H(g(x)) \, dx =$$

$$(H(a) - aH'(a) - (H'(b) - H'(a))b) + \int_\alpha^\beta p(x) \, dx + H'(b) \int_\alpha^\beta p(x) g(x) \, dx$$

$$+ \sum_{j=1}^{r} \sum_{i=0}^{k_j} H^{(i+2)}(a_j) \int_a^b \left(\int_\alpha^\beta p(x) G_4(g(x), s) \, dx \right) H_{ij}(s) \, ds$$

$$+ \int_a^b \int_a^b \left(\int_\alpha^\beta p(x) G_4(g(x), s) \, dx \right) G_{H,n-2}(s, t) H^{(n)}(t) \, dt \, ds.$$

Under the assumptions of Theorem 1.3.7 here we introduce some further linear functional as follows:

$$A_{13}^{[a,b]}([\alpha, \beta], g, p, H) =$$

$$\int_\alpha^\beta p(x) H(g(x)) \, dx - (H(a) - aH'(b)) \int_\alpha^\beta p(x) \, dx - H'(b) \int_\alpha^\beta p(x) g(x) \, dx$$

$$- \sum_{j=1}^{r} \sum_{i=0}^{k_j} H^{(i+2)}(a_j) \int_a^b \left(\int_\alpha^\beta p(x) G_1(g(x), s) \, dx \right) H_{ij}(s) \, ds$$

$$= \int_a^b \int_a^b \left(\int_\alpha^\beta p(x) G_1(g(x), s) \, dx \right) G_{H,n-2}(s, t) H^{(n)}(t) \, dt \, ds, \qquad (1.3.32)$$

$$A_{14}^{[a,b]}([\alpha, \beta], g, p, H) =$$

$$\int_\alpha^\beta p(x) H(g(x)) \, dx - (H(b) - bH'(a)) \int_\alpha^\beta p(x) \, dx + H'(a) \int_\alpha^\beta p(x) g(x) \, dx$$

$$- \sum_{j=1}^{r} \sum_{i=0}^{k_j} H^{(i+2)}(a_j) \int_a^b \left(\int_\alpha^\beta p(x) G_2(g(x), s) \, dx \right) H_{ij}(s) \, ds$$

$$= \int_a^b \int_a^b \left(\int_\alpha^\beta p(x) G_2(g(x), s) \, dx \right) G_{H,n-2}(s, t) H^{(n)}(t) \, dt \, ds. \qquad (1.3.33)$$

$$A_{15}^{[a,b]}([\alpha,\beta],g,p,H) = \int_\alpha^\beta p(x)H(g(x))\,dx - (H(b)-bH'(b))$$

$$+(H'(b)-H'(a))a\int_\alpha^\beta p(x)\,dx - H'(a)\int_\alpha^\beta p(x)g(x)\,dx$$

$$+\sum_{j=1}^r \sum_{i=0}^{k_j} H^{(i+2)}(a_j)\int_a^b \left(\int_\alpha^\beta p(x)G_3(g(x),s)\,dx\right) H_{ij}(s)\,ds$$

$$= \int_a^b \int_a^b \left(\int_\alpha^\beta p(x)G_3(g(x),s)\,dx\right) G_{H,n-2}(s,t)H^{(n)}(t)\,dt\,ds, \quad (1.3.34)$$

$$A_{16}^{[a,b]}([\alpha,\beta],g,p,H) = \int_\alpha^\beta p(x)H(g(x))\,dx$$

$$-(H(a)-aH'(a)-(H'(b)-H'(a))b)$$

$$+\int_\alpha^\beta p(x)\,dx - H'(b)\int_\alpha^\beta p(x)g(x)\,dx$$

$$-\sum_{j=1}^r \sum_{i=0}^{k_j} H^{(i+2)}(a_j)\int_a^b \left(\int_\alpha^\beta p(x)G_4(g(x),s)\,dx\right) H_{ij}(s)\,ds$$

$$= \int_a^b \int_a^b \left(\int_\alpha^\beta p(x)G_4(g(x),s)\,dx\right) G_{H,n-2}(s,t)H^{(n)}(t)\,dt\,ds. \quad (1.3.35)$$

Theorem 1.3.8 *Let all the assumptions of Theorem 1.3.7 be valid. Further let $H: [a,b] \to \mathbb{R}$ be n-convex and for $l \in \{1,2,3,4\}$*

$$\int_\alpha^\beta p(x)G_l(g(x),s)\,dx \geq 0 \quad \text{for all } s \in [a,b], \quad (1.3.36)$$

and consider the inequality

$$A_{l+12}^{[a,b]}([\alpha,\beta],g,p,H) \geq 0 \quad (1.3.37)$$

where functionals A_{l+12} for $l \in \{1,2,3,4\}$ are defined in (1.3.32)–(1.3.35) respectively.

(i) *If k_j for $j \in \{2,\ldots,r\}$ are odd, then inequality (1.3.37) holds for each $l \in \{1,2,3,4\}$.*

(ii) *If k_j for $j \in \{2,\ldots,r-1\}$ are odd and k_r is even, then the reverse inequality holds in (1.3.37) for each $l \in \{1,2,3,4\}$.*

Corollary 1.3.9 *Let τ_i and η_i be given by (1.3.5) and (1.3.6) and let $g : [\alpha, \beta] \to \mathbb{R}$, $p : [\alpha, \beta] \to \mathbb{R}$ be such that (1.3.36) holds. Let $H : [a, b] \to \mathbb{R}$ be n-convex and consider the inequalities*

$$\int_\alpha^\beta p(x)H(g(x))\,dx \geq$$

$$(H(a) - aH'(b))\int_\alpha^\beta p(x)\,dx + H'(b)\int_\alpha^\beta p(x)g(x)\,dx$$

$$+ \int_a^b \left(\int_\alpha^\beta p(x)G_1(g(x), s)\,dx\right)$$

$$\times \left(\sum_{i=0}^{j-1} \tau_i(s)H^{(i+2)}(a) + \sum_{i=0}^{n-j-1} \eta_i(s)H^{(i+2)}(b)\right) ds, \qquad (1.3.38)$$

$$\int_\alpha^\beta p(x)H(g(x))\,dx \geq$$

$$(H(b) - bH'(a))\int_\alpha^\beta p(x)\,dx - H'(a)\int_\alpha^\beta p(x)g(x)\,dx$$

$$+ \int_a^b \left(\int_\alpha^\beta p(x)G_2(g(x), s)\,dx\right)$$

$$\times \left(\sum_{i=0}^{j-1} \tau_i(s)H^{(i+2)}(a) + \sum_{i=0}^{n-j-1} \eta_i(s)H^{(i+2)}(b)\right) ds, \qquad (1.3.39)$$

$$\int_\alpha^\beta p(x)H(g(x))\,dx \geq$$

$$(H(b) - bH'(b) + (H'(b) - H'(a))a)\int_\alpha^\beta p(x)\,dx + H'(a)\int_\alpha^\beta p(x)g(x)\,dx$$

$$+ \int_a^b \left(\int_\alpha^\beta p(x)G_3(g(x), s)\,dx\right)$$

$$\times \left(\sum_{i=0}^{j-1} \tau_i(s)H^{(i+2)}(a) + \sum_{i=0}^{n-j-1} \eta_i(s)H^{(i+2)}(b)\right) ds, \qquad (1.3.40)$$

1.3 Linear Inequalities and Hermite Interpolation with New Green Functions

$$\int_\alpha^\beta p(x)H(g(x))\,dx \geq$$

$$(H(a) - aH'(a) - (H'(b) - H'(a))b)\int_\alpha^\beta p(x)\,dx + H'(b)\int_\alpha^\beta p(x)g(x)\,dx$$

$$+ \int_a^b \left(\int_\alpha^\beta p(x)G_4(g(x),s)\,dx\right)$$

$$\times \left(\sum_{i=0}^{j-1} \tau_i(s)H^{(i+2)}(a) + \sum_{i=0}^{n-j-1} \eta_i(s)H^{(i+2)}(b)\right) ds. \tag{1.3.41}$$

(i) If $n - j$ is even, then (1.3.38)–(1.3.41) hold.
(ii) If $n - j$ is odd, then the reverse of (1.3.38)–(1.3.41) hold.

Corollary 1.3.10 $g : [\alpha, \beta] \to \mathbb{R}$, $p : [\alpha, \beta] \to \mathbb{R}$ be such that (1.3.36) holds. Let $H : [a, b] \to \mathbb{R}$ be n-convex and consider the inequalities

$$\int_\alpha^\beta p(x)H(g(x))\,dx \geq (H(a) - aH'(b))\int_\alpha^\beta p(x)\,dx + H'(b)\int_\alpha^\beta p(x)g(x)\,dx$$

$$+ \int_a^b \left(\int_\alpha^\beta p(x)G_1(g(x),s)\,dx\right)\left(\sum_{i=0}^{j-1}\sum_{k=0}^{j-i-1}\binom{j+k-1}{k}\right)$$

$$\times \left[\frac{(s-a)^i}{i!}\left(\frac{s-b}{a-b}\right)^j\left(\frac{s-a}{b-a}\right)^k H^{(i+2)}(a)\right.$$

$$\left.+ \frac{(s-b)^i}{i!}\left(\frac{s-a}{b-a}\right)^j\left(\frac{s-b}{a-b}\right)^k H^{(i+2)}(b)\right]\right) ds, \tag{1.3.42}$$

$$\int_\alpha^\beta p(x)H(g(x))\,dx \geq (H(b) - bH'(a))\int_\alpha^\beta p(x)\,dx - H'(a)\int_\alpha^\beta p(x)g(x)\,dx$$

$$+ \int_a^b \left(\int_\alpha^\beta p(x)G_2(g(x),s)\,dx\right)\left(\sum_{i=0}^{j-1}\sum_{k=0}^{j-i-1}\binom{j+k-1}{k}\right)$$

$$\times \left[\frac{(s-a)^i}{i!}\left(\frac{s-b}{a-b}\right)^j\left(\frac{s-a}{b-a}\right)^k H^{(i+2)}(a)\right.$$

$$\left.+ \frac{(s-b)^i}{i!}\left(\frac{s-a}{b-a}\right)^j\left(\frac{s-b}{a-b}\right)^k H^{(i+2)}(b)\right]\right) ds, \tag{1.3.43}$$

$$\int_\alpha^\beta p(x)H(g(x))\,dx \geq (H(b) - bH'(b) + (H'(b) - H'(a))a) \int_\alpha^\beta p(x)\,dx$$

$$+ H'(a) \int_\alpha^\beta p(x)g(x)\,dx + \int_a^b \left(\int_\alpha^\beta p(x)G_3(g(x), s)\,dx \right)$$

$$\times \left(\sum_{i=0}^{j-1} \sum_{k=0}^{j-i-1} \binom{j+k-1}{k} \left[\frac{(s-a)^i}{i!} \left(\frac{s-b}{a-b}\right)^j \left(\frac{s-a}{b-a}\right)^k H^{(i+2)}(a) \right. \right.$$

$$\left. \left. + \frac{(s-b)^i}{i!} \left(\frac{s-a}{b-a}\right)^j \left(\frac{s-b}{a-b}\right)^k H^{(i+2)}(b) \right] \right) ds, \tag{1.3.44}$$

$$\int_\alpha^\beta p(x)H(g(x))\,dx \geq (H(a) - aH'(a) - (H'(b) - H'(a))b) \int_\alpha^\beta p(x)\,dx$$

$$+ H'(b) \int_\alpha^\beta p(x)g(x)\,dx + \int_a^b \left(\int_\alpha^\beta p(x)G_4(g(x), s)\,dx \right)$$

$$\times \left(\sum_{i=0}^{j-1} \sum_{k=0}^{j-i-1} \binom{j+k-1}{k} \left[\frac{(s-a)^i}{i!} \left(\frac{s-b}{a-b}\right)^j \left(\frac{s-a}{b-a}\right)^k H^{(i+2)}(a) \right. \right.$$

$$\left. \left. + \frac{(s-b)^i}{i!} \left(\frac{s-a}{b-a}\right)^j \left(\frac{s-b}{a-b}\right)^k H^{(i+2)}(b) \right] \right) ds. \tag{1.3.45}$$

(i) If j is even, then (1.3.42)–(1.3.45) hold.
(ii) If j is odd, then the reverse of (1.3.42)–(1.3.45) hold.

Theorem 1.3.11 *Let all the assumptions of Theorem 1.3.1 be valid and let $g : [\alpha, \beta] \to \mathbb{R}$ and $p : [\alpha, \beta] \to \mathbb{R}$ satisfy (1.0.8). Let $H : [a, b] \to \mathbb{R}$ be n-convex and consider the inequality for $l \in \{1, 2, 3, 4\}$*

$$\int_\alpha^\beta p(x)H(x)\,dx \geq \sum_{j=1}^r \sum_{i=0}^{k_j} H^{(i+2)}(a_j)$$

$$\times \int_a^b \left(\int_\alpha^\beta p(x)G_l(g(x), s)\,dx \right) H_{ij}(s)\,ds \tag{1.3.46}$$

and the function F given by (1.3.31).

(i) *If k_j for $j \in \{2, \ldots, r\}$ are odd, then (1.3.46) holds. Furthermore, if the function F is convex, then inequality $\int_\alpha^\beta p(x)H(g(x))\,dx \geq 0$ holds.*

(ii) If k_j for $j \in \{2,\ldots,r-1\}$ are odd and k_r is even, then the reverse of (1.3.46) holds. Furthermore, if the function F is concave, then inequality $\int_\alpha^\beta p(x)H(g(x))dx \leq 0$ holds.

1.3.2 Inequalities for n-Convex Functions at a Point

Theorem 1.3.12 *Let $\mathbf{x} \in [a,c]^{m_1}$, $\mathbf{p} \in \mathbb{R}^{m_1}$, $\mathbf{y} \in [c,b]^{m_2}$ and $\mathbf{q} \in \mathbb{R}^{m_2}$ be such that for each $l \in \{1,2,3,4\}$*

$$\int_a^c \sum_{k=1}^{m_1} p_k G_l(x_k, s) G_{H,n-2}(s,t)\, ds \geq 0 \quad \text{for all } t \in [a,c], \tag{1.3.47}$$

$$\int_c^b \sum_{k=1}^{m_2} q_k G_l(y_k, s) G_{H,n-2}(s,t)\, ds \geq 0 \quad \text{for all } t \in [c,b], \tag{1.3.48}$$

and

$$A_{l+8}^{[a,b]}(m_1, \mathbf{x}, \mathbf{p}, e_n) = A_{l+8}^{[a,b]}(m_2, \mathbf{y}, \mathbf{q}, e_n) \tag{1.3.49}$$

where G_l are Green functions given by (1.1.6), (1.1.7), (1.1.8) and (1.1.9) respectively, $G_{H,n-2}$ be as defined in (1.3.4), and A_{l+8} be the linear functionals given by (1.3.16) − (1.3.19). If $H : [a,b] \to \mathbb{R}$ is $(n+1)$-convex at point c, then

$$A_{l+8}^{[a,b]}(m_1, \mathbf{x}, \mathbf{p}, H) \leq A_{l+8}^{[c,b]}(m_2, \mathbf{y}, \mathbf{q}, H). \tag{1.3.50}$$

If the inequalities in (1.3.47) and (1.3.48) are reversed, then (1.3.50) holds with the reversed sign of inequality.

Theorem 1.3.13 *Let $g : [\alpha, \beta] \to [a,c]$, $p : [\alpha, \beta] \to \mathbb{R}$, $h : [\gamma, \delta] \to [c,b]$, $q : [\gamma, \delta] \to \mathbb{R}$ be such that for each $l \in \{1,2,3,4\}$*

$$\int_a^c \int_\alpha^\beta p(x) G_l(g(x), s) G_{H,n-2}(s,t)\, dx\, ds \geq 0 \quad \text{for all } t \in [a,c], \tag{1.3.51}$$

$$\int_c^b \int_\gamma^\delta q(x) G_l(h(x), s) G_{H,n-2}(s,t)\, dx\, ds \geq 0 \quad \text{for all } t \in [c,b] \tag{1.3.52}$$

and

$$A_{l+12}^{[a,c]}([\alpha, \beta], g, p, e_n) = A_{l+12}^{[c,b]}([\gamma, \delta], h, q, e_n), \tag{1.3.53}$$

where G_l are Green functions given by (1.1.6), (1.1.7), (1.1.8) and (1.1.9) respectively, $G_{H,n-2}$ be as defined in (1.3.4) and A_{l+12} be the linear functionals given by (1.3.16)–(1.3.19). If $H : [a, b] \to \mathbb{R}$ is $(n + 1)$-convex at point c, then

$$A_{l+12}^{[a,c]}([\alpha, \beta], g, p, H) \leq A_{l+12}^{[c,b]}([\gamma, \delta], h, q, H). \tag{1.3.54}$$

If the inequalities in (1.3.51) and (1.3.52) are reversed, then (1.3.54) holds with the reversed sign of inequality.

1.3.3 Bounds for Remainders and Functionals

Consider the linear functionals $A_9 - A_{16}$ as define in (1.3.16) − (1.3.19) and (1.3.32) − (1.3.35) respectively and $\Omega_9 - \Omega_{12}$ are defined as respectively for $l \in \{9, 10, 11, 12\}$

$$\Omega_l^{[a,b]}(m, \mathbf{x}, \mathbf{p}, t) = \int_a^b \sum_{k=1}^m p_k G_{l-8}(x_k, s) G_{H,n-2}(s, t) ds \geq 0 \quad \text{for all } t \in [a, b], \tag{1.3.55}$$

and $\Omega_{13} - \Omega_{16}$ are defined as respectively for $l \in \{13, 14, 15, 16\}$

$$\Omega_l^{[a,b]}([\alpha, \beta], g, p, t) = \int_a^b \int_\alpha^\beta p(x) G_{l-12}(g(x), s) G_{H,n-2}(s, t) ds\, dx$$

$$\geq 0 \,\forall\, t \in [a, b]. \tag{1.3.56}$$

For the sake of brevity we consider $A_l^{[\cdot,\cdot]}(\cdot, , \cdot, \cdot, H) = A_l(H)$ and $\Omega_l^{[\cdot,\cdot]}(\cdot, \cdot, \cdot, t) = \Omega_l(t)$ We state our next result.

Theorem 1.3.14 *Let* $l \in \{9, \ldots, 16\}$. *Let* $H \in C^{(n)}[a, b]$ *such that for real numbers* γ_1 *and* γ_2 *we have*

$$\gamma_1 \leq H^{(n)}(\eta) \leq \gamma_2 \quad \text{for } \eta \in [a, b].$$

Then in representation

$$A_l(f) = \frac{\left[H^{n-1}(b) - H^{n-1}(a)\right]}{b - a} \int_a^b \Omega_l(x) dx + (b - a) R_n^l, \tag{1.3.57}$$

remainder $R_4^l(f; a, b, n)$ *satisfies estimation*

$$|R_4^l(f; a, b, n)| \leq \frac{1}{2}(\gamma_2 - \gamma_1)\sqrt{T(\Omega_l, \Omega_l)}. \tag{1.3.58}$$

Now we give some Ostrowski-type inequalities related to the generalized linear inequalities.

Theorem 1.3.15 *Let $A_l(F) = A_l^{[\cdot,\cdot]}(\cdot,\cdot,\cdot,F)$ for $(l \in \{9,\ldots,16\})$ as defined in (1.3.16)–(1.3.19) and (1.3.32)–(1.3.35) and $\Omega_l(t) = \Omega_l^{[\cdot,\cdot]}(\cdot,\cdot,\cdot,t)$ for $l \in \{9,\ldots,16\}$ be as defined in (1.3.55) and (1.3.56). Furthermore, let (q,r) be a pair of conjugate exponents, i.e., $1 \leq q,r \leq \infty$, $\frac{1}{q} + \frac{1}{r} = 1$. Let $H^{(n)} \in L_q[a,b]$ for $n \geq 1$. Then we have for $l \in \{9,\ldots,16\}$*

$$|A_l(H)| \leq \|H^{(n)}\|_q \|\Omega_l\|_r . \qquad (1.3.59)$$

The constant on right hand side of (1.3.59) is sharp for $1 < q \leq \infty$ and the best possible for $q = 1$.

Remark 1.3.2 For idea of the proof kindly seek help from previous sections or see [98].

Analogous to the mean value theorems stated in last two sections, similar results for this section as well can be drived. Further, We can construct linear functionals by taking differences of the left and right hand sides of the inequalities from Theorems 1.3.3 and 1.3.8. By using similar methods as in [98, 99] can be constructed new families of exponentially convex functions and Cauchy-type means, furthermore by using some known properties of exponentially convex functions, we can derive new inequalities and prove monotonicity of the obtained Cauchy-type means analogously as in [98, 99].

1.4 Linear Inequalities and the Fink Identity with New Green Functions

In this section we consider positivity of sum $\sum_{i=1}^{n} p_i f(x_i)$ involving convex functions of higher order. Analogous for integral $\int_a^b p(x) f(g(x)) dx$ is also given. Representation of a function f via the Fink identity and the Green's function leads us to identities for which we obtain conditions for positivity of the mentioned sum and integral. We obtain bound for integral remainders which occur in those identities. Some of the results in this section are given without proof. If someone is interested in idea of the proof, previous section can be revisited or see [92].

In [103] we proved various results related to general linear inequalities via Fink identity with and without Green function (see also [101]). Recently, in [29] authors have introduced new Green type functions. Our main objective of present section is to further extend results of [103] using new definitions stated in [29].

Now we recall the Fink identity to prove many useful results. The following result was proved by Fink in [60].

Theorem 1.4.1 *Let $a, b \in \mathbb{R}$, $f : [a, b] \to \mathbb{R}$, $n \geq 1$ and $f^{(n-1)}$ is absolutely continuous on $[a, b]$. Then*

$$f(x) = \frac{n}{b-a} \int_a^b f(t)\,dt - \sum_{k=1}^{n-1} \frac{n-k}{k!}$$

$$\times \left(\frac{f^{(k-1)}(a)(x-a)^k - f^{(k-1)}(b)(x-b)^k}{b-a} \right)$$

$$+ \frac{1}{(n-3)!(b-a)} \int_a^b (x-t)^{n-1} K(t,x) f^{(n)}(t)\,dt, \qquad (1.4.1)$$

where $K(t, x)$ is as defined in (1.1.2).

1.4.1 Results Obtained by the Fink identity and New Green functions

In this section we will put forward some discrete and integral identities and the corresponding linear inequalities using new Green functions and applying the Fink identity.

Theorem 1.4.2 *Fix $l \in \{1, 2, 3, 4\}$. Let $f : [a, b] \to \mathbb{R}$ be such that for $n \geq 3$, $f^{(n-1)}$ is absolutely continuous. Let $x_i, y_i \in [a, b]$, $p_i \in \mathbb{R}$ for $i \in \{1, 2, \ldots, m\}$ be such that $\sum_{i=1}^m p_i = 0$ and $\sum_{i=1}^m p_i x_i = 0$ and let $K(t, x)$ be the same as defined in (1.1.2). If G_l are the Green functions as defined in (1.1.6)–(1.1.9), then we have*

$$\sum_{i=1}^m p_i f(x_i) = \sum_{k=0}^{n-3} \left(\frac{n-k-2}{k!(b-a)} \right) \int_a^b \left(\sum_{i=1}^m p_i G_l(x_i, s) \right)$$

$$\times \left(f^{(k+1)}(b)(s-b)^k - f^{(k+1)}(a)(s-a)^k \right) ds + \frac{1}{(n-3)!(b-a)}$$

$$\times \int_a^b f^{(n)}(t) \left(\int_a^b \sum_{i=1}^m p_i G_l(x_i, s)(s-t)^{n-3} K(t,s)\,ds \right) dt. \qquad (1.4.2)$$

The following theorem is the integral version of Theorem 1.4.2.

Theorem 1.4.3 *Fix $l \in \{1, 2, 3, 4\}$. Let $f : [a, b] \to \mathbb{R}$ be such that for $n \geq 3$, $f^{(n-1)}$ is absolutely continuous on $[a, b]$ and let $p : [\alpha, \beta] \to \mathbb{R}$ and $g : [\alpha, \beta] \to [a, b]$ be integrable functions such that $\int_\alpha^\beta p(x)\,dx = 0$ and $\int_\alpha^\beta p(x)g(x)\,dx = 0$.*

1.4 Linear Inequalities and the Fink Identity with New Green Functions

Let $K(t, x)$ be the same as defined in (1.1.2). If G_l are the Green functions as defined in (1.1.6)–(1.1.9), then we have

$$\int_\alpha^\beta p(x) f(g(x)) dx = \sum_{k=0}^{n-3} \frac{n-k-2}{k!(b-a)} \int_a^b \left(\int_\alpha^\beta p(x) G_l(g(x), s) dx \right)$$

$$\times \left(f^{(k+1)}(b)(s-b)^k - f^{(k+1)}(a)(s-a)^k \right) ds$$

$$+ \frac{1}{(n-3)!(b-a)} \int_a^b f^{(n)}(t)$$

$$\times \left(\int_a^b \left(\int_\alpha^\beta p(x) G_l(g(x), s) dx \right) (s-t)^{n-3} K(t, s) ds \right) dt. \quad (1.4.3)$$

Here we introduce some notations here which will be used in rest of the section:

$$\Omega_{17}^{[a,b]}(m, \mathbf{x}, \mathbf{p}, t) = \int_a^b \sum_{i=1}^m p_i G_l(x_i, s)(s-t)^{n-3} K(t, s) ds, \quad (1.4.4)$$

$$\Omega_{18}^{[a,b]}([\alpha, \beta], g, p, t) = \int_a^b \int_\alpha^\beta p(x) G_l(g(x), s) dx (s-t)^{n-3} K(t, s) ds, \quad (1.4.5)$$

$$A_{17}^{[a,b]}(m, \mathbf{x}, \mathbf{p}, f) = \sum_{i=1}^m p_i f(x_i) - \sum_{k=0}^{n-3} \left(\frac{n-k-2}{k!(b-a)} \right) \int_a^b \sum_{i=1}^m p_i G_l(x_i, s)$$

$$\times \left(f^{(k+1)}(b)(s-b)^k - f^{(k+1)}(a)(s-a)^k \right) ds, \quad (1.4.6)$$

$$A_{18}^{[a,b]}([\alpha, \beta], g, p, f) = \int_\alpha^\beta p(x) f(g(x)) dx$$

$$- \sum_{k=0}^{n-3} \left(\frac{n-k-2}{k!(b-a)} \right) \int_a^b \left(\int_\alpha^\beta p(x) G_l(g(x), s) dx \right)$$

$$\times \left(f^{(k+1)}(b)(s-b)^k - f^{(k+1)}(a)(s-a)^k \right) ds. \quad (1.4.7)$$

The following theorem is our second main result of this section:

Theorem 1.4.4 *Let all the assumptions of Theorem 1.4.2 be satisfied and let for $n \geq 3$, the inequality*

$$\Omega_{17}^{[a,b]}(m, \mathbf{x}, \mathbf{p}, t) \geq 0 \quad (1.4.8)$$

holds. If f is n-convex, then we have

$$A_{17}^{[a,b]}(m, \mathbf{x}, \mathbf{p}, f) \geq 0. \tag{1.4.9}$$

If opposite inequality holds in (1.4.8), *then* (1.4.9) *holds in the reverse direction.*

Proof Since $f^{(n-1)}$ is absolutely continuous on $[a, b]$, $f^{(n)}$ exists almost everywhere. As f is n-convex, applying definition, we have, $f^{(n)}(x) \geq 0$ for all $x \in [a, b]$. Now by using $f^{(n)} \geq 0$ and (1.4.8) in (1.4.2), we have (1.4.9). □

Corollary 1.4.1 *Let all the assumptions of Theorem 1.4.2 be satisfied. In addition we let*

$$\sum_{i=1}^{m} p_i (x_i - x_k)_+ \geq 0 \quad \text{for} \quad k \in \{1, 2, \ldots, m\}.$$

Let n be even and n > 3. If the function $f : [a, b] \to \mathbb{R}$ is n-convex, then inequality (1.4.9) *is satisfied, i.e.*

$$\sum_{i=1}^{m} p_i f(x_i) \geq \sum_{k=0}^{n-3} \frac{n-k-2}{k!(b-a)} \int_a^b \sum_{i=1}^{m} p_i G_l(x_i, s)$$
$$\times \left(f^{(k+1)}(b)(s-b)^k - f^{(k+1)}(a)(s-a)^k \right) ds. \tag{1.4.10}$$

Further if $f^{(k+1)}(a) \leq 0$ and $(-1)^k f^{(k+1)}(b) \geq 0$ for $k \in \{0, 1, \ldots, n-3\}$ then $\sum_{i=1}^{m} p_i f(x_i) \geq 0$.

Proof We fix $l \in \{1, 2, 3, 4\}$ and $n > 3$. As $\mathbf{x}$ and $\mathbf{p}$ are real m-tuples such that they satisfy the assumption (1.0.6), by using the convex function $x \mapsto G_l(x, s)$ in (1.0.3), we obtain

$$\sum_{i=1}^{m} p_i G_l(x_i, s) \geq 0. \tag{1.4.11}$$

For $a \leq s \leq t$, it is easy to see that

$$\int_a^t \sum_{i=1}^{m} p_i G_l(x_i, s)(s-t)^{n-3} K(t, s) ds \geq 0 \tag{1.4.12}$$

holds for even n. Now as f is n-convex for even n, by applying Theorem 1.4.4, we get (1.4.10).

1.4 Linear Inequalities and the Fink Identity with New Green Functions

If $a \leq s \leq b$ and $k \in I_{n-3}$, then from assumptions $f^{(k+1)}(a) \leq 0$ and $(-1)^k f^{(k+1)}(b) \geq 0$ we have that

$$f^{(k+1)}(b)(s-b)^k - f^{(k+1)}(a)(s-a)^k \geq 0. \tag{1.4.13}$$

So, from inequalities (1.4.10), (1.4.11) and (1.4.13) the non-negativity of the right hand side of (1.4.10) is immediate. $\square$

An integral version of our second main result states that:

Theorem 1.4.6 *Let all the assumptions of Theorem 1.4.3 be satisfied and let for $n \geq 3$, the inequality*

$$\Omega_{18}^{[a,b]}([\alpha,\beta],g,p,t) \geq 0 \tag{1.4.14}$$

holds. If f is n-convex, then we have

$$A_{18}^{[a,b]}([\alpha,\beta],g,p,f) \geq 0. \tag{1.4.15}$$

If opposite inequality holds in (1.4.5), then (1.4.15) holds in the reverse direction.

Corollary 1.4.7 *Let all the assumptions of Theorem 1.4.3 be satisfied. In addition we let*

$$\int_\alpha^\beta p(x)(g(x)-t)_+^{n-1}\, dx \geq 0, \quad \text{for every } t \in [a,b].$$

Let n be even and $n > 3$. If the function $f : [a,b] \to \mathbb{R}$ is n-convex, then we have

$$\int_\alpha^\beta p(x) f(g(x))\, dx \geq \sum_{k=0}^{n-3} \frac{n-k-2}{k!(b-a)} \int_a^b \left(\int_\alpha^\beta p(x) G_l(g(x),s)\, dx \right)$$
$$\times \left(f^{(k+1)}(b)(s-b)^k - f^{(k+1)}(a)(s-a)^k \right) ds. \tag{1.4.16}$$

Further if $f^{(k+1)}(a) \leq 0$ and $(-1)^k f^{(k+1)}(b) \geq 0$ for $k \in \{0,1\ldots,n-3\}$, then the right hand side of (1.4.16) is non-negative.

Proof The proof is analogous to the proof of Corollary 1.4.1 but instead of Theorem 1.4.4, we apply Theorem 1.4.6. $\square$

1.4.2 Inequalities for n-Convex Functions at a Point

In this subsection we shall give related results for the class of n-convex functions at a point.

First we state main results for discrete case.

Theorem 1.4.8 *Let $c \in (a,b)$, $\mathbf{x} \in [a,c]^m$, $\mathbf{y} \in [c,b]^l$, $\mathbf{p} \in \mathbb{R}^m$, $\mathbf{q} \in \mathbb{R}^l$ and $f : [a,b] \to \mathbb{R}$ be a function such that $f^{(n-1)}$ is absolutely continuous. Let $\Omega_{17}^{[\cdot,\cdot]}(\cdot,\cdot,\cdot,t)$ and $A_{17}^{[\cdot,\cdot]}(\cdot,\cdot,\cdot,f)$ be defined as in (1.4.4) and (1.4.6) and satisfy the following conditions:*

$$\Omega_{17}^{[a,c]}(m, \mathbf{x}, \mathbf{p}, t) \geq 0 \quad \text{for every} \quad t \in [a,c], \tag{1.4.17}$$

$$\Omega_{17}^{[c,b]}(l, \mathbf{y}, \mathbf{q}, t) \geq 0 \quad \text{for every} \quad t \in [c,b], \tag{1.4.18}$$

and

$$A_{17}^{[a,c]}(m, \mathbf{x}, \mathbf{p}, e_n) = A_{17}^{[c,b]}(l, \mathbf{y}, \mathbf{q}, e_n). \tag{1.4.19}$$

If f is $(n+1)$-convex at point c, then

$$A_{17}^{[a,c]}(m, \mathbf{x}, \mathbf{p}, f) \leq A_{17}^{[c,b]}(l, \mathbf{y}, \mathbf{q}, f). \tag{1.4.20}$$

If inequalities in (1.4.17) and (1.4.18) are reversed, then (1.4.20) holds with the reverse sign of inequality.

Corollary 1.4.9 *Let $j_1, j_2, n \in \mathbb{N}$, $2 \leq j_1, j_2 \leq n$ and let $f : [a,b] \to \mathbb{R}$ be $(n+1)$-convex at point c. Let m-tuples $\mathbf{x} \in [a,c]^m$ and $\mathbf{p} \in \mathbb{R}^m$ satisfy (1.0.4) and (1.0.5) with n replaced by j_1, let l-tuples $\mathbf{y} \in [c,b]^l$ and $\mathbf{q} \in \mathbb{R}^l$ satisfy*

$$\sum_{i=1}^{l} q_i y_i^k = 0, \quad \text{for all } k \in \{0, 1, \ldots, j_2 - 1\};$$

$$\sum_{i=1}^{l} q_i (y_i - t)_+^{j_2 - 1} \geq 0, \quad \text{for every } t \in [y_{(1)}, y_{(l-n+1)}]$$

and let (1.4.19) holds. If $n - j_1$ and $n - j_2$ are even, then (1.4.20) holds.

Remark 1.4.1 For idea of the proof see [101, pp. 171–172].

Integral analogous of previous theorem may be stated as:

Theorem 1.4.10 *Let $c \in (a,b)$ and let $g : [\alpha, \beta] \to [a,c]$, $p : [\alpha, \beta] \to \mathbb{R}$, $h : [\gamma, \delta] \to [c,b]$, $q : [\gamma, \delta] \to \mathbb{R}$ be integrable functions. Let $f : I \to \mathbb{R}$, $[a,b] \subset$*

1.4 Linear Inequalities and the Fink Identity with New Green Functions

I be a function such that $f^{(n-1)}$ is absolutely continuous. Let $\Omega_{18}^{[\cdot,\cdot]}(\cdot,\cdot,\cdot,t)$ and $A_{18}^{[\cdot,\cdot]}(\cdot,\cdot,\cdot,f)$ be defined as in (1.4.5) and (1.4.7) satisfy the following conditions:

$$\Omega_{18}^{[a,c]}([\alpha,\beta],g,p,t) \geq 0 \quad \text{for every} \quad t \in [a,c], \tag{1.4.21}$$

$$\Omega_{18}^{[c,b]}([\gamma,\delta],h,q,t) \geq 0 \quad \text{for every} \quad t \in [c,b], \tag{1.4.22}$$

and

$$A_{18}^{[a,c]}([\alpha,\beta],g,p,e_n) = A_{18}^{[c,b]}([\gamma,\delta],h,q,e_n). \tag{1.4.23}$$

If f is $(n+1)$-convex at point c (for $n \geq 3$), then

$$A_{18}^{[a,c]}([\alpha,\beta],g,p,f) \leq A_{18}^{[c,b]}([\gamma,\delta],h,q,f). \tag{1.4.24}$$

If inequalities in (1.4.21) and (1.4.22) are reversed, then (1.4.24) holds with the reverse sign of inequality.

Corollary 1.4.11 Let $j_1, j_2, n \in \mathbb{N}$, $2 \leq j_1, j_2 \leq n$ and let $f : [a,b] \to \mathbb{R}$ be $(n+1)$-convex at point c. Let integrable functions $g : [\alpha, \beta] \to [a,c]$, $p : [\alpha, \beta] \to \mathbb{R}$ satisfy (1.0.8) with n replaced by j_1, let $h : [\gamma, \delta] \to [c,b]$, $q : [\gamma, \delta] \to \mathbb{R}$ satisfy

$$\int_\gamma^\delta q(x) h(x)^k \, dx = 0, \quad \text{for all } k \in \{0, 1, \ldots, j_2 - 1\};$$

$$\int_\gamma^\delta q(x) (h(x) - t)_+^{j_2 - 1} \, dx \geq 0, \quad \text{for every } t \in [c,b],$$

and let (1.4.23) holds. If $n - j_1$ and $n - j_2$ are even, then (1.4.20) holds.

1.4.3 Bounds for Remainders and Functionals

By using same techniques as used in previous sections we may also proof following results.

Remark 1.4.2 For the sake of brevity, in present and next sections at some places we will use the notations $A_l(f) = A_l^{[\cdot,\cdot]}(\cdot,\cdot,\cdot,f)$ and $\Omega_l(t) = \Omega_l^{[\cdot,\cdot]}(\cdot,\cdot,\cdot,t)$ for $l \in \{17, 18\}$ as defined in Theorems 1.4.4 and 1.4.6.

Now, we are ready to state main results of this section:

Theorem 1.4.12 Let $f : [a, b] \to \mathbb{R}$ be such that $f^{(n)}$ is an absolutely continuous function for $n \in \mathbb{N}$ with $(\cdot - a)(b - \cdot)[f^{(n+1)}]^2 \in L[a, b]$. Then it holds for $l \in \{17, 18\}$

$$A_l(f) = \frac{\left[f^{(n-1)}(b) - f^{(n-1)}(a)\right]}{(n-3)!(b-a)} \int_a^b \Omega_l(s)ds + R_5^l(f; a, b, n),$$

where the remainder $R_5^l(f; a, b, n)$ satisfies the estimation

$$|R_5^l(f; a, b, n)| \leq \frac{1}{(n-3)!}$$
$$\times \left(\frac{(b-a)}{2}\left|T(\Omega_l, \Omega_l)\int_a^b (s-a)(b-s)[f^{(n+1)}(s)]^2 ds\right|\right)^{1/2}. \quad (1.4.25)$$

By using Theorem 1.2.11 we obtain the following Grüss type inequality.

Theorem 1.4.13 Let $f : [a, b] \to \mathbb{R}$ be such that $f^{(n)}$ is an absolutely continuous function for $n \in \mathbb{N}$ with $(\cdot - a)(b - \cdot)[f^{(n+1)}]^2 \in L[a, b]$ with $f^{(n+1)} \geq 0$ on $[a, b]$. Then we have the representation (1.4.25) and the remainder $R_6^l(f; a, b, n)$ satisfies the following condition for $l \in \{17, 18\}$

$$|R_6^l(f; a, b, n)| \leq \frac{1}{(n-3)!}\|\Omega_l'\|_\infty \left\{\frac{b-a}{2}\left[f^{(n-1)}(b) + f^{(n-1)}(a)\right]\right.$$
$$\left. - \left[f^{(n-2)}(b) - f^{(n-2)}(a)\right]\right\}. \quad (1.4.26)$$

Theorem 1.4.14 For $l = 17$ we assume that **x** and **p** satisfy the assumptions of Theorem 1.4.2 and for $l = 18$ we assume that x and p satisfy the assumptions of Theorem 1.4.3.
(i) Let $l \in \{17, 18\}$. Let $f : I \to \mathbb{R}$, $[a, b] \subset I$, be such that $f^{(n)}$ is an absolutely continuous function and

$$\gamma \leq f^{(n)}(x) \leq \Gamma \quad \text{for} \quad x \in [a, b].$$

Then

$$A_l(f) = \frac{\left[f^{(n-1)}(b) - f^{(n-1)}(a)\right]}{(n-3)!(b-a)} \int_a^b \Omega_l(t)dt + R_7^l(f; a, b, n), \quad (1.4.27)$$

where the remainder $R_7^l(f; a, b, n)$ satisfies the estimation

$$|R_7^l(f; a, b, n)| \leq \frac{b-a}{2(n-3)!}(\Gamma - \gamma)\sqrt{T(\Omega_l, \Omega_l)}. \tag{1.4.28}$$

Theorem 1.4.15

(i) *Fix $l \in \{17, 18\}$. Let (q, r) be a pair of conjugate exponents, that is, $1 \leq q, r \leq \infty$, $\frac{1}{q} + \frac{1}{r} = 1$. Let $f^{(n)} \in L_q[a, b]$ for some $n \in \mathbb{N}$, $n > 1$. Further, for $l = 17$ we assume that $\mathbf{x}$ and $\mathbf{p}$ satisfy the assumptions of Theorem 1.4.2 and for $l = 18$ we assume that x and p satisfy the assumptions of Theorem 1.4.3. Then we have*

$$|A_l(f)| \leq \frac{1}{(n-3)!}\|f^{(n)}\|_q \|\Omega_l\|_r. \tag{1.4.29}$$

The constant on the right hand side of (1.4.29) is sharp for $1 < q \leq \infty$ and the best possible for $q = 1$.

Remark 1.4.3 Using the same method as in [98], we can construct new families of exponentially convex functions and Cauchy type means along with mean value theorems, as stated in previous sections.

1.5 Linear Inequalities and the Abel-Gontscharoff's Interpolation Polynomial

The Abel-Gontscharoff interpolation problem in the real case was introduced in 1935 by Whittaker [185] and subsequently by Gontscharoff [63] and Davis [39].

The Abel-Gontscharoff interpolating polynomial for two points with integral remainder is given in [1]:

Theorem 1.5.1 *Let $n, k \in \mathbb{N}$, $n \geq 2$, $0 \leq k \leq n - 1$ and $f \in C^{(n)}[a, b]$; then we have*

$$f(t) = Q_{n-1}(a, b, f, t) + R(f, t), \tag{1.5.1}$$

where Q_{n-1} is the Abel-Gontscharoff interpolating polynomial for two-points of degree $n - 1$, i.e.,

$$Q_{n-1}(a, b, f, t) = \sum_{i=0}^{k} \frac{(t-a)^i}{i!} f^{(i)}(a)$$
$$+ \sum_{j=0}^{n-k-2} \sum_{i=0}^{j} \frac{(t-a)^{k+1+i}(a-b)^{j-i}}{(k+1+i)!(j-i)!} f^{(k+1+j)}(b)$$

and the remainder is given by

$$R(f,t) = \int_a^b G_n(t,s) f^{(n)}(s) ds,$$

where $G_n(t,s)$ is the Green function [12, p. 177] given as

$$G_n(t,s) = \frac{1}{(n-1)!} \begin{cases} \sum_{i=0}^{k} \binom{n-1}{i} (t-a)^i (a-s)^{n-i-1}, & a \le s \le t; \\ -\sum_{i=k+1}^{n-1} \binom{n-1}{i} (t-a)^i (a-s)^{n-i-1}, & t \le s \le b. \end{cases}$$

(1.5.2)

Further, for $a \le s, t \le b$ the following inequality hold

$$(-1)^{n-k-1} \frac{\partial^i G_n(t,s)}{\partial t^i} \ge 0, \quad 0 \le i \le k, \tag{1.5.3}$$

$$(-1)^{n-i} \frac{\partial^i G_n(t,s)}{\partial t^i} \ge 0, \quad k+1 \le i \le n-1. \tag{1.5.4}$$

Before we proceed further it should be noted that all these results related to this section can be found in the article [91].

1.5.1 Results Obtained by the Abel–Gontscharoff's Interpolation

We start this section with the identities of generalizations of Popoviciu type inequality using Abel-Gontscharoff interpolating polynomial for two points.

Theorem 1.5.2 *Let $n, k \in \mathbb{N}$, $n \ge 2$, $0 \le k \le n-1$, $\mathbf{x} \in [a,b]^m$ and $\mathbf{p} \in \mathbb{R}^m$. Let $f \in C^{(n)}[a,b]$ and G_n be the Green function defined as in (1.5.2). Then*

$$\sum_{r=1}^{m} p_r f(x_r) = \theta_1(f) + \int_a^b \left(\sum_{r=1}^{m} p_r G_n(x_r,s) \right) f^{(n)}(s) ds, \tag{1.5.5}$$

1.5 Linear Inequalities and the Abel-Gontscharoff's Interpolation Polynomial

where

$$\theta_1(f) = \sum_{i=0}^{k} \frac{f^{(i)}(a)}{i!} \sum_{r=1}^{m} p_r (x_r - a)^i \qquad (1.5.6)$$

$$+ \sum_{j=0}^{n-k-2} \sum_{i=0}^{j} \left(\sum_{r=1}^{m} p_r (x_r - a)^{k+1+i} \right) \frac{(-1)^{j-i} (b-a)^{j-i}}{(k+1+i)!(j-i)!} f^{(k+1+j)}(b).$$

Proof Consider the expression

$$\sum_{r=1}^{m} p_r f(x_r). \qquad (1.5.7)$$

By using Theorem 1.5.1 we have

$$f(t) = \sum_{i=0}^{k} \frac{(t-a)^i}{i!} f^{(i)}(a)$$

$$+ \sum_{j=0}^{n-k-2} \sum_{i=0}^{j} \frac{(t-a)^{k+1+i} (-1)^{j-i} (b-a)^{j-i}}{(k+1+i)!(j-i)!} f^{(k+1+j)}(b)$$

$$+ \int_a^b G_n(t,s) f^{(n)}(s) ds. \qquad (1.5.8)$$

Substituting this value of f in (1.5.7) and some arrangements, we get (1.5.5). □

Integral version of the above theorem can be stated as:

Theorem 1.5.3 *Let $n, k \in \mathbb{N}$, $n \geq 2$, $0 \leq k \leq n-1$, and $x : [\alpha, \beta] \to [a, b]$, $p : [\alpha, \beta] \to \mathbb{R}$ be continuous functions. Let $f \in C^{(n)}[a, b]$ and G_n be the Green function defined as in (1.5.2). Then*

$$\int_\alpha^\beta p(\tau) f(x(\tau)) d\tau = \theta_2(f) + \int_a^b \left(\int_\alpha^\beta p(\tau) G_n(x(\tau), s) d\tau \right) f^{(n)}(s) ds, \qquad (1.5.9)$$

where

$$\theta_2(f) = \sum_{i=0}^{k} \frac{f^{(i)}(a)}{i!} \int_\alpha^\beta p(\tau) (x(\tau) - a)^i d\tau$$

$$+ \sum_{j=0}^{n-k-2} \sum_{i=0}^{j} \left(\int_\alpha^\beta p(\tau) (x(\tau) - a)^{k+1+i} d\tau \right)$$

$$\times \frac{(-1)^{j-i} (b-a)^{j-i}}{(k+1+i)!(j-i)!} f^{(k+1+j)}(b). \qquad (1.5.10)$$

If **x** and **p** satisfy additional conditions, then we get generalization of Popoviciu type inequality for n-convex functions, i. e., we give a lower bound for the sum $\sum p_r f(x_r)$ which depends only on nodes $x_1, \ldots, x_m$, weights $p_1, \ldots, p_m$ and values of higher derivatives of a function f at points a and b.

Theorem 1.5.4 *Let all the assumptions of Theorem* 1.5.2 *be valid. In addition, if for all* $s \in [a, b]$

$$\sum_{r=1}^{m} p_r G_n(x_r, s) \geq 0, \qquad (1.5.11)$$

then for every n-convex function $f : [a, b] \to \mathbb{R}$, *following inequality holds*

$$\sum_{r=1}^{m} p_r f(x_r) \geq \theta_1(f), \qquad (1.5.12)$$

where $\theta_1(f)$ *is given in* (1.5.6).

If the reverse inequality in (1.5.11) *holds, then also the reverse inequality in* (1.5.12) *holds.*

Proof Since the function f is n-convex, therefore without loss of generality we can assume that f is n-times differentiable and $f^{(n)}(x) \geq 0$, for all $x \in [a, b]$. Hence we can apply Theorem 1.5.2 to get (1.5.12). □

Integral version of the above theorem can be stated as:

Theorem 1.5.5 *Let all the assumptions of Theorem* 1.5.3 *be valid. In addition, if for all* $s \in [a, b]$

$$\int_{\alpha}^{\beta} p(\tau) G_n(x(\tau), s) d\tau \geq 0, \qquad (1.5.13)$$

then for every n-convex function $f : [a, b] \to \mathbb{R}$, *it holds*

$$\int_{\alpha}^{\beta} p(\tau) f(x(\tau)) d\tau \geq \theta_2(f), \qquad (1.5.14)$$

where $\theta_2(f)$ *is defined in* (1.5.10).

If the reverse inequality in (1.5.13) *holds, then also the reverse inequality in* (1.5.14) *holds.*

1.5 Linear Inequalities and the Abel-Gontscharoff's Interpolation Polynomial

In some cases the assumption $\sum_{r=1}^{m} p_r G_n(x_r, s) \geq 0$, $s \in [a, b]$ can be replaced with more simpler conditions in which we recognize assumptions from Popoviciu's theorem about positivity of sum $\sum p_r f(x_r)$ for a convex function f. Namely we have the following statement.

Theorem 1.5.6 *Let $n, k \in \mathbb{N}$, $n \geq 2$, $1 \leq k \leq n-1$, $\mathbf{x} \in [a, b]^m$ $\mathbf{p} \in \mathbb{R}^m$ be m-tuples such that condition (1.0.6) and let G_n be the Green function defined as in (1.5.2).*

(i) If k is odd and n is even or k is even and n is odd, then for every n-convex function $f : [a, b] \to \mathbb{R}$, it holds

$$\sum_{r=1}^{m} p_r f(x_r) \geq \theta_1(f), \qquad (1.5.15)$$

where $\theta_1(f)$ is given in (1.5.6).

Moreover, if $f^{(i)}(a) \geq 0$ for $i \in \{2, \ldots, k\}$ and $f^{(k+1+j)}(b) \geq 0$ if $j - i$ is even and $f^{(k+1+j)}(b) \leq 0$ if $j - i$ is odd for $i \in \{0, \ldots, j\}$ and $j \in \{0, \ldots, n-k-2\}$, then $\sum_{r=1}^{m} p_r f(x_r) \geq 0$.

(ii) If k and n both are even or odd, then for every n-convex function $f : [a, b] \to \mathbb{R}$, the reverse inequality in (1.5.15) holds.

Moreover, if $f^{(i)}(a) \leq 0$ for $i \in \{0, \ldots, k\}$ and $f^{(k+1+j)}(b) \leq 0$ if $j - i$ is even, and $f^{(k+1+j)}(b) \geq 0$ if $j - i$ is odd for $i \in \{0, \ldots, j\}$ and $j \in \{0, \ldots, n-k-2\}$, then $\sum_{r=1}^{m} p_r f(x_r) \leq 0$.

Proof (i) Let us consider properties (1.5.3) and (1.5.4) for $i = 2$. If k is odd and n is even, then for $k = 1$ we get $(-1)^{n-2} \frac{\partial^2 G_n(t, s)}{\partial t^2} \geq 0$ from (1.5.4), i.e. $\frac{\partial^2 G_n(t, s)}{\partial t^2} \geq 0$, i. e. G_n is convex. For $k > 1$, from (1.5.3) we get the same inequality. If k is even and n is odd, then $k \geq 2$ and from (1.5.3) we get that G_n is convex in the first variable. By Remark 1.0.4, applied on the function G_n we get

$$\sum_{r=1}^{m} p_r G_n(x_r, s) \geq 0,$$

i.e., the assumptions of Theorem 1.5.4 are fullfilled and inequality (1.5.15) holds. If further assumptions on $f^{(i)}(a)$ and $f^{(k+1+j)}(b)$ are valid, then the right-hand side of (1.5.15) is nonnegative.

The case (ii) is proved in a similar manner. $\square$

An integral analogue of the previous theorem is the following theorem.

Theorem 1.5.7 Let $n, k \in \mathbb{N}$, $n \geq 2$, $1 \leq k \leq n-1$, $x : [\alpha, \beta] \to [a, b]$ and $p : [\alpha, \beta] \to \mathbb{R}$ be continuous functions satisfying

$$\int_\alpha^\beta p(\tau) d\tau = 0, \quad \int_\alpha^\beta p(\tau) x(\tau) d\tau = 0,$$

and

$$\int_\alpha^\beta p(\tau)(x(\tau) - s)_+ d\tau \geq 0 \quad \text{for } s \in [a, b],$$

and let G_n be the Green function defined as in (1.5.2).

(i) If k is odd and n is even or k is even and n is odd, then for every n-convex function $f : [a, b] \to \mathbb{R}$, then

$$\int_\alpha^\beta p(\tau) f(x(\tau)) d\tau \geq \theta_2(f). \tag{1.5.16}$$

Moreover, if $f^{(i)}(a) \geq 0$ for $i \in \{0, \ldots, k\}$ and $f^{(k+1+j)}(b) \geq 0$ if $j - i$ is even and $f^{(k+1+j)}(b) \leq 0$ if $j - i$ is odd for $i \in \{0, \ldots, j\}$ and $j \in \{0, \ldots, n-k-2\}$, then $\int_\alpha^\beta p(t) f(x(t)) dt \geq 0$.

(ii) If k and n both are even or odd, then for every n-convex function $f : [a, b] \to \mathbb{R}$, then the reverse inequality holds in (1.5.16).

Moreover, if $f^{(i)}(a) \leq 0$ for $i \in \{0, \ldots, k\}$ and $f^{(k+1+j)}(b) \leq 0$ if $j - i$ is even, and $f^{(k+1+j)}(b) \geq 0$ if $j - i$ is odd for $i \in \{0, \ldots, j\}$ and $j \in \{0, \ldots, n-k-2\}$, then $\int_\alpha^\beta p(t) f(x(t)) dt \leq 0$.

1.5.2 Results Obtained by the Abel–Gontscharoff's Interpolation Polynomial and Green Functions

Now we obtain results using the Green function G, (1.5.17), together with the Abel-Gontscharoff polynomials. Here it is worth mentioning that we would use G_0 for Green function G defined in (1.5.17).

Now we recall the definition of Green function G which would be used in some of our results. The function $G : [a, b] \times [a, b]$ is defined by

$$G(s, t) = \begin{cases} \dfrac{(s-b)(t-a)}{b-a} & \text{for } a \leq t \leq s; \\ \dfrac{(t-b)(s-a)}{b-a} & \text{for } s \leq t \leq b. \end{cases} \tag{1.5.17}$$

The function G is convex and continuous with respect to both s and t.

1.5 Linear Inequalities and the Abel-Gontscharoff's Interpolation Polynomial

For any function $f : [a, b] \to \mathbb{R}$, $f \in C^{(2)}[a, b]$, we can obtain the following integral identity by simply using integration by parts

$$f(x) = \frac{b-x}{b-a} f(a) + \frac{x-a}{b-a} f(b) + \int_a^b G(x, s) f''(s) ds, \quad (1.5.18)$$

where the function G is defined as above in (1.5.17) (see also [187]).

We begin with some identities related to generalizations of Popoviciu type inequality.

Theorem 1.5.8 Let $n, k \in \mathbb{N}$, $n \geq 4$, $0 \leq k \leq n-1$, $f \in C^{(n)}[a, b]$ and $\mathbf{x} \in [a, b]^m$, $\mathbf{p} \in \mathbb{R}^m$. Also let G and G_n be defined by (1.5.17) and (1.5.2) respectively. Then

$$\sum_{r=1}^m p_r f(x_r) = \theta_3(f, G_0) + \int_a^b \int_a^b \left(\sum_{r=1}^m p_r G(x_r, s) \right) G_{n-2}(s, t) f^{(n)}(t) dt ds,$$

where $\theta_3(f, G_0)$ is defined as

$$\theta_3(f, G_0) = \frac{f(b) - f(a)}{b-a} \sum_{r=1}^m p_r x_r + \frac{bf(a) - af(b)}{b-a} \sum_{r=1}^m p_r$$

$$+ \sum_{i=0}^k \frac{f^{(i+2)}(a)}{i!} \int_a^b \sum_{r=1}^m p_r G(x_r, s) (s-a)^i ds \quad (1.5.19)$$

$$+ \sum_{j=0}^{n-k-4} \sum_{i=0}^j \frac{(-1)^{j-i} (b-a)^{j-i} f^{(k+3+j)}(b)}{(k+1+i)! (j-i)!}$$

$$\times \int_a^b \sum_{r=1}^m p_r G(x_r, s) (s-a)^{k+1+i} ds.$$

In similar manner we can state further results related to other Green functions $G_1 - G_4$ as follows:

Theorem 1.5.9 Let $n, k \in \mathbb{N}$, $n \geq 4$, $0 \leq k \leq n-1$, $f \in C^{(n)}[a, b]$ and $\mathbf{x} \in [a, b]^m$, $\mathbf{p} \in \mathbb{R}^m$. Also let $G_1 - G_4$ and G_n be defined by (1.1.6)–(1.1.9) and (1.5.2) respectively. Then for $l \in \{4, 5, 6, 7\}$

$$\sum_{r=1}^m p_r f(x_r) = \theta_l(f, G_{l-3})$$

$$+ \int_a^b \int_a^b \left(\sum_{r=1}^m p_r G_{l-3}(x_r, s) \right) G_{n-2}(s, t) f^{(n)}(t) dt ds,$$

where

$$\theta_4(f, G_1) = (f(a) - af'(b))\sum_{r=1}^{m} p_r + f'(b)\sum_{r=1}^{m} p_r x_r$$

$$+ \sum_{i=0}^{k} \frac{f^{(i+2)}(a)}{i!} \int_a^b \sum_{r=1}^{m} p_r G_1(x_r, s)(s-a)^i \, ds$$

$$+ \sum_{j=0}^{n-k-4} \sum_{i=0}^{j} \frac{(-1)^{j-i}(b-a)^{j-i} f^{(k+3+j)}(b)}{(k+1+i)!(j-i)!}$$

$$\times \int_a^b \sum_{r=1}^{m} p_r G_1(x_r, s)(s-a)^{k+1+i} \, ds, \qquad (1.5.20)$$

$$\theta_5(f, G_2) = (f(b) - bf'(a))\sum_{r=1}^{m} p_r - f'(a)\sum_{r=1}^{m} p_r x_r$$

$$+ \sum_{i=0}^{k} \frac{f^{(i+2)}(a)}{i!} \int_a^b \sum_{r=1}^{m} p_r G_2(x_r, s)(s-a)^i \, ds$$

$$+ \sum_{j=0}^{n-k-4} \sum_{i=0}^{j} \frac{(-1)^{j-i}(b-a)^{j-i} f^{(k+3+j)}(b)}{(k+1+i)!(j-i)!}$$

$$\times \int_a^b \sum_{r=1}^{m} p_r G_2(x_r, s)(s-a)^{k+1+i} \, ds, \qquad (1.5.21)$$

$$\theta_6(f, G_3) = (f(b) - bf'(b) + (f'(b) - f'(a))a)\sum_{r=1}^{m} p_r + f'(a)\sum_{r=1}^{m} p_r x_r$$

$$+ \sum_{i=0}^{k} \frac{f^{(i+2)}(a)}{i!} \int_a^b \sum_{r=1}^{m} p_r G_3(x_r, s)(s-a)^i \, ds$$

$$+ \sum_{j=0}^{n-k-4} \sum_{i=0}^{j} \frac{(-1)^{j-i}(b-a)^{j-i} f^{(k+3+j)}(b)}{(k+1+i)!(j-i)!}$$

$$\times \int_a^b \sum_{r=1}^{m} p_r G_3(x_r, s)(s-a)^{k+1+i} \, ds, \qquad (1.5.22)$$

$$\theta_7(f, G_4) = (f(a) - af'(a) - (f'(b) - f'(a))b) \sum_{r=1}^{m} p_r + f'(b) \sum_{r=1}^{m} p_r x_r$$

$$+ \sum_{i=0}^{k} \frac{f^{(i+2)}(a)}{i!} \int_a^b \sum_{r=1}^{m} p_r G_4(x_r, s) (s - a)^i \, ds$$

$$+ \sum_{j=0}^{n-k-4} \sum_{i=0}^{j} \frac{(-1)^{j-i} (b - a)^{j-i} f^{(k+3+j)}(b)}{(k + 1 + i)! (j - i)!}$$

$$\times \int_a^b \sum_{r=1}^{m} p_r G_4(x_r, s) (s - a)^{k+1+i} \, ds. \tag{1.5.23}$$

Theorem 1.5.10 *Let $n, k \in \mathbb{N}$, $n \geq 4$, $0 \leq k \leq n - 1$, $f \in C^{(n)}[a, b]$, and let $x : [\alpha, \beta] \to [a, b]$, $p : [\alpha, \beta] \to \mathbb{R}$ be continuous functions. Also let $G_1 - G_4$ and G_n be defined by (1.1.6)–(1.1.9) and (1.5.2) respectively. Then*

$$\int_\alpha^\beta p(\tau) f(x(\tau)) d\tau = \theta_8(f, G_0)$$

$$+ \int_a^b \int_a^b \int_\alpha^\beta p(\tau) G(x(\tau), s) G_{n-2}(s, t) f^{(n)}(t) d\tau \, ds \, dt,$$

where

$$\theta_8(f, G_0) = \frac{f(b) - f(a)}{b - a} \int_\alpha^\beta p(\tau) x(\tau) d\tau + \frac{bf(a) - af(b)}{b - a} \int_\alpha^\beta p(\tau) d\tau$$

$$+ \sum_{i=0}^{k} \frac{f^{(i+2)}(a)}{i!} \int_a^b \int_\alpha^\beta p(\tau) G(x(\tau), s) d\tau (s - a)^i \, ds \tag{1.5.24}$$

$$+ \sum_{j=0}^{n-k-4} \sum_{i=0}^{j} \frac{(-1)^{j-i} (b - a)^{j-i} f^{(k+3+j)}(b)}{(k + 1 + i)! (j - i)!}$$

$$\times \int_a^b \int_\alpha^\beta p(\tau) G(x(\tau), s) (s - a)^{k+1+i} \, d\tau \, ds.$$

Theorem 1.5.11 Let $n, k \in \mathbb{N}$, $n \geq 4$, $0 \leq k \leq n-1$, $f \in C^{(n)}[a,b]$, and let $x : [\alpha, \beta] \to [a,b]$, $p : [\alpha, \beta] \to \mathbb{R}$ be continuous functions and $G_0 - G_4$, G_n be defined by (1.5.17), (1.1.6)–(1.1.9) and (1.5.2) respectively. Then for $l \in \{9, 10, 11, 12\}$

$$\int_\alpha^\beta p(\tau) f(x(\tau)) \, d\tau = \theta_l(f, G_{l-8})$$

$$+ \int_a^b \int_a^b \int_\alpha^\beta p(\tau) G_{l-8}(x(\tau), s) G_{n-2}(s, t) f^{(n)}(t) \, d\tau \, ds \, dt,$$

$$\theta_9(f, G_1) = (f(a) - af'(b)) \int_\alpha^\beta p(\tau) \, d\tau + f'(b) \int_\alpha^\beta p(\tau) x(\tau) \, d\tau$$

$$+ \sum_{i=0}^k \frac{f^{(i+2)}(a)}{i!} \int_a^b \int_\alpha^\beta p(\tau) G_1(x(\tau), s) \, d\tau (s-a)^i \, ds$$

$$+ \sum_{j=0}^{n-k-4} \sum_{i=0}^j \frac{(-1)^{j-i} (b-a)^{j-i} f^{(k+3+j)}(b)}{(k+1+i)! (j-i)!}$$

$$\times \int_a^b \int_\alpha^\beta p(\tau) G_1(x(\tau), s) (s-a)^{k+1+i} \, d\tau \, ds,$$

$$\theta_{10}(f, G_2) = (f(b) - bf'(a)) \int_\alpha^\beta p(\tau) \, d\tau - f'(a) \int_\alpha^\beta p(\tau) x(\tau) \, d\tau$$

$$+ \sum_{i=0}^k \frac{f^{(i+2)}(a)}{i!} \int_a^b \int_\alpha^\beta p(\tau) G_2(x(\tau), s) \, d\tau (s-a)^i \, ds$$

$$+ \sum_{j=0}^{n-k-4} \sum_{i=0}^j \frac{(-1)^{j-i} (b-a)^{j-i} f^{(k+3+j)}(b)}{(k+1+i)! (j-i)!}$$

$$\times \int_a^b \int_\alpha^\beta p(\tau) G_2(x(\tau), s) (s-a)^{k+1+i} \, d\tau \, ds,$$

$$\theta_{11}(f, G_3) = (f(b) - bf'(b) + (f'(b) - f'(a))a) \int_\alpha^\beta p(\tau) \, d\tau$$

$$+ f'(a) \int_\alpha^\beta p(\tau) x(\tau) \, d\tau$$

1.5 Linear Inequalities and the Abel-Gontscharoff's Interpolation Polynomial

$$+ \sum_{i=0}^{k} \frac{f^{(i+2)}(a)}{i!} \int_a^b \int_\alpha^\beta p(\tau) G_3(x(\tau), s) d\tau (s-a)^i ds$$

$$+ \sum_{j=0}^{n-k-4} \sum_{i=0}^{j} \frac{(-1)^{j-i} (b-a)^{j-i} f^{(k+3+j)}(b)}{(k+1+i)! (j-i)!}$$

$$\times \int_a^b \int_\alpha^\beta p(\tau) G_3(x(\tau), s) (s-a)^{k+1+i} d\tau ds,$$

$$\theta_{12}(f, G_4) = (f(a) - af'(a)) - (f'(b) - f'(a))b) + \int_\alpha^\beta p(x) dx$$

$$+ f'(b) \int_\alpha^\beta p(x) g(x) dx$$

$$+ \sum_{i=0}^{k} \frac{f^{(i+2)}(a)}{i!} \int_a^b \int_\alpha^\beta p(\tau) G_4(x(\tau), s) d\tau (s-a)^i ds$$

$$+ \sum_{j=0}^{n-k-4} \sum_{i=0}^{j} \frac{(-1)^{j-i} (b-a)^{j-i} f^{(k+3+j)}(b)}{(k+1+i)! (j-i)!}$$

$$\times \int_a^b \int_\alpha^\beta p(\tau) G_4(x(\tau), s) (s-a)^{k+1+i} d\tau ds,$$

Theorem 1.5.12 *Let $n, k \in \mathbb{N}$, $n \geq 4$, $0 \leq k \leq n-1$, $\mathbf{x} \in [a, b]^m$ and $\mathbf{p} \in \mathbb{R}^m$. Also let $G_0, G_1 - G_4, G_n$ be defined by (1.5.17), (1.1.6)–(1.1.9) and (1.5.2) respectively. If $f : [a, b] \to \mathbb{R}$ is n-convex, and*

$$\int_a^b \left(\sum_{r=1}^{m} p_r G_{l-3}(x_r, s) \right) G_{n-2}(s, t) ds \geq 0, \qquad t \in [a, b], \qquad (1.5.25)$$

then for $l \in \{3, 4, 5, 6, 7\}$

$$\sum_{r=1}^{m} p_r f(x_r) \geq \theta_l(f, G_{l-3}). \qquad (1.5.26)$$

If the reverse inequality in (1.5.25) holds, then also the reverse inequality in (1.5.26) holds.

Proof It follows from n-convexity of a function f and from Theorem 1.5.8. □

As from (1.5.3) we have $(-1)^{n-k-3}G_{n-2}(s,t) \geq 0$, therefore for the case when n is even and k is odd or n is odd and k is even, it is enough to assume that $\sum_{r=1}^{m} p_r G_{l-3}(x_r, s) \geq 0$, $s \in [\alpha, \beta]$, $\{3, 4, 5, 6, 7\}$, instead of the assumption (1.5.25) in Theorem 1.5.12. Similarly we can discuss for the reverse inequality in (1.5.26).

Integral version of the above theorem can be stated as:

Theorem 1.5.13 *Let $n, k \in \mathbb{N}$, $n \geq 4$, $0 \leq k \leq n-1$, $x : [\alpha, \beta] \to [a, b]$, $p : [\alpha, \beta] \to \mathbb{R}$ be continuous functions and $G_0, G_1 - G_4$, G_n be defined by (1.5.17), (1.1.6)–(1.1.9) and (1.5.2) respectively. If $f : [a, b] \to \mathbb{R}$ is n-convex, and*

$$\int_a^b \int_\alpha^\beta p(\tau) G_{l-8}(x(\tau), s) G_{n-2}(s, t) d\tau \, ds \geq 0, \tag{1.5.27}$$

then $l \in \{8, 9, 10, 11, 12\}$

$$\int_\alpha^\beta p(\tau) f(x(\tau)) d\tau \geq \theta_l(f, G_{l-8}). \tag{1.5.28}$$

If the reverse inequality in (1.5.27) holds, then also the reverse inequality in (1.5.28) holds.

As from (1.5.3) we have $(-1)^{n-k-3}G_{n-2}(s,t) \geq 0$, therefore for the case when n is even and k is odd or n is odd and k is even, it is enough to assume that $\int_a^b p(\tau) G_{l-8}(x(\tau), s) d\tau \geq 0$, $s \in [\alpha, \beta]$, $l \in \{8, 9, 10, 11, 12\}$, instead of the assumption (1.5.27) in Theorem 1.5.13. Similarly we can discuss for the reverse inequality in (1.5.28).

If we deal with assumptions from Remark 1.0.4, which are equivalent to Popoviciu's conditions for positivity of sum involving convex function f, then for some combinations of n and k we get result for n-convex function f. Precisely, we get the following theorem.

Theorem 1.5.14 *Let $n, k \in \mathbb{N}$, $n \geq 4$, $0 \leq k \leq n-1$. Let G_0 and $G_1 - G_4$ be defined by (1.5.17), (1.1.6)–(1.1.9) and let $f : [a, b] \to \mathbb{R}$ be n-convex. Let $\mathbf{x} \in [a, b]^m$ and $\mathbf{p} \in \mathbb{R}$ satisfy (1.0.4).*

(i) *If n is even and k is odd or n is odd and k is even, then for $l \in \{0, 1, 2, 3, 4\}$*

$$\sum_{r=1}^{m} p_l f(x_r) \geq \sum_{i=0}^{k} \frac{f^{(i+2)}(a)}{i!} \int_a^b \sum_{r=1}^{m} p_r G_l(x_r, s)(s-a)^i \, ds$$

$$+ \sum_{j=0}^{n-k-4} \sum_{i=0}^{j} \frac{(-1)^{j-i}(b-a)^{j-i} f^{(k+3+j)}(b)}{(k+1+i)!(j-i)!}$$

$$\times \int_a^b \sum_{r=1}^{m} p_r G_l(x_r, s)(s-a)^{k+1+i} \, ds. \tag{1.5.29}$$

1.5 Linear Inequalities and the Abel-Gontscharoff's Interpolation Polynomial

Moreover if $f^{(i+2)}(a) \geq 0$ *for* $i \in \{0, \ldots, k\}$ *and* $f^{(k+3+j)}(b) \geq 0$ *if* $j - i$ *is even and* $f^{(k+3+j)}(b) \leq 0$ *if* $j - i$ *is odd for* $i \in \{0, \ldots, j\}$ *and* $j \in \{0, \ldots, n - k - 4\}$, *then* $\sum_{r=1}^{m} p_r f(x_r) \geq 0$.

(ii) *If n and k both are even or both are odd, then reverse inequality holds in* (1.5.29).

Moreover if $f^{(i+2)}(a) \leq 0$ *for* $i \in \{0, \ldots, k\}$ *and* $f^{(k+3+j)}(b) \leq 0$ *if* $j - i$ *is even and* $f^{(k+3+j)}(b) \geq 0$ *if* $j - i$ *is odd for* $i \in \{0, \ldots, j\}$ *and* $j \in \{0, \ldots, n - k - 4\}$, *then* $\sum_{r=1}^{m} p_r f(x_r) \leq 0$.

Proof (i) By using (1.5.3) we have $(-1)^{n-k-3} G_{n-2}(s, t) \geq 0$, $s, t \in [a, b]$, therefore if n is even and k is odd or n is odd and k is even then $G_{n-2}(s, t) \geq 0$. Since G_l are convex for $\{0, 1, 2, 3, 4\}$ and G_{n-2} is nonnegative, the inequality (1.5.25) holds. Hence by Theorem 1.5.12 the inequality (1.5.29) holds. By using the other conditions the nonnegativity of the right-hand side of (1.5.29) is obvious. Similarly we prove (ii).

□

The integral version of Theorem 1.5.14 can be stated as:

Theorem 1.5.15 *Let* $n, k \in \mathbb{N}$, $n \geq 4$, $0 \leq k \leq n - 1$ *and let* $x : [\alpha, \beta] \to [a, b]$ *and* $p : [\alpha, \beta] \to \mathbb{R}$ *be any continuous functions satisfy:*

$$\int_\alpha^\beta p(\tau)\, d\tau \geq 0, \qquad \int_\alpha^\beta p(\tau)(x(\tau) - t)_+\, d\tau \geq 0 \text{ for } t \in [a, b]. \quad (1.5.30)$$

Also let G, $G_1 - G_4$ be defined by (1.5.17) and (1.1.6)–(1.1.9). Consider $f : [a, b] \to \mathbb{R}$ is n-convex.

(i) *If n is even and k is odd or n is odd and k is even, then for $l \in \{0, 1, 2, 3, 4\}$*

$$\int_\alpha^\beta p(\tau) f(x(\tau))\, d\tau \geq \sum_{i=0}^{k} \frac{f^{(i+2)}(a)}{i!} \int_a^b \int_\alpha^\beta p(\tau) G_l(x(\tau), s)(s-a)^i\, d\tau\, ds$$
$$+ \sum_{j=0}^{n-k-4} \sum_{i=0}^{j} \frac{(-1)^{j-i}(b-a)^{j-i} f^{(k+3+j)}(b)}{(k+1+i)!\,(j-i)!}$$
$$\times \int_a^b \int_\alpha^\beta p(\tau) G_l(x(\tau), s)(s-a)^{k+1+i}\, d\tau\, ds. \quad (1.5.31)$$

Moreover if $f^{(i+2)}(a) \geq 0$ *for* $i \in \{0, \ldots, k\}$ *and* $f^{(k+3+j)}(b) \geq 0$ *if* $j - i$ *is even and* $f^{(k+3+j)}(b) \leq 0$ *if* $j - i$ *is odd for* $i \in \{0, \ldots, j\}$ *and* $j \in \{0, \ldots, n - k - 4\}$, *then the right-hand side of (1.5.31) is nonnegative, that is integral version of (1.0.3) holds.*

(ii) *If n and k both are even or both are odd, then reverse inequality holds in* (1.5.31).

Moreover if $f^{(i+2)}(a) \leq 0$ for $i \in \{0, \ldots, k\}$ and $f^{(k+3+j)}(b) \leq 0$ if $j-i$ is even and $f^{(k+3+j)}(b) \geq 0$ if $j-i$ is odd for $i \in \{0, \ldots, j\}$ and $j \in \{0, \ldots, n-k-4\}$, then the right hand side of the reverse inequality in (1.5.31) is nonpositive, that is the reverse inequality in the integral version of (1.0.3) holds.

1.5.3 Inequalities for n-Convex Functions at a Point

Under the assumptions of Theorems 1.5.2, 1.5.8 and 1.5.9 here we define some linear functional as follows:

$$A_{19}^{[a,b]}(m, \mathbf{x}, \mathbf{p}, f) = \sum_{r=1}^{m} p_r f(x_r) - \theta_1(f). \quad (1.5.32)$$

For $l \in \{4, 5, 6, 7, 8\}$

$$A_{l+16}^{[a,b]}(m, \mathbf{x}, \mathbf{p}, f) = \sum_{r=1}^{m} p_r f(x_r) - \theta_{l-1}(f, G_{l-4}). \quad (1.5.33)$$

Similarly, under the assumptions of Theorem 1.5.3, 1.5.10 and 1.5.11 here we introduce some further linear functional as follows:

$$A_{25}^{[a,b]}([\alpha, \beta], x, p, f) = \int_{\alpha}^{\beta} p(\tau) f(x(\tau)) \, d\tau - \theta_2(f). \quad (1.5.34)$$

For $l \in \{10, 11, 12, 13, 14\}$ we have

$$A_{l+16}^{[a,b]}([\alpha, \beta], x, p, f) = \int_{\alpha}^{\beta} p(\tau) f(x(\tau)) \, d\tau - \theta_{l-1}(f, G_{l-10}). \quad (1.5.35)$$

Here we also define some new functionals Ω_{19} and Ω_{l+16} for $l \in \{4, 5, 6, 7, 8\}$ as follows:

$$\Omega_{19}^{[a,b]}(m, \mathbf{x}, \mathbf{p}, t) = \sum_{r=1}^{m} p_r G_n(x_r, t) \text{ for all } t \in [a, b], \quad (1.5.36)$$

$$\Omega_{l+16}^{[a,b]}(m, \mathbf{x}, \mathbf{p}, t) = \int_{a}^{b} \sum_{r=1}^{m} p_r G_{l-4}(x_r, s) G_{n-2}(s, t) ds \geq 0 \quad \text{for all } t \in [a, b],$$

$$(1.5.37)$$

1.5 Linear Inequalities and the Abel-Gontscharoff's Interpolation Polynomial

and Ω_{25} and Ω_{l+16} for $l \in \{10, 11, 12, 13, 14\}$ are defined as

$$\Omega_{25}^{[a,b]}([\alpha, \beta], x, p, t) = \int_\alpha^\beta p(\tau) G_n(x(\tau), t) \, d\tau \, dx \geq 0 \quad \forall \; t \in [a, b], \tag{1.5.38}$$

$$\Omega_{l+16}^{[a,b]}([\alpha, \beta], x, p, t) = \int_a^b \int_\alpha^\beta p(\tau) G_{l-10}(x(\tau), s) G_{n-2}(s, t) d\tau \, ds$$
$$\geq 0 \quad \forall \; t \in [a, b]. \tag{1.5.39}$$

For the sake of brevity we consider $A_l^{[\cdot,\cdot]}(\cdot, \cdot, \cdot, f) = A_l(f)$ and $\Omega_l^{[\cdot,\cdot]}(\cdot, \cdot, \cdot, t) = \Omega_l(t)$. We state our next result.

Theorem 1.5.16 Let $\mathbf{x} \in [a, c]^{m_1}$, $\mathbf{p} \in \mathbb{R}^{m_1}$, $\mathbf{y} \in [c, b]^{m_2}$ and $\mathbf{q} \in \mathbb{R}^{m_2}$ be such that for each $l \in \{19, 20, 21, 22, 23, 24\}$

$$\Omega_l^{[a,c]}(m_1, \mathbf{x}, \mathbf{p}, t) \geq 0 \quad \text{for all } t \in [a, c], \tag{1.5.40}$$

$$\Omega_l^{[c,]}(m_2, \mathbf{y}, \mathbf{q}, t) \geq 0 \quad \text{for all } t \in [c, b], \tag{1.5.41}$$

and

$$A_l^{[a,c]}(m_1, \mathbf{x}, \mathbf{p}, e_n) = A_l^{[c,b]}(m_2, \mathbf{y}, \mathbf{q}, e_n). \tag{1.5.42}$$

where A_l and Ω_l be the linear functionals given by (1.5.32)–(1.5.35) and (1.5.36)–(1.5.39). If $f : [a, b] \to \mathbb{R}$ is $(n+1)$-convex at point c, then

$$A_l^{[a,c]}(m_1, \mathbf{x}, \mathbf{p}, f) \leq A_l^{[c,b]}(m_2, \mathbf{y}, \mathbf{q}, f). \tag{1.5.43}$$

If the inequalities in (1.5.40) and (1.5.41) are reversed, then (1.5.43) holds with the reversed sign of inequality.

Theorem 1.5.17 Let $x : [\alpha, \beta] \to [a, c]$, $p : [\alpha, \beta] \to \mathbb{R}$, $y : [\gamma, \delta] \to [c, b]$, $q : [\gamma, \delta] \to \mathbb{R}$ be such that for each $l \in \{25, 26, 27, 28, 29, 30\}$

$$\Omega_l^{[a,c]}([\alpha, \beta], x, p, t) \geq 0 \quad \text{for all } t \in [a, c], \tag{1.5.44}$$

$$\Omega_l^{[c,b]}([\gamma, \delta], y, q, t) \geq 0 \quad \text{for all } t \in [c, b] \tag{1.5.45}$$

and

$$A_l^{[a,c]}([\alpha, \beta], x, p, e_n) = A_l^{[c,b]}([\gamma, \delta], y, p, e_n) \tag{1.5.46}$$

Then

$$A_l^{[a,c]}([\alpha, \beta], x, p, f) \leq A_l^{[c,b]}([\gamma, \delta], y, p, f), \quad (1.5.47)$$

where A_l and Ω_l be the linear functionals given by (1.5.32)–(1.5.35) and (1.5.36)–(1.5.39) respectively.

If the inequalities in (1.5.44) and (1.5.45) are reversed, then (1.5.47) holds with the reversed sign of inequality.

1.5.4 Bounds for Remainders and Functionals

For the sake of brevity we consider $A_l^{[\cdot,\cdot]}(\cdot, \cdot, \cdot, f) = A_l(f)$ and $\Omega_l^{[\cdot,\cdot]}(\cdot, \cdot, \cdot, t) = \Omega_l(t)$ which was defined in previous subsection. Now we state our next result.

Theorem 1.5.18 *Let $l \in \{19, \ldots, 30\}$. Let $f \in C^{(n)}[a, b]$ such that for real numbers γ_1 and γ_2 we have*

$$\gamma_1 \leq f^{(n)}(\eta) \leq \gamma_2 \quad \text{for } \eta \in [a, b].$$

Then in representation

$$A_l(f) = \frac{[f^{n-1}(b) - f^{n-1}(a)]}{b-a} \int_a^b \Omega_l(\xi) d\xi + (b-a) R_8^l(f; a, b, n), \quad (1.5.48)$$

remainder $R_8^l(f; a, b, n)$ satisfies estimation

$$|R_8^l(f; a, b, n)| \leq \frac{1}{2}(\gamma_2 - \gamma_1)\sqrt{T(\Omega_l, \Omega_l)}. \quad (1.5.49)$$

Now we give some Ostrowski-type inequalities related to the generalized linear inequalities.

Theorem 1.5.19 *Let for $l \in \{19, \ldots, 30\}$ A_l and Ω_l be linear functionals as defined in previous subsection. Furthermore, let (q, r) be a pair of conjugate exponents, i.e., $1 \leq q, r \leq \infty$, $\frac{1}{q} + \frac{1}{r} = 1$. Let $f^{(n)} \in L_q[a, b]$ for $n \geq 1$. Then we have for $l \in \{19, \ldots, 30\}$*

$$|A_l(f)| \leq \|f^{(n)}\|_q \|\Omega_l\|_r. \quad (1.5.50)$$

The constant on right hand side of (1.5.50) is sharp for $1 < q \leq \infty$ and the best possible for $q = 1$.

Remark 1.5.1 For idea of the proof kindly see [98].

Using the same method as given in [5] we can state mean value theorems and results related to exponentially convexity.

Chapter 2
Ostrowski Inequality

> *The art of doing mathematics consists in finding that special case which contains all the germs of generality.*
>
> —David Hilbert

According to Hardy's famous statement "Behind every theorem lies an inequality," and it is true that many significant and well-known results in mathematics are inequalities, new types of interesting inequalities are discovered every year and number of researchers improve, extend, refine and generalize various mathematical inequalities. Several books on inequalities have been published, interested readers can see [20, 27, 48, 61, 69]. While the topic of inequalities covers many branches of mathematics but this chapter focuses on those associated with Ostrowski inequality.

Firstly, we would like to recall the classical Ostrowski inequality.

Ostrowski Inequality

In 1937, Ostrowski established an inequality in his paper [141]. It is a known fact that this type of inequality could be used to approximate the absolute deviation of functional value from its integral mean. This result is now a days known as Ostrowski inequality. The database of MathSciNet includes more than 360 papers with the keywords "Ostrowski" and "Inequality". Various generalizations and extensions of the Ostrowski inequality have appeared in the literature and detailed history on Ostrowski inequality, see [6, 8, 17, 19, 36, 52] and references therein. For further literature we can see unpublished doctoral thesis [75, 192] and Dragomir's monograph [48].

In following theorem, we give this inequality from [48].

Theorem 2.0.1 *Let $f : I = [a,b] \subseteq \mathbb{R} \to \mathbb{R}$ be a differentiable function on I^o (where I^o is the interior of I) such that $f \in L[a,b]$, where $a, b \in I$ and $a < b$.*

If $|f'(t)| \leq M$ valid $\forall\, t \in (a,b)$ where M is positive real constant. Then

$$\left| f(x) - \frac{1}{b-a}\int_a^b f(t)dt \right| \leq M(b-a)\left[\frac{1}{4} + \frac{\left(x - \frac{a+b}{2}\right)^2}{(b-a)^2} \right]$$

$$= \left[\frac{(x-a)^2 + (b-x)^2}{2(b-a)} \right] M. \qquad (2.0.1)$$

Proof We have Montgomery identity from (1.1.1)

$$(b-a)f(x) - \int_a^b f(t)dt = \int_a^x (t-a)f'(t)dt + \int_x^b (t-b)f'(t)dt.$$

Taking the modulus, we deduce

$$\left| (b-a)f(x) - \int_a^b f(t)dt \right| \leq \left| \int_a^x (t-a)f'(t)dt \right| + \left| \int_x^b (t-b)f'(t)dt \right|$$

$$\leq \int_a^x (t-a)\left|f'(t)\right|dt + \int_x^b (t-b)\left|f'(t)\right|dt$$

$$\leq M\int_a^x (t-a)dt + M\int_x^b (t-b)dt$$

$$= \frac{M}{2}\left[(x-a)^2 + (b-x)^2\right]$$

$$= M(b-a)^2\left[\frac{1}{4} + \left(\frac{x - \frac{a+b}{2}}{b-a}\right) \right].$$

This produces our classical Ostrowski inequality (2.0.1). $\square$

Remark 2.0.2

(1) Here the constant $\frac{1}{4}$ in first inequality is the best possible in the sense that it cannot be replaced by smaller one.
(2) In latest versions M is usually replaced by $\|f'\|_\infty = \operatorname{ess\,sup}_{t\in(a,b)} |f'(t)| < \infty$.
(3) Since f' is bounded so the result is also valid for functions of bounded variation.
(4) In this result some assumptions may be relaxed by using absolutely continuous functions instead of differentiable functions.
(5) This result may be proved in variety of ways by using different techniques including Lagrange mean value theorem and Montgomery identity, and by using direct calculation etc.
(6) Ostrowski's inequality helps us to estimate the bound of Hermite-Hadamard's left inequality.

2 Ostrowski Inequality

(7) This inequality may be interpreted in the following manners:
   (a) Estimation of deviation of functional values of function with bounded derivative from the integral mean.
   (b) It measures the estimate of approximating area under the curve by rectangle.
(8) The celebrated inequality has vast applications in statistics, numerical integration, probability theory and special mean(s) among many others.
(9) It has close connection with other celebrated inequalities including Ostrowski-Grüss, Grüss and Hermite-Hadamard inequalities.

Hermite-Hadamard Inequality

In [72], Hermite presented the following result known as Hermite-Hadamard dual inequality for convex function.

Theorem 2.0.2 *Let $f : I \to \mathbb{R}$ be a convex function. Then*

$$f\left(\frac{a+b}{2}\right) \leq \frac{1}{b-a}\int_a^b f(x)dx \leq \frac{f(a)+f(b)}{2}.$$

It is worth mentioning that, for concave function f, both inequalities would be in reverse order.

Standard and Non-Standard Quadrature Rules

From [52] and [115], we have some standard quadrature rules of midpoint, trapezoidal, $\frac{1}{3}$ Simpson's, $\frac{3}{8}$ Simpson's and average of midpoint and trapezoidal rule as given in $I_1(f)$, $I_2(f)$, $I_3(f)$, $I_4(f)$ and $I_5(f)$ respectively. We will also capture some non-standard quadrature rules from our findings as given in $I_6(f) - I_9(f)$.

$$I_1(f) : \frac{1}{b-a}\int_a^b f(x)dx \cong f\left(\frac{a+b}{2}\right),$$

$$I_2(f) : \frac{1}{b-a}\int_a^b f(x)dx \cong \frac{f(a)+f(b)}{2},$$

$$I_3(f) : \frac{1}{b-a}\int_a^b f(x)dx \cong \frac{1}{3}\left[\frac{f(a)+f(b)}{2} + 2f\left(\frac{a+b}{2}\right)\right],$$

$$I_4(f) : \frac{1}{b-a}\int_a^b f(x)dx \cong \frac{3}{8}\left[\frac{f(a)+f(b)}{3} + f\left(\frac{2a+b}{3}\right) + f\left(\frac{a+2b}{3}\right)\right],$$

$$I_5(f) : \frac{1}{b-a}\int_a^b f(x)dx \cong \frac{1}{2}\left[\frac{f(a)+f(b)}{2} + f\left(\frac{a+b}{2}\right)\right],$$

$$I_6(f) : \frac{1}{b-a}\int_a^b f(x)dx \cong \frac{1}{2}\left[-f(a) + 2f\left(\frac{a+b}{2}\right) + f(b)\right],$$

$$I_7(f) : \frac{1}{b-a}\int_a^b f(x)dx \cong \frac{1}{2}\left[f(a) + 2f\left(\frac{a+b}{2}\right) - f(b)\right],$$

$$I_8(f) : \frac{1}{b-a}\int_a^b f(x)dx \cong f(a),$$

$$I_9(f) : \frac{1}{b-a}\int_a^b f(x)dx \cong f(b).$$

2.1 Generalized Ostrowski Type Inequalities with Parameter

Generalized Ostrowski type inequalities for bounded, bounded below only and bounded above only differentiable functions are given in this section. Some applications to find error bounds of midpoint, trapezoidal, $\frac{1}{3}$ Simpson's and $\frac{3}{8}$ Simpson's quadrature rules are discussed. Error bounds are also provided for some non-standard quadratures rules. Our proposed findings would generalize the results of Masjed-Jamei and Dragomir in [116] and [115] and Ujević in [184]. The results of this section have been obtained in [77].

Ostrowski inequality for differentiable functions have been generalized many times as stated in [53, 79, 115, 116, 184]. To establish our main results, we need these two lemmas from [7] as given below.

Lemma 2.1.1 *Let $f: I \to \mathbb{R}$, be absolutely continuous function in I^0 such that $x \in [a, b]$ with*

$$a + \epsilon\frac{b-a}{2} \leq x \leq \frac{a+b}{2}.$$

Then

$$\int_a^b P(x,t)f'(t)dt = (b-a)\left[\epsilon\frac{f(a)+f(b)}{2} + (1-\epsilon)\frac{f(x)+f(a+b-x)}{2}\right]$$
$$-\int_a^b f(t)dt, \qquad (2.1.1)$$

$\forall \epsilon \in [0, 1]$, *where Peano kernel $P(x, t)$ is defined as*

$$P(x,t) = \begin{cases} t - \left(a + \epsilon\frac{b-a}{2}\right), & t \in [a, x]; \\ t - \frac{a+b}{2}, & t \in (x, a+b-x]; \\ t - \left(b - \epsilon\frac{b-a}{2}\right), & t \in (a+b-x, b]. \end{cases}$$

2.1 Generalized Ostrowski Type Inequalities with Parameter

Lemma 2.1.2 *Let $f: I \to \mathbb{R}$, be a differentiable function and if $\phi(t) \leq f'(t) \leq \Phi(t)$ such that $\phi, \Phi \in C[a,b]$ and $t \in [a,b]$. Then we have*

$$\left| f'(t) - \frac{\phi(t) + \Phi(t)}{2} \right| \leq \frac{\Phi(t) - \phi(t)}{2}. \tag{2.1.2}$$

Now we are going to present our main results of this section.

2.1.1 Ostrowski Type Inequality for Bounded Differentiable Functions

Theorem 2.1.1 *Let $f: I \to \mathbb{R}$, be a differentiable function in I^0. If $\phi(x) \leq f'(x) \leq \Phi(x)$ for any $\phi, \Phi \in C[a,b]$ and $x \in [a,b]$, then the following inequality holds*

$$m(x, \epsilon) \leq \epsilon \frac{f(a) + f(b)}{2} + (1-\epsilon) \frac{f(x) + f(a+b-x)}{2} - \frac{1}{b-a} \int_a^b f(t)dt$$

$$\leq M(x, \epsilon), \tag{2.1.3}$$

where

$$m(x, \epsilon) = \frac{1}{b-a} \left[\int_{-\epsilon \frac{b-a}{2}}^{x-\left(a+\epsilon \frac{b-a}{2}\right)} \left(\frac{\lambda + |\lambda|}{2} \phi \left(\lambda + a + \epsilon \frac{b-a}{2} \right) \right. \right.$$

$$+ \frac{\lambda - |\lambda|}{2} \Phi \left(\lambda + a + \epsilon \frac{b-a}{2} \right) \right) d\lambda + \int_{x-\frac{a+b}{2}}^{\frac{a+b}{2}-x} \left(\frac{\lambda + |\lambda|}{2} \phi \left(\lambda + \frac{a+b}{2} \right) \right.$$

$$+ \frac{\lambda - |\lambda|}{2} \Phi \left(\lambda + \frac{a+b}{2} \right) \right) d\lambda + \int_{a+\epsilon \frac{b-a}{2}-x}^{\epsilon \frac{b-a}{2}} \left(\frac{\lambda + |\lambda|}{2} \phi \left(\lambda + b - \epsilon \frac{b-a}{2} \right) \right.$$

$$\left. \left. + \frac{\lambda - |\lambda|}{2} \Phi \left(\lambda + b - \epsilon \frac{b-a}{2} \right) \right) d\lambda \right]$$

and

$$M(x, \epsilon) = \frac{1}{b-a} \left[\int_{-\epsilon \frac{b-a}{2}}^{x-\left(a+\epsilon \frac{b-a}{2}\right)} \left(\frac{\lambda - |\lambda|}{2} \phi \left(\lambda + a + \epsilon \frac{b-a}{2} \right) \right. \right.$$

$$+ \frac{\lambda + |\lambda|}{2} \Phi \left(\lambda + a + \epsilon \frac{b-a}{2} \right) \right) d\lambda + \int_{x-\frac{a+b}{2}}^{\frac{a+b}{2}-x} \left(\frac{\lambda - |\lambda|}{2} \phi \left(\lambda + \frac{a+b}{2} \right) \right.$$

$$+ \frac{\lambda + |\lambda|}{2} \Phi\left(\lambda + \frac{a+b}{2}\right)\right) d\lambda + \int_{a+\epsilon\frac{b-a}{2}-x}^{\epsilon\frac{b-a}{2}} \left(\frac{\lambda - |\lambda|}{2} \phi\left(\lambda + b - \epsilon\frac{b-a}{2}\right)\right.$$
$$\left.+ \frac{\lambda + |\lambda|}{2} \Phi\left(\lambda + b - \epsilon\frac{b-a}{2}\right)\right) d\lambda \Bigg].$$

Proof Replacing $f'(t)$ by $f'(t) - \dfrac{\phi(t) + \Phi(t)}{2}$ in (2.1.1), we get

$$\int_a^b P(x,t) \left(f'(t) - \frac{\phi(t) + \Phi(t)}{2} \right) dt$$

$$= \int_a^b P(x,t) f'(t) dt - \frac{1}{2} \left(\int_a^b P(x,t) (\phi(t) + \Phi(t)) dt \right)$$

$$= (b-a) \left[\epsilon \frac{f(a) + f(b)}{2} + (1-\epsilon) \frac{f(x) + f(a+b-x)}{2} \right] - \int_a^b f(t) dt$$

$$- \frac{1}{2} \left[\int_a^x \left(t - \left(a + \epsilon \frac{b-a}{2} \right) \right) (\phi(t) + \Phi(t)) dt \right.$$

$$+ \int_x^{a+b-x} \left(t - \frac{a+b}{2} \right) (\phi(t) + \Phi(t)) dt$$

$$\left. + \int_{a+b-x}^b \left(t - \left(b - \epsilon \frac{b-a}{2} \right) \right) (\phi(t) + \Phi(t)) dt \right]. \tag{2.1.4}$$

From (2.1.2) and (2.1.4), we get

$$\left| (b-a) \left[\epsilon \frac{f(a) + f(b)}{2} + (1-\epsilon) \frac{f(x) + f(a+b-x)}{2} \right] - \int_a^b f(t) dt \right.$$

$$- \frac{1}{2} \left[\int_a^x \left(t - \left(a + \epsilon \frac{b-a}{2} \right) \right) (\phi(t) + \Phi(t)) dt + \int_x^{a+b-x} \left(t - \frac{a+b}{2} \right) \right.$$

$$\left. \times (\phi(t) + \Phi(t)) dt + \int_{a+b-x}^b \left(t - \left(b - \epsilon \frac{b-a}{2} \right) \right) (\phi(t) + \Phi(t)) dt \right] \Bigg|$$

$$= \left| \int_a^b P(x,t) \left(f'(t) - \frac{\phi(t) + \Phi(t)}{2} \right) dt \right|$$

$$\leq \int_a^b |P(x,t)| \left| \left(f'(t) - \frac{\phi(t) + \Phi(t)}{2} \right) dt \right|$$

$$\leq \int_a^b |P(x,t)| \left(\frac{\Phi(t) - \phi(t)}{2} \right) dt$$

2.1 Generalized Ostrowski Type Inequalities with Parameter

$$= \frac{1}{2}\left[\int_a^x \left|t - \left(a + \epsilon\frac{b-a}{2}\right)\right| (\Phi(t) - \phi(t)) \, dt\right.$$
$$+ \int_x^{a+b-x} \left|t - \frac{a+b}{2}\right| (\Phi(t) - \phi(t)) \, dt$$
$$+ \left.\int_{a+b-x}^b \left|t - \left(b - \epsilon\frac{b-a}{2}\right)\right| (\Phi(t) - \phi(t)) \, dt\right]. \tag{2.1.5}$$

After rearranging the terms of (2.1.5), we get

$$m(x,\epsilon) \leq \epsilon \frac{f(a) + f(b)}{2} + (1-\epsilon)\frac{f(x) + f(a+b-x)}{2} - \frac{1}{b-a}\int_a^b f(t)\,dt$$
$$\leq M(x,\epsilon),$$

where

$$m(x,\epsilon) = \frac{1}{b-a}\left[\int_a^x \left(\left(t - \left(a + \epsilon\frac{b-a}{2}\right)\right) - \left|t - \left(a + \epsilon\frac{b-a}{2}\right)\right|\right)\frac{\Phi(t)}{2}\right.$$
$$+ \left(t - \left(a + \epsilon\frac{b-a}{2}\right) + \left|t - \left(a + \epsilon\frac{b-a}{2}\right)\right|\right)\frac{\phi(t)}{2}\,dt$$
$$+ \int_x^{a+b-x}\left(\left(t - \frac{a+b}{2} - \left|t - \frac{a+b}{2}\right|\right)\frac{\Phi(t)}{2} + \left(t - \frac{a+b}{2}\right.\right.$$
$$+ \left.\left|t - \frac{a+b}{2}\right|\right)\frac{\phi(t)}{2}\,dt + \int_{a+b-x}^b \left(\left(t - \left(b - \epsilon\frac{b-a}{2}\right)\right)\right.$$
$$- \left|t - \left(b - \epsilon\frac{b-a}{2}\right)\right|\right)\frac{\Phi(t)}{2} + \left(t - \left(b - \epsilon\frac{b-a}{2}\right)\right.$$
$$+ \left.\left.\left|t - \left(b - \epsilon\frac{b-a}{2}\right)\right|\right)\frac{\phi(t)}{2}\,dt\right]$$
$$= \frac{1}{b-a}\left[\int_{-\epsilon\frac{b-a}{2}}^{x-\left(a+\epsilon\frac{b-a}{2}\right)}\left(\frac{\lambda + |\lambda|}{2}\phi\left(\lambda + a + \epsilon\frac{b-a}{2}\right) + \frac{\lambda - |\lambda|}{2}\right.\right.$$
$$\times \Phi\left(\lambda + a + \epsilon\frac{b-a}{2}\right)\right)d\lambda + \int_{x-\frac{a+b}{2}}^{\frac{a+b}{2}-x}\left(\frac{\lambda + |\lambda|}{2}\phi\left(\lambda + \frac{a+b}{2}\right)\right.$$
$$+ \left.\frac{\lambda - |\lambda|}{2}\Phi\left(\lambda + \frac{a+b}{2}\right)\right)d\lambda + \int_{a+\epsilon\frac{b-a}{2}-x}^{\epsilon\frac{b-a}{2}}\left(\frac{\lambda + |\lambda|}{2}\phi\left(\lambda + b - \epsilon\frac{b-a}{2}\right)\right.$$
$$+ \left.\left.\frac{\lambda - |\lambda|}{2}\Phi\left(\lambda + b - \epsilon\frac{b-a}{2}\right)\right)d\lambda\right]$$

and

$$M(x,\epsilon) = \frac{1}{b-a}\left[\int_a^x \left(\left(\left(t-\left(a+\epsilon\frac{b-a}{2}\right)\right)+\left|t-\left(a+\epsilon\frac{b-a}{2}\right)\right|\right)\frac{\Phi(t)}{2}\right.\right.$$
$$+\left(\left(t-\left(a+\epsilon\frac{b-a}{2}\right)\right)-\left|t-\left(a+\epsilon\frac{b-a}{2}\right)\right|\right)\frac{\phi(t)}{2}\right)dt$$
$$+\int_x^{a+b-x}\left(\left(\left(t-\frac{a+b}{2}\right)+\left|t-\frac{a+b}{2}\right|\right)\frac{\Phi(t)}{2}+\left(t-\frac{a+b}{2}\right.\right.$$
$$\left.\left.-\left|t-\frac{a+b}{2}\right|\right)\frac{\phi(t)}{2}\right)dt+\int_{a+b-x}^b\left(\left(\left(t-\left(b-\epsilon\frac{b-a}{2}\right)\right)\right.\right.$$
$$\left.+\left|t-\left(b-\epsilon\frac{b-a}{2}\right)\right|\right)\frac{\Phi(t)}{2}$$
$$\left.+\left(\left(t-\left(b-\epsilon\frac{b-a}{2}\right)\right)-\left|t-\left(b-\epsilon\frac{b-a}{2}\right)\right|\right)\frac{\phi(t)}{2}\right)dt\right]$$
$$=\frac{1}{b-a}\left[\int_{-\epsilon\frac{b-a}{2}}^{x-\left(a+\epsilon\frac{b-a}{2}\right)}\left(\frac{\lambda-|\lambda|}{2}\phi\left(\lambda+a+\epsilon\frac{b-a}{2}\right)+\frac{\lambda+|\lambda|}{2}\right.\right.$$
$$\left.\times\Phi\left(\lambda+a+\epsilon\frac{b-a}{2}\right)\right)d\lambda+\int_{x-\frac{a+b}{2}}^{\frac{a+b}{2}-x}\left(\frac{\lambda-|\lambda|}{2}\phi\left(\lambda+\frac{a+b}{2}\right)\right.$$
$$\left.+\frac{\lambda+|\lambda|}{2}\Phi\left(\lambda+\frac{a+b}{2}\right)\right)d\lambda+\int_{a+\epsilon\frac{b-a}{2}-x}^{\epsilon\frac{b-a}{2}}\left(\frac{\lambda-|\lambda|}{2}\phi\left(\lambda+b-\epsilon\frac{b-a}{2}\right)\right.$$
$$\left.\left.+\frac{\lambda+|\lambda|}{2}\Phi\left(\lambda+b-\epsilon\frac{b-a}{2}\right)\right)d\lambda\right].$$

□

Remark 2.1.1 By choosing $\epsilon = 1$ in (2.1.3), it would be independent of x, we get the bound for trapezoidal rule (Hermite–Hadamard right bound) in the following corollary.

Corollary 2.1.2 *Let all assumptions of Theorem 2.1.1 be valid. Then*

$$m_1 \leq \frac{f(a)+f(b)}{2} - \frac{1}{b-a}\int_a^b f(t)dt \leq M_1, \qquad (2.1.6)$$

where

$$m_1 = \frac{1}{b-a}\left[\int_{-\frac{b-a}{2}}^{\frac{b-a}{2}}\left(\frac{\lambda+|\lambda|}{2}\phi\left(\lambda+\frac{a+b}{2}\right)+\frac{\lambda-|\lambda|}{2}\Phi\left(\lambda+\frac{a+b}{2}\right)\right)d\lambda\right]$$

2.1 Generalized Ostrowski Type Inequalities with Parameter

and

$$M_1 = \frac{1}{b-a}\left[\int_{-\frac{b-a}{2}}^{\frac{b-a}{2}}\left(\frac{\lambda-|\lambda|}{2}\phi\left(\lambda+\frac{a+b}{2}\right)+\frac{\lambda+|\lambda|}{2}\Phi\left(\lambda+\frac{a+b}{2}\right)\right)d\lambda\right],$$

which is Corollary 2 of [116].

From now onwards, throughout the section ϕ_0, ϕ_1, Φ_0 and Φ_1 are real constants.

Special Case 2.1.2.(a) If we take, $\phi(x) = \phi_0 \neq 0$ and $\Phi(x) = \Phi_0 \neq 0$ in (2.1.6), then

$$\frac{(b-a)}{8}(\phi_0 - \Phi_0) \le \frac{f(a)+f(b)}{2} - \frac{1}{b-a}\int_a^b f(t)dt \le \frac{(b-a)}{8}(\Phi_0 - \phi_0),$$

which is Corollary 2 of [184].

Special Case 2.1.2.(b) If we take, $\phi(x) = \phi_1 x + \phi_0 \neq 0$ and $\Phi(x) = \Phi_1 x + \Phi_0 \neq 0$ in (2.1.6), then

$$m_2 \le \frac{f(a)+f(b)}{2} - \frac{1}{b-a}\int_a^b f(t)dt \le M_2,$$

where

$$m_2 = \frac{(b-a)}{8}\left[\frac{(b-a)}{3}(\phi_1 + \Phi_1) + \frac{(a+b)}{2}(\phi_1 - \Phi_1) + \phi_0 - \Phi_0\right]$$

and

$$M_2 = \frac{(b-a)}{8}\left[\frac{(b-a)}{3}(\phi_1 + \Phi_1) + \frac{(a+b)}{2}(\Phi_1 - \phi_1) + \Phi_0 - \phi_0\right].$$

Remark 2.1.2 By choosing $x = \frac{a+b}{2}$ in (2.1.3), we get the inequality in the following corollary.

Corollary 2.1.3 *Let all assumptions of Theorem 2.1.1 be valid. Then*

$$m_3(\epsilon) \le \epsilon\frac{f(a)+f(b)}{2} + (1-\epsilon)f\left(\frac{a+b}{2}\right) - \frac{1}{b-a}\int_a^b f(t)dt \le M_3(\epsilon),$$
(2.1.7)

where

$$m_3(\epsilon) = \frac{1}{b-a}\left[\int_{-\epsilon\frac{b-a}{2}}^{(1-\epsilon)\frac{b-a}{2}} \left(\frac{\lambda+|\lambda|}{2}\phi\left(\lambda+a+\epsilon\frac{b-a}{2}\right)\right.\right.$$
$$\left.+\frac{\lambda-|\lambda|}{2}\Phi\left(\lambda+a+\epsilon\frac{b-a}{2}\right)\right)d\lambda$$
$$+\int_{(\epsilon-1)\frac{b-a}{2}}^{\epsilon\frac{b-a}{2}} \left(\frac{\lambda+|\lambda|}{2}\times\phi\left(\lambda+b-\epsilon\frac{b-a}{2}\right)\right.$$
$$\left.\left.+\frac{\lambda-|\lambda|}{2}\Phi\left(\lambda+b-\epsilon\frac{b-a}{2}\right)\right)d\lambda\right]$$

and

$$M_3(\epsilon) = \frac{1}{b-a}\left[\int_{-\epsilon\frac{b-a}{2}}^{(1-\epsilon)\frac{b-a}{2}} \left(\frac{\lambda-|\lambda|}{2}\phi\left(\lambda+a+\epsilon\frac{b-a}{2}\right)\right.\right.$$
$$\left.+\frac{\lambda+|\lambda|}{2}\Phi\left(\lambda+a+\epsilon\frac{b-a}{2}\right)\right)d\lambda$$
$$+\int_{(\epsilon-1)\frac{b-a}{2}}^{\epsilon\frac{b-a}{2}} \left(\frac{\lambda-|\lambda|}{2}\phi\left(\lambda+b-\epsilon\frac{b-a}{2}\right)+\frac{\lambda+|\lambda|}{2}\Phi\left(\lambda+b-\epsilon\frac{b-a}{2}\right)\right)d\lambda\right].$$

The Corollary 2.1.3 could be more useful to get different quadrature bounds as under.

Remark 2.1.3 By choosing $\epsilon = 0$ in (2.1.7), we get the bound for midpoint rule (Hermite–Hadamard left bound) in the following corollary.

Corollary 2.1.4 *Let all assumptions of Theorem 2.1.1 be valid. Then*

$$m_4 \leq f\left(\frac{a+b}{2}\right) - \frac{1}{b-a}\int_a^b f(t)dt \leq M_4, \qquad (2.1.8)$$

where

$$m_4 = \frac{1}{b-a}\left[\int_{-\frac{b-a}{2}}^{0} \left(\frac{\lambda+|\lambda|}{2}\phi(\lambda+b) + \frac{\lambda-|\lambda|}{2}\Phi(\lambda+b)\right)d\lambda\right.$$
$$\left.+\int_{0}^{\frac{b-a}{2}} \left(\frac{\lambda+|\lambda|}{2}\phi(\lambda+a) + \frac{\lambda-|\lambda|}{2}\Phi(\lambda+a)\right)d\lambda\right]$$

2.1 Generalized Ostrowski Type Inequalities with Parameter

and

$$M_4 = \frac{1}{b-a} \left[\int_{-\frac{b-a}{2}}^{0} \left(\frac{\lambda - |\lambda|}{2} \phi(\lambda+b) + \frac{\lambda + |\lambda|}{2} \Phi(\lambda+b) \right) d\lambda \right.$$

$$\left. + \int_{0}^{\frac{b-a}{2}} \left(\frac{\lambda - |\lambda|}{2} \phi(\lambda+a) + \frac{\lambda + |\lambda|}{2} \Phi(\lambda+a) \right) d\lambda \right],$$

which is Corollary 1 of [116].

Special Case 2.1.4.(a) If we take, $\phi(x) = \phi_0 \neq 0$ and $\Phi(x) = \Phi_0 \neq 0$ in (2.1.8), then

$$\frac{(b-a)}{8}(\phi_0 - \Phi_0) \leq f\left(\frac{a+b}{2}\right) - \frac{1}{b-a}\int_a^b f(t)dt \leq \frac{(b-a)}{8}(\Phi_0 - \phi_0),$$

which is in fact the Special Case 1 of Theorem 1 and Corollary 1 as cited in [116] and [184] respectively.

Special Case 2.1.4.(b) If we take, $\phi(x) = \phi_1 x + \phi_0 \neq 0$ and $\Phi(x) = \Phi_1 x + \Phi_0 \neq 0$ in (2.1.8), then

$$m_5 \leq f\left(\frac{a+b}{2}\right) - \frac{1}{b-a}\int_a^b f(t)dt \leq M_5,$$

where

$$m_5 = \frac{(b-a)}{8}\left(\frac{b-a}{3}(\phi_1 + \Phi_1) + a\phi_1 - b\Phi_1 + \phi_0 - \Phi_0\right)$$

and

$$M_5 = \frac{(b-a)}{8}\left(\frac{b-a}{3}(\phi_1 + \Phi_1) + a\Phi_1 - b\phi_1 + \Phi_0 - \phi_0\right),$$

which is the result of Corollary 1 of [116].

Remark 2.1.4 By choosing $\epsilon = \frac{1}{3}$ in (2.1.7), we obtain bounds for $\frac{1}{3}$ Simpson's rule in the following corollary.

Corollary 2.1.5 *Let all the assumptions of Theorem 2.1.1 be valid. Then*

$$m_6 \leq \frac{1}{3}\left[\frac{f(a)+f(b)}{2} + 2f\left(\frac{a+b}{2}\right)\right] - \frac{1}{b-a}\int_a^b f(t)dt \leq M_6, \quad (2.1.9)$$

where

$$m_6 = \frac{1}{b-a}\left[\int_{-\frac{b-a}{6}}^{\frac{b-a}{3}}\left(\frac{\lambda+|\lambda|}{2}\phi\left(\lambda+\frac{5a+b}{6}\right)+\frac{\lambda-|\lambda|}{2}\Phi\left(\lambda+\frac{5a+b}{6}\right)\right)d\lambda\right.$$
$$\left.+\int_{-\frac{b-a}{3}}^{\frac{b-a}{6}}\left(\frac{\lambda+|\lambda|}{2}\phi\left(\lambda+\frac{a+5b}{6}\right)+\frac{\lambda-|\lambda|}{2}\Phi\left(\lambda+\frac{a+5b}{6}\right)\right)d\lambda\right]$$

and

$$M_6 = \frac{1}{b-a}\left[\int_{-\frac{b-a}{6}}^{\frac{b-a}{3}}\left(\frac{\lambda-|\lambda|}{2}\phi\left(\lambda+\frac{5a+b}{6}\right)+\frac{\lambda+|\lambda|}{2}\Phi\left(\lambda+\frac{5a+b}{6}\right)\right)d\lambda\right.$$
$$\left.+\int_{-\frac{b-a}{3}}^{\frac{b-a}{6}}\left(\frac{\lambda-|\lambda|}{2}\phi\left(\lambda+\frac{a+5b}{6}\right)+\frac{\lambda+|\lambda|}{2}\Phi\left(\lambda+\frac{a+5b}{6}\right)\right)d\lambda\right].$$

Special Case 2.1.5.(a) If we take, $\phi(x) = \phi_0 \neq 0$ and $\Phi(x) = \Phi_0 \neq 0$ in (2.1.9), then

$$\frac{5(b-a)}{72}(\phi_0 - \Phi_0) \leq \frac{1}{3}\left[\frac{f(a)+f(b)}{2}+2f\left(\frac{a+b}{2}\right)\right] - \frac{1}{b-a}\int_a^b f(t)dt$$
$$\leq \frac{5(b-a)}{72}(\Phi_0 - \phi_0),$$

which is Corollay 4 of [184].

Special Case 2.1.5.(b) If we take, $\phi(x) = \phi_1 x + \phi_0 \neq 0$ and $\Phi(x) = \Phi_1 x + \Phi_0 \neq 0$ in (2.1.9), then

$$m_7 \leq \frac{1}{3}\left[\frac{f(a)+f(b)}{2}+2f\left(\frac{a+b}{2}\right)\right] - \frac{1}{b-a}\int_a^b f(t)dt \leq M_7,$$

where

$$m_7 = \frac{(b-a)}{72}\left[(b-a)(\phi_1+\Phi_1)+\frac{a}{2}(7\phi_1-3\Phi_1)+\frac{b}{2}(3\phi_1-7\Phi_1)+5(\phi_0-\Phi_0)\right]$$

and

$$M_7 = \frac{(b-a)}{72}\left[(b-a)(\phi_1+\Phi_1)+\frac{a}{2}(7\phi_1-3\Phi_1)+\frac{b}{2}(3\Phi_1-7\phi_1)+5(\Phi_0-\phi_0)\right].$$

Remark 2.1.5 By choosing $\epsilon = \frac{1}{2}$ in (2.1.7), we get the bound of average midpoint and trapezoidal rule in the following corollary.

2.1 Generalized Ostrowski Type Inequalities with Parameter

Corollary 2.1.6 *Let all the assumptions of Theorem 2.1.1 be valid. Then*

$$m_8 \leq \frac{1}{2}\left[\frac{f(a)+f(b)}{2} + f\left(\frac{a+b}{2}\right)\right] - \frac{1}{b-a}\int_a^b f(t)dt \leq M_8, \quad (2.1.10)$$

where

$$m_8 = \frac{1}{b-a}\left[\int_{-\frac{b-a}{4}}^{\frac{b-a}{4}} \left(\frac{\lambda+|\lambda|}{2}\phi\left(\lambda + \frac{3a+b}{4}\right) + \frac{\lambda-|\lambda|}{2}\Phi\left(\lambda + \frac{3a+b}{4}\right)\right)d\lambda\right.$$
$$\left. + \int_{-\frac{b-a}{4}}^{\frac{b-a}{4}} \left(\frac{\lambda+|\lambda|}{2}\phi\left(\lambda + \frac{a+3b}{4}\right) + \frac{\lambda-|\lambda|}{2}\Phi\left(\lambda + \frac{a+3b}{4}\right)\right)d\lambda\right]$$

and

$$M_8 = \frac{1}{b-a}\left[\int_{-\frac{b-a}{4}}^{\frac{b-a}{4}} \left(\frac{\lambda-|\lambda|}{2}\phi\left(\lambda + \frac{3a+b}{4}\right) + \frac{\lambda+|\lambda|}{2}\Phi\left(\lambda + \frac{3a+b}{4}\right)\right)d\lambda\right.$$
$$\left. + \int_{-\frac{b-a}{4}}^{\frac{b-a}{4}} \left(\frac{\lambda-|\lambda|}{2}\phi\left(\lambda + \frac{a+3b}{4}\right) + \frac{\lambda+|\lambda|}{2}\Phi\left(\lambda + \frac{a+3b}{4}\right)\right)d\lambda\right].$$

Special Case 2.1.6.(a) If we take, $\phi(x) = \phi_0 \neq 0$ and $\Phi(x) = \Phi_0 \neq 0$ in (2.1.10), then

$$\frac{(b-a)}{16}(\phi_0 - \Phi_0) \leq \frac{1}{2}\left[\frac{f(a)+f(b)}{2} + f\left(\frac{a+b}{2}\right)\right] - \frac{1}{b-a}\int_a^b f(t)dt$$
$$\leq \frac{(b-a)}{16}(\Phi_0 - \phi_0),$$

which is Corollary 3 of [184].

Special Case 2.1.6.(b) If we take, $\phi(x) = \phi_1 x + \phi_0 \neq 0$ and $\Phi(x) = \Phi_1 x + \Phi_0 \neq 0$ in (2.1.10), then

$$m_9 \leq \frac{1}{2}\left[\frac{f(a)+f(b)}{2} + f\left(\frac{a+b}{2}\right)\right] - \frac{1}{b-a}\int_a^b f(t)dt \leq M_9,$$

where

$$m_9 = \frac{(b-a)}{16}\left[\frac{(b-a)}{6}(\phi_1 + \Phi_1) + \frac{a}{2}(\phi_1 - \Phi_1) + \frac{b}{2}(\phi_1 - \Phi_1) + \phi_0 - \Phi_0\right]$$

and

$$M_9 = \frac{(b-a)}{16}\left[\frac{(b-a)}{6}(\phi_1 + \Phi_1) + \frac{a}{2}(\Phi_1 - \phi_1) + \frac{b}{2}(\Phi_1 - \phi_1) + \Phi_0 - \phi_0\right].$$

Remark 2.1.6 By choosing $x = a$ in (2.1.3), then for any value of $\epsilon \in [0, 1]$, we get the bound of trapezoidal rule (Hermite–Hadamard right bound) in the following corollary.

Corollary 2.1.7 *Let all the assumptions of Theorem 2.1.1 be valid. Then*

$$m_{10} \leq \frac{f(a) + f(b)}{2} - \frac{1}{b-a}\int_a^b f(t)dt \leq M_{10}, \qquad (2.1.11)$$

where

$$m_{10} = \frac{1}{b-a}\left[\int_{-\frac{b-a}{2}}^{\frac{b-a}{2}}\left(\frac{\lambda + |\lambda|}{2}\phi\left(\lambda + \frac{a+b}{2}\right) + \frac{\lambda - |\lambda|}{2}\Phi\left(\lambda + \frac{a+b}{2}\right)\right)d\lambda\right]$$

and

$$M_{10} = \frac{1}{b-a}\left[\int_{-\frac{b-a}{2}}^{\frac{b-a}{2}}\left(\frac{\lambda - |\lambda|}{2}\phi\left(\lambda + \frac{a+b}{2}\right) + \frac{\lambda + |\lambda|}{2}\Phi\left(\lambda + \frac{a+b}{2}\right)\right)d\lambda\right],$$

which is Corollary 2 cited in [116].

If we take, $\phi(x) = \phi_0 \neq 0$ and $\Phi(x) = \Phi_0 \neq 0$, and $\phi(x) = \phi_1 x + \phi_0 \neq 0$ and $\Phi(x) = \Phi_1 x + \Phi_0 \neq 0$ in (2.1.11), then we obtain the results as stated in Special Case 2.1.2.(a) and Special Case 2.1.2.(b) respectively.

Remark 2.1.7 By choosing $x = b$ and $\epsilon = 0$ in (2.1.3), we get the bound for trapezoidal rule (Hermite–Hadamard right bound) in following corollary.

Corollary 2.1.8 *Let all the assumptions of Theorem 2.1.1 be valid. Then*

$$m_{11} \leq \frac{f(a) + f(b)}{2} - \frac{1}{b-a}\int_a^b f(t)dt \leq M_{11}, \qquad (2.1.12)$$

where

$$m_{11} = \frac{1}{b-a}\left[\int_{-(b-a)}^{0}\left(\frac{\lambda + |\lambda|}{2}\phi(\lambda + b) + \frac{\lambda - |\lambda|}{2}\Phi(\lambda + b)\right)d\lambda\right.$$
$$-\int_{-\frac{b-a}{2}}^{\frac{b-a}{2}}\left(\frac{\lambda + |\lambda|}{2}\phi\left(\lambda + \frac{a+b}{2}\right) + \frac{\lambda - |\lambda|}{2}\Phi\left(\lambda + \frac{a+b}{2}\right)\right)d\lambda$$
$$\left.+\int_{0}^{b-a}\left(\frac{\lambda + |\lambda|}{2}\phi(\lambda + a) + \frac{\lambda - |\lambda|}{2}\Phi(\lambda + a)\right)d\lambda\right]$$

2.1 Generalized Ostrowski Type Inequalities with Parameter

and

$$M_{11} = \frac{1}{b-a}\left[\int_{-(b-a)}^{0}\left(\frac{\lambda-|\lambda|}{2}\phi(\lambda+b)+\frac{\lambda+|\lambda|}{2}\Phi(\lambda+b)\right)d\lambda\right.$$
$$-\int_{-\frac{b-a}{2}}^{\frac{b-a}{2}}\left(\frac{\lambda-|\lambda|}{2}\phi\left(\lambda+\frac{a+b}{2}\right)+\frac{\lambda+|\lambda|}{2}\Phi\left(\lambda+\frac{a+b}{2}\right)\right)d\lambda$$
$$\left.+\int_{0}^{b-a}\left(\frac{\lambda-|\lambda|}{2}\phi(\lambda+a)+\frac{\lambda+|\lambda|}{2}\Phi(\lambda+a)\right)d\lambda\right].$$

Special Case 2.2.8.(a) If we take, $\phi(x) = \phi_0 \neq 0$ and $\Phi(x) = \Phi_0 \neq 0$ in (2.1.12), then

$$\frac{3(b-a)}{8}(\phi_0-\Phi_0) \leq \left[\frac{f(a)+f(b)}{2}\right] - \frac{1}{b-a}\int_{a}^{b}f(t)dt \leq \frac{3(b-a)}{8}(\Phi_0-\phi_0).$$

Special Case 2.2.8.(b) If we take, $\phi(x) = \phi_1 x + \phi_0 \neq 0$ and $\Phi(x) = \Phi_1 x + \Phi_0 \neq 0$ in (2.1.12), then

$$m_{12} \leq \left[\frac{f(a)+f(b)}{2}\right] - \frac{1}{b-a}\int_{a}^{b}f(t)dt \leq M_{12},$$

where

$$m_{12} = \frac{(b-a)}{2}\left[\frac{7(b-a)}{12}(\phi_1+\Phi_1)+\frac{a}{8}(7\phi_1+\Phi_1)-\frac{b}{8}(\phi_1+7\Phi_1)+\frac{3}{4}(\phi_0-\Phi_0)\right]$$

and

$$M_{12} = \frac{(b-a)}{2}\left[\frac{7(b-a)}{12}(\phi_1+\Phi_1)+\frac{a}{8}(\phi_1+7\Phi_1)-\frac{b}{8}(7\phi_1+\Phi_1)+\frac{3}{4}(\Phi_0-\phi_0)\right].$$

Remark 2.1.8 By choosing $x = b$ and $\epsilon = \frac{1}{2}$ in (2.1.3), we get the bound for trapezoidal rule (Hermite–Hadamard right bound) in the following corollary.

Corollary 2.1.9 *Let all the assumptions of Theorem* 2.1.1 *be valid. Then*

$$m_{13} \leq \left[\frac{f(a)+f(b)}{2}\right] - \frac{1}{b-a}\int_{a}^{b}f(t)dt \leq M_{13}, \qquad (2.1.13)$$

where

$$m_{13} = \frac{1}{b-a}\left[\int_{-\frac{(b-a)}{4}}^{\frac{3(b-a)}{4}}\left(\frac{\lambda+|\lambda|}{2}\phi\left(\lambda+\frac{3a+b}{4}\right)+\frac{\lambda-|\lambda|}{2}\Phi\left(\lambda+\frac{3a+b}{4}\right)\right)d\lambda\right.$$

$$-\int_{-\frac{b-a}{2}}^{\frac{b-a}{2}}\left(\frac{\lambda+|\lambda|}{2}\phi\left(\lambda+\frac{a+b}{2}\right)+\frac{\lambda-|\lambda|}{2}\Phi\left(\lambda+\frac{a+b}{2}\right)\right)d\lambda$$

$$\left.+\int_{-\frac{3(b-a)}{4}}^{\frac{b-a}{4}}\left(\frac{\lambda+|\lambda|}{2}\phi\left(\lambda+\frac{a+3b}{4}\right)+\frac{\lambda-|\lambda|}{2}\Phi\left(\lambda+\frac{a+3b}{4}\right)\right)d\lambda\right]$$

and

$$M_{13} = \frac{1}{b-a}\left[\int_{-\frac{(b-a)}{4}}^{\frac{3(b-a)}{4}}\left(\frac{\lambda-|\lambda|}{2}\phi\left(\lambda+\frac{3a+b}{4}\right)+\frac{\lambda+|\lambda|}{2}\Phi\left(\lambda+\frac{3a+b}{4}\right)\right)d\lambda\right.$$

$$-\int_{-\frac{b-a}{2}}^{\frac{b-a}{2}}\left(\frac{\lambda-|\lambda|}{2}\phi\left(\lambda+\frac{a+b}{2}\right)+\frac{\lambda+|\lambda|}{2}\Phi\left(\lambda+\frac{a+b}{2}\right)\right)d\lambda$$

$$\left.+\int_{-\frac{3(b-a)}{4}}^{\frac{b-a}{4}}\left(\frac{\lambda-|\lambda|}{2}\phi\left(\lambda+\frac{a+3b}{4}\right)+\frac{\lambda+|\lambda|}{2}\Phi\left(\lambda+\frac{a+3b}{4}\right)\right)d\lambda\right].$$

Special Case 2.1.9.(a) If we take, $\phi(x) = \phi_0 \neq 0$ and $\Phi(x) = \Phi_0 \neq 0$ in (2.1.13), then

$$\frac{3(b-a)}{16}(\phi_0 - \Phi_0) \leq \left[\frac{f(a)+f(b)}{2}\right] - \frac{1}{b-a}\int_a^b f(t)dt \leq \frac{3(b-a)}{16}(\Phi_0 - \phi_0).$$

Special Case 2.1.9.(b) If we take, $\phi(x) = \phi_1 x + \phi_0 \neq 0$ and $\Phi(x) = \Phi_1 x + \Phi_0 \neq 0$ in (2.1.13), then

$$m_{14} \leq \left[\frac{f(a)+f(b)}{2}\right] - \frac{1}{b-a}\int_a^b f(t)dt \leq M_{14},$$

where

$$m_{14} = \frac{(b-a)}{16}\left[\frac{5}{3}(b-a)(\phi_1+\Phi_1)+\frac{a}{2}(5\phi_1-\Phi_1)+\frac{b}{2}(\phi_1-5\Phi_1)+3(\phi_0-\Phi_0)\right]$$

and

$$M_{14} = \frac{(b-a)}{16}\left[\frac{5}{3}(b-a)(\phi_1+\Phi_1)+\frac{a}{2}(5\Phi_1-\phi_1)+\frac{b}{2}(\Phi_1-5\phi_1)+3(\Phi_0-\phi_0)\right].$$

2.1 Generalized Ostrowski Type Inequalities with Parameter

Remark 2.1.9 By choosing $x = b$ and $\epsilon = \dfrac{1}{3}$ in (2.1.3), we get the bound for trapezoidal rule (Hermite–Hadamard right bound) in the following corollary.

Corollary 2.1.10 *Let all the assumptions of Theorem 2.1.1 be valid. Then*

$$m_{15} \leq \left[\frac{f(a)+f(b)}{2}\right] - \frac{1}{b-a}\int_a^b f(t)dt \leq M_{15}, \qquad (2.1.14)$$

where

$$m_{15} = \frac{1}{b-a}\left[\int_{-\frac{(b-a)}{6}}^{\frac{5(b-a)}{6}} \left(\frac{\lambda+|\lambda|}{2}\phi\left(\lambda+\frac{5a+b}{6}\right) + \frac{\lambda-|\lambda|}{2}\Phi\left(\lambda+\frac{5a+b}{6}\right)\right)d\lambda\right.$$
$$- \int_{-\frac{b-a}{2}}^{\frac{b-a}{2}} \left(\frac{\lambda+|\lambda|}{2}\phi\left(\lambda+\frac{a+b}{2}\right) + \frac{\lambda-|\lambda|}{2}\Phi\left(\lambda+\frac{a+b}{2}\right)\right)d\lambda$$
$$\left. + \int_{-\frac{5(b-a)}{6}}^{\frac{b-a}{6}} \left(\frac{\lambda+|\lambda|}{2}\phi\left(\lambda+\frac{a+5b}{6}\right) + \frac{\lambda-|\lambda|}{2}\Phi\left(\lambda+\frac{a+5b}{6}\right)\right)d\lambda\right]$$

and

$$M_{15} = \frac{1}{b-a}\left[\int_{-\frac{(b-a)}{6}}^{\frac{5(b-a)}{6}} \left(\frac{\lambda-|\lambda|}{2}\phi\left(\lambda+\frac{5a+b}{4}\right) + \frac{\lambda+|\lambda|}{2}\Phi\left(\lambda+\frac{5a+b}{6}\right)\right)d\lambda\right.$$
$$- \int_{-\frac{b-a}{2}}^{\frac{b-a}{2}} \left(\frac{\lambda-|\lambda|}{2}\phi\left(\lambda+\frac{a+b}{2}\right) + \frac{\lambda+|\lambda|}{2}\Phi\left(\lambda+\frac{a+b}{2}\right)\right)d\lambda$$
$$\left. + \int_{-\frac{5(b-a)}{6}}^{\frac{b-a}{6}} \left(\frac{\lambda-|\lambda|}{2}\phi\left(\lambda+\frac{a+5b}{6}\right) + \frac{\lambda+|\lambda|}{2}\Phi\left(\lambda+\frac{a+5b}{6}\right)\right)d\lambda\right].$$

Special Case 2.1.10.(a) If we take, $\phi(x) = \phi_0 \neq 0$ and $\Phi(x) = \Phi_0 \neq 0$ in (2.1.14), then

$$\frac{17(b-a)}{72}(\phi_0 - \Phi_0) \leq \left[\frac{f(a)+f(b)}{2}\right] - \frac{1}{b-a}\int_a^b f(t)dt \leq \frac{17(b-a)}{72}(\Phi_0 - \phi_0).$$

Special Case 2.1.10.(b) If we take, $\phi(x) = \phi_1 x + \phi_0 \neq 0$ and $\Phi(x) = \Phi_1 x + \Phi_0 \neq 0$ in (2.1.14), then

$$m_{16} \leq \left[\frac{f(a)+f(b)}{2}\right] - \frac{1}{b-a}\int_a^b f(t)dt \leq M_{16},$$

where

$$m_{16} = \frac{(b-a)}{72}\left[11(b-a)(\phi_1 + \Phi_1) + \frac{3a}{2}(33\phi_1 - \Phi_1) + \frac{3b}{2}(\phi_1 - 33\Phi_1)\right.$$
$$\left. + 17(\phi_0 - \Phi_0)\right]$$

and

$$M_{16} = \frac{(b-a)}{72}\left[(b-a)(\phi_1 + \Phi_1) + \frac{3a}{2}(33\Phi_1 - \phi_1) + \frac{3b}{2}(\Phi_1 - 33\phi_1)\right.$$
$$\left. + 17(\Phi_0 - \phi_0)\right].$$

Remark 2.1.10 By choosing $x = b$ and $\epsilon = \frac{1}{4}$ in (2.1.3), we get the bound for trapezoidal rule (Hermite–Hadamard right bound) in following corollary.

Corollary 2.1.11 *Let all the assumptions of Theorem 2.1.1 be valid. Then*

$$m_{17} \leq \left[\frac{f(a)+f(b)}{2}\right] - \frac{1}{b-a}\int_a^b f(t)dt \leq M_{17}, \qquad (2.1.15)$$

where

$$m_{17} = \frac{1}{b-a}\left[\int_{-\frac{(b-a)}{8}}^{\frac{7(b-a)}{8}}\left(\frac{\lambda+|\lambda|}{2}\phi\left(\lambda+\frac{7a+b}{8}\right) + \frac{\lambda-|\lambda|}{2}\Phi\left(\lambda+\frac{7a+b}{8}\right)\right)d\lambda\right.$$
$$- \int_{-\frac{b-a}{2}}^{\frac{b-a}{2}}\left(\frac{\lambda+|\lambda|}{2}\phi\left(\lambda+\frac{a+b}{2}\right) + \frac{\lambda-|\lambda|}{2}\Phi\left(\lambda+\frac{a+b}{2}\right)\right)d\lambda$$
$$\left. + \int_{-\frac{7(b-a)}{8}}^{\frac{b-a}{8}}\left(\frac{\lambda+|\lambda|}{2}\phi\left(\lambda+\frac{a+7b}{8}\right) + \frac{\lambda-|\lambda|}{2}\Phi\left(\lambda+\frac{a+7b}{8}\right)\right)d\lambda\right]$$

and

$$M_{17} = \frac{1}{b-a}\left[\int_{-\frac{(b-a)}{8}}^{\frac{7(b-a)}{8}}\left(\frac{\lambda-|\lambda|}{2}\phi\left(\lambda+\frac{7a+b}{8}\right) + \frac{\lambda+|\lambda|}{2}\Phi\left(\lambda+\frac{7a+b}{8}\right)\right)d\lambda\right.$$
$$- \int_{-\frac{b-a}{2}}^{\frac{b-a}{2}}\left(\frac{\lambda-|\lambda|}{2}\phi\left(\lambda+\frac{a+b}{2}\right) + \frac{\lambda+|\lambda|}{2}\Phi\left(\lambda+\frac{a+b}{2}\right)\right)d\lambda$$
$$\left. + \int_{-\frac{7(b-a)}{8}}^{\frac{b-a}{8}}\left(\frac{\lambda-|\lambda|}{2}\phi\left(\lambda+\frac{a+7b}{8}\right) + \frac{\lambda+|\lambda|}{2}\Phi\left(\lambda+\frac{a+7b}{8}\right)\right)d\lambda\right].$$

2.1 Generalized Ostrowski Type Inequalities with Parameter

Special Case 2.1.11.(a) If we take, $\phi(x) = \phi_0 \neq 0$ and $\Phi(x) = \Phi_0 \neq 0$ in (2.1.15), then

$$\frac{17(b-a)}{64}(\phi_0 - \Phi_0) \leq \left[\frac{f(a)+f(b)}{2}\right] - \frac{1}{b-a}\int_a^b f(t)dt \leq \frac{17(b-a)}{64}(\Phi_0 - \phi_0).$$

Special Case 2.1.11.(b) If we take, $\phi(x) = \phi_1 x + \phi_0 \neq 0$ and $\Phi(x) = \Phi_1 x + \Phi_0 \neq 0$ in (2.1.15), then

$$m_{18} \leq \left[\frac{f(a)+f(b)}{2}\right] - \frac{1}{b-a}\int_a^b f(t)dt \leq M_{18},$$

where

$$m_{18} = \frac{(b-a)}{64}\left[\frac{35}{3}(b-a)(\phi_1 + \Phi_1) + \frac{a}{2}(35\phi_1 + \Phi_1) - \frac{b}{2}(\phi_1 + 35\Phi_1)\right.$$
$$\left. + 17(\phi_0 - \Phi_0)\right]$$

and

$$M_{18} = \frac{(b-a)}{64}\left[\frac{35}{3}(b-a)(\phi_1 + \Phi_1) + \frac{a}{2}(35\Phi_1 + \phi_1+) - \frac{b}{2}(\Phi_1 + 35\phi_1)\right.$$
$$\left. + 17(\Phi_0 - \phi_0)\right].$$

Remark 2.1.11 If we choose $x = \frac{3a+b}{4}$ and $\epsilon = 0$ in (2.1.3), we get the bound for trapezoidal type rule in the following corollary.

Corollary 2.1.12 Let all the assumptions of Theorem 2.1.1 be valid. Then

$$m_{19} \leq \frac{1}{2}\left[f\left(\frac{3a+b}{4}\right) + f\left(\frac{a+3b}{4}\right)\right] - \frac{1}{b-a}\int_a^b f(t)dt \leq M_{19}, \quad (2.1.16)$$

where

$$m_{19} = \frac{1}{b-a}\left[\int_{-\frac{b-a}{4}}^0 \left(\frac{\lambda + |\lambda|}{2}\phi(\lambda + b) + \frac{\lambda - |\lambda|}{2}\Phi(\lambda + b)\right)d\lambda\right.$$
$$+ \int_{-\frac{b-a}{4}}^{\frac{b-a}{4}} \left(\frac{\lambda + |\lambda|}{2}\phi\left(\lambda + \frac{a+b}{2}\right) + \frac{\lambda - |\lambda|}{2}\Phi\left(\lambda + \frac{a+b}{2}\right)\right)d\lambda$$
$$\left. + \int_0^{\frac{b-a}{4}} \left(\frac{\lambda + |\lambda|}{2}\phi(\lambda + a) + \frac{\lambda - |\lambda|}{2}\Phi(\lambda + a)\right)d\lambda\right]$$

and

$$M_{19} = \frac{1}{b-a}\left[\int_{-\frac{b-a}{4}}^{0}\left(\frac{\lambda-|\lambda|}{2}\phi(\lambda+b) + \frac{\lambda+|\lambda|}{2}\Phi(\lambda+b)\right)d\lambda\right.$$

$$+ \int_{-\frac{b-a}{4}}^{\frac{b-a}{4}}\left(\frac{\lambda-|\lambda|}{2}\phi\left(\lambda+\frac{a+b}{2}\right) + \frac{\lambda+|\lambda|}{2}\Phi\left(\lambda+\frac{a+b}{2}\right)\right)d\lambda$$

$$+ \left.\int_{0}^{\frac{b-a}{4}}\left(\frac{\lambda-|\lambda|}{2}\phi(\lambda+a) + \frac{\lambda+|\lambda|}{2}\Phi(\lambda+a)\right)d\lambda\right].$$

Special Case 2.1.12.(a) If we take, $\phi(x) = \phi_0 \neq 0$ and $\Phi(x) = \Phi_0 \neq 0$ in (2.1.16), then

$$\frac{(b-a)}{16}(\phi_0 - \Phi_0) \leq \frac{1}{2}\left[f\left(\frac{3a+b}{4}\right) + f\left(\frac{a+3b}{4}\right)\right] - \frac{1}{b-a}\int_a^b f(t)dt$$

$$\leq \frac{(b-a)}{16}(\Phi_0 - \phi_0).$$

Special Case 2.1.12.(b) If we take, $\phi(x) = \phi_1 x + \phi_0 \neq 0$ and $\Phi(x) = \Phi_1 x + \Phi_0 \neq 0$ in (2.1.16), then

$$m_{20} \leq \frac{1}{2}\left[f\left(\frac{3a+b}{4}\right) + f\left(\frac{a+3b}{4}\right)\right] - \frac{1}{b-a}\int_a^b f(t)dt \leq M_{20},$$

where

$$m_{20} = \frac{(b-a)}{16}\left[\frac{(b-a)}{6}(\phi_1 + \Phi_1) + \frac{a}{4}(3\phi_1 - \Phi_1) + \frac{b}{4}(\phi_1 - 3\Phi_1) + \phi_0 - \Phi_0\right]$$

and

$$M_{20} = \frac{(b-a)}{16}\left[\frac{(b-a)}{6}(\phi_1 + \Phi_1) + \frac{a}{4}(3\Phi_1 - \phi_1) + \frac{b}{4}(\Phi_1 - 3\phi_1) + \Phi_0 - \phi_0\right].$$

Remark 2.1.12 If we choose $x = \frac{2a+b}{3}$ and $\epsilon = \frac{1}{4}$ in (2.1.3), then we get the bound of $\frac{3}{8}$ Simpson's rule in the following corollary.

Corollary 2.1.13 *Let all the assumptions of Theorem 2.1.1 be valid. Then*

$$m_{21} \leq \frac{3}{8}\left[\frac{f(a)+f(b)}{3} + f\left(\frac{2a+b}{3}\right) + f\left(\frac{a+2b}{3}\right)\right] - \frac{1}{b-a}\int_a^b f(t)dt \leq M_{21},$$

$$(2.1.17)$$

2.1 Generalized Ostrowski Type Inequalities with Parameter

where

$$m_{21} = \frac{1}{b-a}\left[\int_{-\frac{b-a}{8}}^{\frac{5(b-a)}{24}}\left(\frac{\lambda+|\lambda|}{2}\phi\left(\lambda+\frac{7a+b}{8}\right)+\frac{\lambda-|\lambda|}{2}\Phi\left(\lambda+\frac{7a+b}{8}\right)\right)d\lambda\right.$$

$$+\int_{-\frac{b-a}{6}}^{\frac{b-a}{6}}\left(\frac{\lambda+|\lambda|}{2}\phi\left(\lambda+\frac{a+b}{2}\right)+\frac{\lambda-|\lambda|}{2}\Phi\left(\lambda+\frac{a+b}{2}\right)\right)d\lambda$$

$$+\int_{-\frac{5(b-a)}{24}}^{\frac{b-a}{8}}\left(\frac{\lambda+|\lambda|}{2}\phi\left(\lambda+\frac{a+7b}{8}\right)+\frac{\lambda-|\lambda|}{2}\Phi\left(\lambda+\frac{a+7b}{8}\right)\right)d\lambda\right]$$

and

$$M_{21} = \frac{1}{b-a}\left[\int_{-\frac{b-a}{8}}^{\frac{5(b-a)}{24}}\left(\frac{\lambda-|\lambda|}{2}\phi\left(\lambda+\frac{7a+b}{8}\right)+\frac{\lambda+|\lambda|}{2}\Phi\left(\lambda+\frac{7a+b}{8}\right)\right)d\lambda\right.$$

$$+\int_{-\frac{b-a}{6}}^{\frac{b-a}{6}}\left(\frac{\lambda-|\lambda|}{2}\phi\left(\lambda+\frac{a+b}{2}\right)+\frac{\lambda+|\lambda|}{2}\Phi\left(\lambda+\frac{a+b}{2}\right)\right)d\lambda$$

$$+\int_{-\frac{5(b-a)}{24}}^{\frac{b-a}{8}}\left(\frac{\lambda-|\lambda|}{2}\phi\left(\lambda+\frac{a+7b}{8}\right)+\frac{\lambda+|\lambda|}{2}\Phi\left(\lambda+\frac{a+7b}{8}\right)\right)d\lambda\right].$$

Special Case 2.1.13.(a) If we take, $\phi(x) = \phi_0 \neq 0$ and $\Phi(x) = \Phi_0 \neq 0$ in (2.1.17), then

$$\frac{25(b-a)}{576}(\Phi_0 - \phi_0)$$

$$\leq \frac{3}{8}\left[\frac{f(a)+f(b)}{3}+f\left(\frac{2a+b}{3}\right)+f\left(\frac{a+2b}{3}\right)\right]-\frac{1}{b-a}\int_a^b f(t)dt$$

$$\leq \frac{25(b-a)}{576}(\phi_0 - \Phi_0).$$

Special Case 2.1.13.(b) If we take, $\phi(x) = \phi_1 x + \phi_0 \neq 0$ and $\Phi(x) = \Phi_1 x + \Phi_0 \neq 0$ in (2.1.17), then

$$m_{22} \leq \frac{3}{8}\left[\frac{f(a)+f(b)}{3}+f\left(\frac{2a+b}{3}\right)+f\left(\frac{a+2b}{3}\right)\right]-\frac{1}{b-a}\int_a^b f(t)dt \leq M_{22},$$

where

$$m_{22} = \frac{(b-a)}{192}\left[(b-a)(\phi_1+\Phi_1)+\frac{31}{6}(a\phi_1-b\Phi_1)+\frac{19}{6}(b\phi_1-a\Phi_1)\right.$$

$$\left.+\frac{25}{3}(\phi_0-\Phi_0)\right]$$

and

$$M_{22} = \frac{(b-a)}{192}\left[(b-a)(\phi_1 + \Phi_1) + \frac{31}{6}(b\Phi_1 - a\phi_1) + \frac{19}{6}(a\Phi_1 - b\phi_1)\right.$$
$$\left. + \frac{25}{3}(\Phi_0 - \phi_0)\right].$$

2.1.2 Ostrowski Type Inequalities for Bounded Below Only and Bounded Above Only Differentiable Functions

The condition $\phi(x) \leq f'(x) \leq \Phi(x)$ is true in Theorem 2.1.1, but sometimes we are not able to find both bounds of a function. Now we define two theorems. The first one would be helpful when f' is bounded from above and the second one would be helpful when f' is bounded from below.

Theorem 2.1.14 *Let $f : I \to \mathbb{R}$ be differentiable function in I^0. If f' is bounded from below, then $\phi(x) \leq f'(x)$ such that $\phi \in C[a,b]$, $x \in [a,b]$, $\forall \epsilon \in [0,1]$ we have*

$$m_{23}(x,\epsilon) \leq \left[\epsilon \frac{f(a)+f(b)}{2} + (1-\epsilon)\frac{f(x)+f(a+b-x)}{2}\right] - \frac{1}{b-a}\int_a^b f(t)dt$$
$$\leq M_{23}(x,\epsilon), \qquad (2.1.18)$$

where

$$m_{23}(x,\epsilon) = \frac{1}{b-a}\left[\int_a^b \left(t - \frac{a+b}{2}\right)\phi(t)dt\right.$$
$$+ \frac{b-a}{2}\left(\int_a^x (1-\epsilon)\phi(t)dt - \int_{a+b-x}^b (1-\epsilon)\phi(t)dt\right)$$
$$- \max\left\{\epsilon\frac{b-a}{2}, \left(x - \frac{(2-\epsilon)a + \epsilon b}{2}\right),\right.$$
$$\left.\left.\left(\frac{a+b}{2} - x\right)\right\}\left(f(b) - f(a) - \int_a^b \phi(t)dt\right)\right]$$

2.1 Generalized Ostrowski Type Inequalities with Parameter

and

$$M_{23}(x, \epsilon) = \frac{1}{b-a}\left[\int_a^b \left(t - \frac{a+b}{2}\right)\phi(t)dt \right.$$
$$+ \frac{b-a}{2}\left(\int_a^x (1-\epsilon)\phi(t)dt\right.$$
$$\left.- \int_{a+b-x}^b (1-\epsilon)\phi(t)dt\right)$$
$$+ \max\left\{\epsilon\frac{b-a}{2}, \left(x - \frac{(2-\epsilon)a + \epsilon b}{2}\right),\right.$$
$$\left.\left.\left(\frac{a+b}{2} - x\right)\right\}\left(f(b) - f(a) - \int_a^b \phi(t)dt\right)\right].$$

Proof Since

$$\int_a^b P(x,t)\left(f'(t) - \phi(t)\right)dt = (b-a)\left[\epsilon\frac{f(a) + f(b)}{2} + (1-\epsilon)\frac{f(x) + f(a+b-x)}{2}\right]$$
$$- \int_a^b f(t)dt - \int_a^b P(x,t)\phi(t)dt$$
$$= (b-a)\left[\epsilon\frac{f(a) + f(b)}{2} + (1-\epsilon)\frac{f(x) + f(a+b-x)}{2}\right]$$
$$- \int_a^b f(t)dt - \left[\int_a^x \left(t - \left(a + \epsilon\frac{b-a}{2}\right)\right)\phi(t)dt\right.$$
$$+ \int_x^{a+b-x} \left(t - \frac{a+b}{2}\right)\phi(t)dt$$
$$\left.+ \int_{a+b-x}^b \left(t - \left(b - \epsilon\frac{b-a}{2}\right)\right)\phi(t)dt\right].$$

Applying modulus property, we get

$$\left|(b-a)\left[\epsilon\frac{f(a) + f(b)}{2} + (1-\epsilon)\frac{f(x) + f(a+b-x)}{2}\right] - \int_a^b f(t)dt\right.$$
$$- \left[\int_a^x \left(t - \left(a + \epsilon\frac{b-a}{2}\right)\right)\phi(t)dt + \int_x^{a+b-x}\left(t - \frac{a+b}{2}\right)\phi(t)dt\right.$$

$$+ \int_{a+b-x}^{b}\left(t-\left(b-\epsilon\frac{b-a}{2}\right)\right)\phi(t)dt\Bigg]\Bigg|$$

$$= \left|\int_{a}^{b} P(x,t)\left(f'(t)-\phi(t)\right)dt\right|$$

$$\leq \int_{a}^{b} |P(x,t)|\left(f'(t)-\phi(t)\right)dt$$

$$\leq \max_{t\in[a,b]} |P(x,t)| \int_{a}^{b}\left(f'(t)-\phi(t)\right)dt$$

$$= \max\left\{\epsilon\frac{b-a}{2},\left(x-\frac{(2-\epsilon)a+\epsilon b}{2}\right),\left(\frac{a+b}{2}-x\right)\right\}$$

$$\times \left(f(b)-f(a)-\int_{a}^{b}\phi(t)dt\right). \tag{2.1.19}$$

After re-arranging (2.1.19), we get the required inequality (2.1.18). □

Remark 2.1.13 The inequality (2.1.18) is the generalized case of Theorem 2 which is presented in [116] and Theorem 2 which is presented in [115].

Remark 2.1.14 If we select $x = \frac{a+b}{2}$ in (2.1.18), then we get the following corollary.

Corollary 2.1.15 *Let all the assumptions of Theorem 2.1.14 be valid. Then*

$$m_{24}(\epsilon) \leq \left[\epsilon\frac{f(a)+f(b)}{2}+(1-\epsilon)f\left(\frac{a+b}{2}\right)\right]-\frac{1}{b-a}\int_{a}^{b} f(t)dt \leq M_{24}(\epsilon),$$

$$\tag{2.1.20}$$

where

$$m_{24}(\epsilon) = \frac{1}{b-a}\Bigg[\int_{a}^{b}\left(t-\frac{a+b}{2}\right)\phi(t)dt$$

$$+\frac{b-a}{2}\left(\int_{a}^{\frac{a+b}{2}}(1-\epsilon)\phi(t)dt - \int_{\frac{a+b}{2}}^{b}(1-\epsilon)\phi(t)dt\right)$$

$$-\max\left\{\epsilon\frac{b-a}{2},(1-\epsilon)\frac{b-a}{2}\right\}$$

$$\left(f(b)-f(a)-\int_{a}^{b}\phi(t)dt\right)\Bigg]$$

2.1 Generalized Ostrowski Type Inequalities with Parameter

and

$$M_{24}(\epsilon) = \frac{1}{b-a}\left[\int_a^b \left(t - \frac{a+b}{2}\right)\phi(t)dt\right.$$
$$+ \frac{b-a}{2}\left(\int_a^{\frac{a+b}{2}}(1-\epsilon)\phi(t)dt - \int_{\frac{a+b}{2}}^b(1-\epsilon)\phi(t)dt\right)$$
$$+ \max\left\{\epsilon\frac{b-a}{2}, (1-\epsilon)\frac{b-a}{2}\right\}$$
$$\left.\left(f(b) - f(a) - \int_a^b \phi(t)dt\right)\right].$$

Theorem 2.1.16 *Let $f : I \to \mathbb{R}$ be differentiable function in I^0. If f' is bounded above, i.e., $f'(x) \leq \Phi(x)$ such that $\Phi \in C[a, b]$, $x \in [a, b]$, $\forall \epsilon \in [0, 1]$ we have*

$$m_{25}(x,\epsilon) \leq \left[\epsilon\frac{f(a)+f(b)}{2} + (1-\epsilon)\frac{f(x)+f(a+b-x)}{2}\right] - \frac{1}{b-a}\int_a^b f(t)dt$$
$$\leq M_{25}(x,\epsilon), \qquad (2.1.21)$$

where

$$m_{25}(x,\epsilon) = \frac{1}{b-a}\left[\int_a^b \left(t - \frac{a+b}{2}\right)\Phi(t)dt\right.$$
$$+ \frac{b-a}{2}\left(\int_a^x (1-\epsilon)\Phi(t)dt - \int_{a+b-x}^b (1-\epsilon)\Phi(t)dt\right)$$
$$- \max\left\{\epsilon\frac{b-a}{2}, \left(x - \frac{(2-\epsilon)a + \epsilon b}{2}\right), \left(\frac{a+b}{2} - x\right)\right\}$$
$$\left.\times \left(\int_a^b \Phi(t)dt - f(b) + f(a)\right)\right]$$

and

$$M_{25}(x,\epsilon) = \frac{1}{b-a}\left[\int_a^b \left(t - \frac{a+b}{2}\right)\Phi(t)dt\right.$$
$$+ \frac{b-a}{2}\left(\int_a^x (1-\epsilon)\Phi(t)dt - \int_{a+b-x}^b (1-\epsilon)\Phi(t)dt\right)$$

$$+ \max\left\{\epsilon\frac{b-a}{2}, \left(x - \frac{(2-\epsilon)a + \epsilon b}{2}\right), \left(\frac{a+b}{2} - x\right)\right\}$$
$$\times \left(\int_a^b \Phi(t)dt - f(b) + f(a)\right)\Bigg].$$

Proof Since

$$\int_a^b P(x,t)\left(f'(t) - \Phi(t)\right)dt$$

$$= (b-a)\left[\epsilon\frac{f(a) + f(b)}{2} + (1-\epsilon)\frac{f(x) + f(a+b-x)}{2}\right]$$

$$- \int_a^b f(t)dt - \int_a^b P(x,t)\Phi(t)dt$$

$$= (b-a)\left[\epsilon\frac{f(a) + f(b)}{2} + (1-\epsilon)\frac{f(x) + f(a+b-x)}{2}\right] - \int_a^b f(t)dt$$

$$- \left[\int_a^x \left(t - \left(a + \epsilon\frac{b-a}{2}\right)\right)\Phi(t)dt + \int_x^{a+b-x}\left(t - \frac{a+b}{2}\right)\Phi(t)dt\right.$$

$$\left.+ \int_{a+b-x}^b \left(t - \left(b - \epsilon\frac{b-a}{2}\right)\right)\Phi(t)dt\right],$$

so we get

$$\Bigg|(b-a)\left[\epsilon\frac{f(a) + f(b)}{2} + (1-\epsilon)\frac{f(x) + f(a+b-x)}{2}\right] - \int_a^b f(t)dt$$

$$- \left[\int_a^x \left(t - \left(a + \epsilon\frac{b-a}{2}\right)\right)\Phi(t)dt + \int_x^{a+b-x}\left(t - \frac{a+b}{2}\right)\Phi(t)dt\right.$$

$$\left.+ \int_{a+b-x}^b \left(t - \left(b - \epsilon\frac{b-a}{2}\right)\right)\Phi(t)dt\right]\Bigg|$$

$$= \left|\int_a^b P(x,t)\left(f'(t) - \Phi(t)\right)dt\right|$$

$$\leq \int_a^b |P(x,t)|\left(\Phi(t) - f'(t)\right)dt$$

$$\leq \max_{t\in[a,b]} |P(x,t)| \int_a^b \left(\Phi(t) - f'(t)\right)dt$$

$$= \max\left\{\epsilon\frac{b-a}{2}, \left(x - \frac{(2-\epsilon)a + \epsilon b}{2}\right), \left(\frac{a+b}{2} - x\right)\right\}\left(\int_a^b \Phi(t)dt - f(b) + f(a)\right).$$
$$\tag{2.1.22}$$

2.1 Generalized Ostrowski Type Inequalities with Parameter

After rearranging (2.1.22), we get the inequality (2.1.21). □

Remark 2.1.15 The inequality (2.1.21) is the special case of Theorem 3 which is presented in [116] and Theorem 3 which is presented in [115].

Remark 2.1.16 By choosing $x = \frac{a+b}{2}$ in (2.1.21), then we achieve following corollary.

Corollary 2.1.17 *Let all the assumptions of Theorem 2.1.16 be valid. Then*

$$m_{26}(\epsilon) \le \left[\epsilon\frac{f(a)+f(b)}{2} + (1-\epsilon)f\left(\frac{a+b}{2}\right)\right] - \frac{1}{b-a}\int_a^b f(t)dt \le M_{26}(\epsilon), \quad (2.1.23)$$

where

$$m_{26}(\epsilon) = \frac{1}{b-a}\left[\int_a^b \left(t - \frac{a+b}{2}\right)\Phi(t)dt \right.$$

$$+ \frac{b-a}{2}\left(\int_a^{\frac{a+b}{2}}(1-\epsilon)\,\Phi(t)dt - \int_{\frac{a+b}{2}}^b (1-\epsilon)\,\Phi(t)dt\right)$$

$$\left. - \max\left\{\epsilon\frac{b-a}{2}, (1-\epsilon)\frac{b-a}{2}\right\}\left(\int_a^b \Phi(t)dt - f(b)+f(a)\right)\right]$$

and

$$M_{26}(\epsilon) = \frac{1}{b-a}\left[\int_a^b \left(t - \frac{a+b}{2}\right)\Phi(t)dt \right.$$

$$+ \frac{b-a}{2}\left(\int_a^{\frac{a+b}{2}}(1-\epsilon)\,\Phi(t)dt - \int_{\frac{a+b}{2}}^b (1-\epsilon)\,\Phi(t)dt\right)$$

$$\left. + \max\left\{\epsilon\frac{b-a}{2}, (1-\epsilon)\frac{b-a}{2}\right\}\left(\int_a^b \Phi(t)dt - f(b)+f(a)\right)\right].$$

Remark 2.1.17 If $\phi(x) \le f'(x) \le \Phi(x)$ such that $x \in [a,b]$ and $\phi, \Phi \in C[a,b]$, and if we select $\epsilon = 0$, then $I_6(f)$ can be bounded as

$$m_{27} \le \frac{1}{2}\left[-f(a) + 2f\left(\frac{a+b}{2}\right) + f(b)\right] - \frac{1}{b-a}\int_a^b f(t)dt \le M_{27}, \quad (2.1.24)$$

where

$$m_{27} = \frac{1}{b-a}\left[\int_a^{\frac{a+b}{2}}(t-a)\,\phi(t)dt + \int_{\frac{a+b}{2}}^b (t-b)\phi(t)dt + \frac{(b-a)}{2}\int_a^b \phi(t)dt\right]$$

and

$$M_{27} = \frac{1}{b-a} \left[\int_a^{\frac{a+b}{2}} (t-a)\Phi(t)dt + \int_{\frac{a+b}{2}}^b (t-b)\Phi(t)dt + \frac{(b-a)}{2} \int_a^b \Phi(t)dt \right],$$

which is in fact Corollary 3 and Corollary 4 of [116] and [115] respectively.

Proof In order to prove (2.1.24) both the Corollary 2.1.15 and Corollary 2.1.17 results should be used at the same time. First of all by selecting $\epsilon = 0$ in (2.1.20), we get

$$\frac{1}{b-a} \left[\int_a^b (t-a)\phi(t)dt + \frac{b-a}{2} \left(\int_a^{\frac{a+b}{2}} \phi(t)dt - \int_{\frac{a+b}{2}}^b \phi(t)dt \right) \right]$$

$$\leq \frac{1}{2} \left[-f(a) + 2f\left(\frac{a+b}{2}\right) + f(b) \right] - \frac{1}{b-a} \int_a^b f(t)dt, \quad (2.1.25)$$

provided that $\phi(t) \leq f'(t) \; \forall \, t \in [a,b]$.

On the other hand, by assuming $\epsilon = 0$ in (2.1.23), we obtain

$$\frac{1}{2} \left[-f(a) + 2f\left(\frac{a+b}{2}\right) + f(b) \right] - \frac{1}{b-a} \int_a^b f(t)dt$$

$$\leq \frac{1}{b-a} \left[\int_a^b (t-a)\Phi(t)dt + \frac{b-a}{2} \left(\int_a^{\frac{a+b}{2}} \Phi(t)dt - \int_{\frac{a+b}{2}}^b \Phi(t)dt \right) \right], \quad (2.1.26)$$

provided that $f'(t) \leq \Phi(t) \; \forall \, t \in [a,b]$.

Now by combining the above two results (2.1.25) and (2.1.26), the result (2.1.24) is derived. □

Remark 2.1.18 If $\phi(x) \leq f'(x) \leq \Phi(x)$ such that $x \in [a,b]$ and $\phi, \Phi \in C[a,b]$, and if we select $\epsilon = 0$, then $I_7(f)$ can be bounded as

$$m_{28} \leq \frac{1}{2} \left[f(a) + 2f\left(\frac{a+b}{2}\right) - f(b) \right] - \frac{1}{b-a} \int_a^b f(t)dt \leq M_{28}, \quad (2.1.27)$$

where

$$m_{28} = \frac{1}{b-a} \left[\int_a^{\frac{a+b}{2}} (t-a)\Phi(t)dt + \int_{\frac{a+b}{2}}^b (t-b)\Phi(t)dt - \frac{(b-a)}{2} \int_a^b \Phi(t)dt \right]$$

and

$$M_{28} = \frac{1}{b-a}\left[\int_a^{\frac{a+b}{2}} (t-a)\phi(t)dt + \int_{\frac{a+b}{2}}^b (t-b)\phi(t)dt - \frac{(b-a)}{2}\int_a^b \phi(t)dt\right],$$

which is in fact Corollary 4 and Corollary 5 of [116] and [115] respectively.

Proof The proof of (2.1.27) is similar to Remark 2.1.17. $\square$

Remark 2.1.19 If $\phi(x) \leq f'(x) \leq \Phi(x)$ for any $x \in [a, b]$ and $\phi, \Phi \in C[a, b]$ then by replacing $x = b$, and $\epsilon = 0$ in (2.1.20) and (2.1.23), respectively, then $I_8(f)$ can be bounded as

$$\frac{1}{(b-a)}\int_a^b (t-b)\Phi(t)dt \leq f(a) - \frac{1}{(b-a)}\int_a^b f(t)dt \leq \frac{1}{(b-a)}\int_a^b (t-b)\phi(t)dt,$$

which is in fact Corollary 5 and Corollary 2 of [116] and [115] respectively.

Remark 2.1.20 If $\phi(x) \leq f'(x) \leq \Phi(x)$ such that $x \in [a, b]$ and $\phi, \Phi \in C[a, b]$ then by replacing $x = a$ and $\epsilon = 0$ in (2.1.20) and (2.1.23), respectively, then $I_9(f)$ can be bounded as

$$\frac{1}{(b-a)}\int_a^b (t-a)\phi(t)dt \leq f(b) - \frac{1}{(b-a)}\int_a^b f(t)dt \leq \frac{1}{(b-a)}\int_a^b (t-a)\Phi(t)dt,$$

which is in fact Corollary 6 and Corollary 3 of [116] and [115] respectively.

Now, we will discuss certain applications in numerical quadrature rules that can be used to get some sharp bounds.

2.1.3 Applications to Numerical Integration

Let $I_n : a = \zeta_0 < \zeta_1 < \cdots < \zeta_n = b$ be a partition of the interval $[a, b]$ and $\Delta \zeta_k = \zeta_{k+1} - \zeta_k, k \in \{0, 1, 2, \ldots, n-1\}$. Then

$$\int_a^b f(t)dt = Q_n(f, I_n) + R_n(f, I_n), \tag{2.1.28}$$

where $Q_n(f, I_n)$ is defined as

$$Q_n(f, I_n) := \sum_{k=0}^{n-1}\left[\epsilon \frac{f(\zeta_k) + f(\zeta_{k+1})}{2} + (1-\epsilon)f(\xi_k) + f\left(\frac{\zeta_k + \zeta_{k+1} - \xi_k}{2}\right)\right]\Delta\zeta_k,$$

$$\tag{2.1.29}$$

∀ ϵ ∈ [0, 1] and

$$\zeta_k + \epsilon \frac{\Delta \zeta_k}{2} \leq \xi_k \leq \frac{\zeta_k + \zeta_{k+1}}{2}.$$

Theorem 2.1.18 *Let all the assumptions of Theorem 2.1.1 be valid. Then (2.1.28) is valid and $Q_n(f, I_n)$ is given in the form of (2.1.29) and the remainder $R_n(f, I_n)$ becomes*

$$|R_n(f, I_n)| \leq \sup\{|R|, |R'|\}, \qquad (2.1.30)$$

where

$$R = \int_{-\epsilon \frac{\Delta \zeta_k}{2}}^{\xi_k - \left(\zeta_k + \epsilon \frac{\Delta \zeta_k}{2}\right)} \left(\frac{\lambda_k + |\lambda_k|}{2} \phi \left(\lambda_k + \zeta_k + \epsilon \frac{\Delta \zeta_k}{2} \right) \right.$$

$$\left. + \frac{\lambda_k - |\lambda_k|}{2} \Phi \left(\lambda_k + \zeta_k + \epsilon \frac{\Delta \zeta_k}{2} \right) \right) d\lambda_k$$

$$+ \int_{\xi_k - \frac{\zeta_k + \zeta_{k+1}}{2}}^{\frac{\zeta_k + \zeta_{k+1}}{2} - \xi_k} \left(\frac{\lambda_k + |\lambda_k|}{2} \phi \left(\lambda_k + \frac{\zeta_k + \zeta_{k+1}}{2} \right) \right.$$

$$\left. + \frac{\lambda_k - |\lambda_k|}{2} \Phi \left(\lambda_k + \frac{\zeta_k + \zeta_{k+1}}{2} \right) \right) d\lambda_k$$

$$+ \int_{\zeta_k + \epsilon \frac{\Delta \zeta_k}{2} - \xi_k}^{\epsilon \frac{\Delta \zeta_k}{2}} \left(\frac{\lambda_k + |\lambda_k|}{2} \phi \left(\lambda_k + \zeta_{k+1} - \epsilon \frac{\Delta \zeta_k}{2} \right) \right.$$

$$\left. + \frac{\lambda_k - |\lambda_k|}{2} \Phi \left(\lambda_k + \zeta_{k+1} - \epsilon \frac{\Delta \zeta_k}{2} \right) \right) d\lambda_k$$

and

$$R' = \int_{-\epsilon \frac{\Delta \zeta_k}{2}}^{\xi_k - \left(\zeta_k + \epsilon \frac{\Delta \zeta_k}{2}\right)} \left(\frac{\lambda_k - |\lambda_k|}{2} \phi \left(\lambda_k + \zeta_k + \epsilon \frac{\Delta \zeta_k}{2} \right) \right.$$

$$\left. + \frac{\lambda_k + |\lambda_k|}{2} \Phi \left(\lambda_k + \zeta_k + \epsilon \frac{\Delta \zeta_k}{2} \right) \right) d\lambda_k$$

$$+ \int_{\xi_k - \frac{\zeta_k + \zeta_{k+1}}{2}}^{\frac{\zeta_k + \zeta_{k+1}}{2} - \xi_k} \left(\frac{\lambda_k - |\lambda_k|}{2} \phi \left(\lambda_k + \frac{\zeta_k + \zeta_{k+1}}{2} \right) \right.$$

$$\left. + \frac{\lambda_k + |\lambda_k|}{2} \Phi \left(\lambda_k + \frac{\zeta_k + \zeta_{k+1}}{2} \right) \right) d\lambda_k$$

2.1 Generalized Ostrowski Type Inequalities with Parameter

$$+ \int_{\zeta_k+\epsilon\frac{\Delta\zeta_k}{2}-\xi_k}^{\epsilon\frac{\Delta\zeta_k}{2}} \left(\frac{\lambda_k - |\lambda_k|}{2} \phi \left(\lambda_k + \zeta_{k+1} - \epsilon\frac{\Delta\zeta_k}{2} \right) \right.$$
$$\left. + \frac{\lambda_k + |\lambda_k|}{2} \Phi \left(\lambda_k + \zeta_{k+1} - \epsilon\frac{\Delta\zeta_k}{2} \right) \right) d\lambda_k,$$

$\forall \xi_k \in [\zeta_k, \zeta_{k+1}]$.

Proof If inequality (2.1.3) is applied on $[\zeta_k, \zeta_{k+1}]$, we have

$$R_k(f, I_k) = \int_{\zeta_k}^{\zeta_{k+1}} f(t)dt$$
$$- \left[\epsilon \frac{f(\zeta_k) + f(\zeta_{k+1})}{2} + (1-\epsilon)f(\xi_k) + f\left(\frac{\zeta_k + \zeta_{k+1} - \xi_k}{2} \right) \right] \Delta\zeta_k,$$

we sum up $R_k(f, I_k)$ from 0 to $n-1$ over k. This produces

$$R_n(f, I_n) = \int_a^b f(t)dt$$
$$- \sum_{k=0}^{n-1} \left[\epsilon \frac{f(\zeta_k) + f(\zeta_{k+1})}{2} + (1-\epsilon)f(\xi_k) + f\left(\frac{\zeta_k + \zeta_{k+1} - \xi_k}{2} \right) \right] \Delta\zeta_k.$$

It follows from (2.1.3) that

$$|R_n(f, I_n)|$$
$$= \left| \int_a^b f(t)dt - \sum_{k=0}^{n-1} \left[\epsilon \frac{f(\zeta_k) + f(\zeta_{k+1})}{2} + (1-\epsilon)f(\xi_k) + f\left(\frac{\zeta_k + \zeta_{k+1} - \xi_k}{2} \right) \right] \right|$$
$$\leq \sup \left\{ \left| \int_{-\epsilon\frac{\Delta\zeta_k}{2}}^{\xi_k - \left(\zeta_k + \epsilon\frac{\Delta\zeta_k}{2}\right)} \left(\frac{\lambda_k + |\lambda_k|}{2} \phi \left(\lambda_k + \zeta_k + \epsilon\frac{\Delta\zeta_k}{2} \right) \right. \right. \right.$$
$$\left. + \frac{\lambda_k - |\lambda_k|}{2} \Phi \left(\lambda_k + \zeta_k + \epsilon\frac{\Delta\zeta_k}{2} \right) \right) d\lambda_k$$
$$+ \int_{\xi_k - \frac{\zeta_k+\zeta_{k+1}}{2}}^{\frac{\zeta_k+\zeta_{k+1}}{2} - \xi_k} \left(\frac{\lambda_k + |\lambda_k|}{2} \phi \left(\lambda_k + \frac{\zeta_k + \zeta_{k+1}}{2} \right) \right.$$
$$\left. + \frac{\lambda_k - |\lambda_k|}{2} \Phi \left(\lambda_k + \frac{\zeta_k + \zeta_{k+1}}{2} \right) \right) d\lambda_k$$
$$+ \int_{\zeta_k+\epsilon\frac{\Delta\zeta_k}{2}-\xi_k}^{\epsilon\frac{\Delta\zeta_k}{2}} \left(\frac{\lambda_k + |\lambda_k|}{2} \phi \left(\lambda_k + \zeta_{k+1} - \epsilon\frac{\Delta\zeta_k}{2} \right) \right.$$

$$+ \frac{\lambda_k - |\lambda_k|}{2} \Phi\left(\lambda_k + \zeta_{k+1} - \epsilon \frac{\Delta \zeta_k}{2}\right)\right) d\lambda_k\bigg|,$$

$$\left|\int_{-\epsilon \frac{\Delta \zeta_k}{2}}^{\xi_k - \left(\zeta_k + \epsilon \frac{\Delta \zeta_k}{2}\right)} \left(\frac{\lambda_k - |\lambda_k|}{2} \phi\left(\lambda_k + \zeta_k + \epsilon \frac{\Delta \zeta_k}{2}\right)\right.\right.$$

$$+ \frac{\lambda_k + |\lambda_k|}{2} \Phi\left(\lambda_k + \zeta_k + \epsilon \frac{\Delta \zeta_k}{2}\right)\right) d\lambda_k$$

$$+ \int_{\xi_k - \frac{\zeta_k + \zeta_{k+1}}{2}}^{\frac{\zeta_k + \zeta_{k+1}}{2} - \xi_k} \left(\frac{\lambda_k - |\lambda_k|}{2} \phi\left(\lambda_k + \frac{\zeta_k + \zeta_{k+1}}{2}\right)\right.$$

$$\left.\frac{\lambda_k + |\lambda_k|}{2} \Phi\left(\lambda_k + \frac{\zeta_k + \zeta_{k+1}}{2}\right)\right) d\lambda_k$$

$$+ \int_{\zeta_k + \epsilon \frac{\Delta \zeta_k}{2} - \xi_k}^{\epsilon \frac{\Delta \zeta_k}{2}} \left(\frac{\lambda_k - |\lambda_k|}{2} \phi\left(\lambda_k + \zeta_{k+1} - \epsilon \frac{\Delta \zeta_k}{2}\right)\right.$$

$$\left.+ \frac{\lambda_k + |\lambda_k|}{2} \Phi\left(\lambda_k + \zeta_{k+1} - \epsilon \frac{\Delta \zeta_k}{2}\right)\right) d\lambda_k\bigg|\bigg\}.$$

□

Remark 2.1.21 In similar manner, we can state applications of all corollaries, remarks and special cases presented in the previous section.

2.2 Generalized Ostrowski Type Inequalities for Functions of L_p Spaces and Bounded Variation

In current section, some Ostrowski type inequalities for L_p spaces and functions of bounded variations are stated. Applications are also given for obtaining error bounds of some composite quadrature formulae. Result of this section can be found in [76].

2.2.1 Ostrowski Type Inequality for Functions of L_p Spaces

Lets recall the result of Hölder's inequality from Theorem 1.1.13 that will be useful in our next results.

In 2010, a generalization of Ostrowski inequality [141] discussed by Milovanović and Cvetković in [129], which follows:

2.2 Generalized Ostrowski Type Inequalities for Functions of L_p Spaces and...

Theorem 2.2.1 *Let $f : [a, b] \to \mathbb{R}$ be a function such that f' is bounded on (a, b), i.e.,*

$$\|f'\|_\infty = \sup_{t \in (a,b)} |f'(t)| < \infty.$$

Then $\forall\, x \in [a, b]$, following inequality holds

$$\left| f(x) - \frac{1}{b-a} \int_a^b f(t)dt \right| \leq \left(\frac{(x-a)^2 + (b-x)^2}{2(b-a)} \right) \|f'\|_\infty$$

$$= \left[\frac{1}{4} + \frac{\left(x - \frac{a+b}{2}\right)^2}{(b-a)^2} \right] (b-a)\|f'\|_\infty. \quad (2.2.1)$$

To prove our main results in this section, we need the following lemma extracted from [116].

Lemma 2.2.1 *Let $f : [a, b] \to \mathbb{R}$ be absolutely continuous function where kernel $P_1(x, t)$ given as*

$$P_1(x, t) = \begin{cases} t - x + \dfrac{b-a}{2}, & t \in [a, x]; \\[2mm] t - x - \dfrac{b-a}{2}, & t \in (x, b]. \end{cases} \quad (2.2.2)$$

Then

$$\frac{1}{b-a} \int_a^b P_1(x,t) f'(t) dt = f(x) - \frac{f(b) - f(a)}{b-a}\left(x - \frac{a+b}{2}\right) - \frac{1}{b-a} \int_a^b f(t)dt. \quad (2.2.3)$$

Proof Using kernel (2.2.2), after some computation, we obtain

$$\int_a^x \left(t - x + \frac{b-a}{2} \right) f'(t) dt = \frac{b-a}{2} f(x) + \left(x - \frac{a+b}{2} \right) f(a) - \int_a^x f(t)dt$$

and

$$\int_x^b \left(t - x - \frac{b-a}{2} \right) f'(t) dt = \frac{b-a}{2} f(x) - \left(x - \frac{a+b}{2} \right) f(b) - \int_x^b f(t)dt.$$

By adding above two equations and after some simplifications, we get (2.2.3).

Now we shall use the above identity to obtain our results for first derivative bounded functions and prove inequalities for absolutely continuous functions in which $f' \in L_q[a, b]$ for $q \geq 1$.

Theorem 2.2.2 Let $f : [a, b] \to \mathbb{R}$ be an absolutely continuous function on I^o where $a, b \in I$ and $a < b$. If $f' \in L_q[a, b]$, for $q \geq 1$ and $\frac{1}{q} + \frac{1}{r} = 1$, then for any $x \in [a, b]$, this inequality holds

$$\left| f(x) - \frac{f(b) - f(a)}{b - a} \left(x - \frac{a+b}{2} \right) - \frac{1}{b-a} \int_a^b f(t) dt \right|$$
$$\leq \frac{\|f'\|_q}{(b-a)(r+1)^{\frac{1}{r}}} \left[\left(\frac{a+b}{2} - x \right)^{r+1} + \left(x - \frac{a+b}{2} \right)^{r+1} + 2 \left(\frac{b-a}{2} \right)^{r+1} \right]^{\frac{1}{r}}.$$
(2.2.4)

Proof Applying absolute on (2.2.3) and then applying Hölder inequality, we obtain

$$\left| f(x) - \frac{f(b) - f(a)}{b - a} \left(x - \frac{a+b}{2} \right) - \frac{1}{b-a} \int_a^b f(t) dt \right|$$
$$= \frac{1}{b-a} \left| \int_a^b P_1(x, t) f'(t) dt \right|$$
$$\leq \frac{1}{b-a} \left(\int_a^b |P_1(x,t)|^r dt \right)^{\frac{1}{r}} \left(\int_a^b |f'(t)|^q dt \right)^{\frac{1}{q}}$$
$$= \frac{1}{b-a} \left[\int_a^x \left| t - \left(x - \frac{b-a}{2} \right) \right|^r dt + \int_x^b \left| t - \left(x + \frac{b-a}{2} \right) \right|^r dt \right]^{\frac{1}{r}} \|f'\|_q$$
$$= \frac{\|f'\|_q}{(b-a)(r+1)^{\frac{1}{r}}} \left[\left(\frac{a+b}{2} - x \right)^{r+1} + \left(x - \frac{a+b}{2} \right)^{r+1} + 2 \left(\frac{b-a}{2} \right)^{r+1} \right]^{\frac{1}{r}}.$$

□

Remark 2.2.1 If we substitute $r = 1$ (and $q \to \infty$) in (2.2.4), then we get the following result.

Corollary 2.2.3 Let $f : [a, b] \to \mathbb{R}$ be an absolutely continuous function on I^o for $a, b \in I$ where $a < b$. If $f' \in L_\infty[a, b]$, then for any $x \in [a, b]$, this inequality holds

$$\left| f(x) - \frac{f(b) - f(a)}{b - a} \left(x - \frac{a+b}{2} \right) - \frac{1}{b-a} \int_a^b f(t) dt \right|$$
$$\leq \left[\frac{1}{4} + \frac{(x - \frac{a+b}{2})^2}{(b-a)^2} \right] (b-a) \|f'\|_\infty.$$
(2.2.5)

2.2 Generalized Ostrowski Type Inequalities for Functions of L_p Spaces and...

Remark 2.2.2 The inequality (2.2.5) is the generalization of Ostrowski inequality, i.e., by replacing $f(a) = f(b)$ in (2.2.5) we get (2.2.1) and also by choosing $\|f'\|_\infty = M$ we get (2.0.1).

Remark 2.2.3 If we replace $x = \frac{a+b}{2}$ in (2.2.4), then we get the midpoint inequality (Hermite-Hadamard left bound) in the following corollary.

Corollary 2.2.4 *Let all the assumptions of Theorem 2.2.2 be valid. Then*

$$\left| f\left(\frac{a+b}{2}\right) - \frac{1}{b-a}\int_a^b f(t)dt \right| \leq \frac{1}{2}\left(\frac{b-a}{r+1}\right)^{\frac{1}{r}} \|f'\|_q. \qquad (2.2.6)$$

Remark 2.2.4 If we replace $x = a$ or $x = b$ in (2.2.4), we get the trapezoidal inequality (Hermite-Hadamard right bound) in the following corollary.

Corollary 2.2.5 *Let all the assumptions of Theorem 2.2.2 be valid. Then*

$$\left| \frac{f(a)+f(b)}{2} - \frac{1}{b-a}\int_a^b f(t)dt \right|$$

$$\leq \frac{1}{(b-a)(r+1)^{\frac{1}{r}}} \left[\left(\frac{a-b}{2}\right)^{r+1} + 3\left(\frac{b-a}{2}\right)^{r+1} \right]^{\frac{1}{r}} \|f'\|_q. \qquad (2.2.7)$$

1. *If r is odd, then*

$$\left| \frac{f(a)+f(b)}{2} - \frac{1}{b-a}\int_a^b f(t)dt \right| \leq \frac{1}{2}\left(\frac{2(b-a)}{r+1}\right)^{\frac{1}{r}} \|f'\|_q. \qquad (2.2.8)$$

2. *If r is even, then*

$$\left| \frac{f(a)+f(b)}{2} - \frac{1}{b-a}\int_a^b f(t)dt \right| \leq \frac{1}{2}\left(\frac{b-a}{r+1}\right)^{\frac{1}{r}} \|f'\|_q. \qquad (2.2.9)$$

Remark 2.2.5 If we replace $x = \frac{a+b}{2}$ in (2.2.5), then we get midpoint inequality (Hermite Hadamard left bound) in the following corollary.

Corollary 2.2.6 *Let all the assumptions of Corollary 2.2.3 be valid. Then*

$$\left| f\left(\frac{a+b}{2}\right) - \frac{1}{b-a}\int_a^b f(t)dt \right| \leq \frac{1}{4}(b-a)\|f'\|_\infty. \qquad (2.2.10)$$

Remark 2.2.6 By replacing $x = a$ or $x = b$ in (2.2.5), we get trapezoidal inequality (Hermite Hadamard right bound) in the following corollary.

Corollary 2.2.7 *Let all the assumptions of Corollary 2.2.3 be valid. Then*

$$\left| \frac{f(a) + f(b)}{2} - \frac{1}{b-a} \int_a^b f(t) dt \right| \leq \frac{1}{2}(b-a) \|f'\|_\infty, \qquad (2.2.11)$$

where the constant $\frac{1}{2}$ is better one.

Remark 2.2.7 From the inequality (2.2.10) and (2.2.6), we retrieve the result of Corollary 5 and Corollary 8 of [7], respectively.

2.2.2 Ostrowski Type Inequality for Functions of Bounded Variation

The main purpose of present subsection is to obtain some Ostrowski type inequalities for functions of bounded variation and to give some special cases of the obtained results.

In start, we need a useful definition stated as under:

Definition 2.2.1 Total variation of a continuous real-valued function f on $[a, b] \subseteq \mathbb{R}$ is

$$\bigvee_a^b (f) = \sup_{P \in \mathcal{P}[a,b]} \sum_{i=0}^{n_P - 1} |f(x_{i+1}) - f(x_i)|,$$

where $P = \{x_0, \cdots, x_{n_P}\}$, be a partition of $[a, b]$ satisfying $x_i \leq x_{i+1}$ for $0 \leq i \leq n_P - 1$ and supremum is taken over $\mathcal{P}[a, b] = \{P | P \text{ is partition of } [a, b]\}$ of all partitions of $[a, b]$.

Definition 2.2.2 Continue real-valued function ρ on $\mathbb{R}$ is of bounded variation on $[a, b] \subset \mathbb{R}$ if its total variation is finite, i. e.,

$$f \in BV[a, b] \iff \bigvee_a^b (f) < +\infty.$$

Function of bounded variation is a real-valued function in which total variation is bounded. Functions of bounded variation of single variable were first presented by Camille Jordan in 1881 dealing with convergence of fourier series. For further details see [8].

First of all, we would like to recall some useful results related to bounded variation.

2.2 Generalized Ostrowski Type Inequalities for Functions of L_p Spaces and...

Lemma 2.2.2 ([7]) Let $f : [a, b] \to \mathbb{R}$ be absolutely continuous function and $g : [a, b] \to \mathbb{R}$ be function of bounded variation, the following inequality holds:

$$\left| \int_a^b f(t) dg(t) \right| \leq \sup_{t \in [a,b]} |f(t)| \bigvee_a^b (g). \qquad (2.2.12)$$

Because of its numerous applications, the Ostrowski inequality is one of the fundamental results in mathematical inequalities. In 2001, Dragomir presented the result of the Ostrowski inequality for bounded variation [43] in the following theorem.

Theorem 2.2.8 Let $f : [a, b] \to \mathbb{R}$ be a function of bounded variation on $[a, b]$ holds for $x \in [a, b]$. Then

$$\left| f(x) - \frac{1}{b-a} \int_a^b f(t) dt \right| \leq \left[\frac{1}{2} + \left| \frac{x - \frac{a+b}{2}}{b-a} \right| \right] \bigvee_a^b (f).$$

The constant $\frac{1}{2}$ is the best possible in such a way that it cannot be replaced by the smaller one.

In [46], Dragomir established another result for functions of bounded variation.

Theorem 2.2.9 Let all the assumptions of Theorem 2.2.8 be valid. Then

$$\left| \frac{f(x + f(a+b-x))}{2} - \frac{1}{b-a} \int_a^b f(t) dt \right| \leq \left[\frac{1}{4} + \left| \frac{x - \frac{3a+b}{4}}{b-a} \right| \right] \bigvee_a^b (f).$$

The constant $\frac{1}{4}$ is the best possible in such a way that it cannot be replaced by the smaller one.

In [37, 40, 44, 111, 155], many authors generalized and improved similar type of inequalities. Now we will discuss generalization of Ostrowski type inequality using the same technique.

Theorem 2.2.10 Let all the assumptions of Theorem 2.2.8 be valid. Then

$$\left| f(x) - \frac{f(b) - f(a)}{b-a} \left(x - \frac{a+b}{2} \right) - \frac{1}{b-a} \int_a^b f(t) dt \right|$$

$$\leq \frac{1}{2} \max \left\{ \left| \frac{a+b-2x}{b-a} \right|, 1 \right\} \bigvee_a^b (f), \qquad (2.2.13)$$

where the constant $\frac{1}{2}$ is sharp.

Proof Using Lemma 2.2.1 with the result (2.2.12) for $f(t) = P(x, t)$, and $g(t) = f(t)$, $t \in [a, b]$, we get

$$\left| \frac{1}{b-a} \int_a^b P_1(x,t) df(t) \right|$$

$$\leq \frac{1}{b-a} \left| \int_a^x P_1(x,t) df(t) \right| + \frac{1}{b-a} \left| \int_x^b P_1(x,t) df(t) \right|$$

$$\leq \frac{1}{b-a} \sup_{t \in [a,x]} |P_1(x,t)| \bigvee_a^x (f) + \frac{1}{b-a} \sup_{t \in (x,b]} |P_1(x,t)| \bigvee_x^b (f)$$

$$= \frac{1}{b-a} \max\left\{ \left|\frac{a+b}{2} - x\right|, \frac{b-a}{2} \right\} \bigvee_a^x (f) + \frac{1}{b-a} \max\left\{ \left|\frac{a+b}{2} - x\right| \right\} \bigvee_x^b (f)$$

$$:= M(x).$$

We notice that

$$M(x) \leq \frac{1}{b-a} \max\left\{ \left|\frac{a+b}{2} - x\right|, \frac{b-a}{2} \right\} \left[\bigvee_a^x (f) + \bigvee_x^b (f) \right]$$

$$= \frac{1}{2} \max\left\{ \left|\frac{a+b-2x}{b-a}\right|, 1 \right\} \bigvee_a^b (f),$$

which proves the inequality (2.2.13).

To show that the $\frac{1}{2}$ constant is sharp in (2.2.13), let (2.2.13) is valid for constant $C > 0$, we get

$$\left| f(x) - \frac{f(b) - f(a)}{b-a} \left(x - \frac{a+b}{2} \right) - \frac{1}{b-a} \int_a^b f(t) dt \right|$$

$$\leq C \max\left\{ \left|\frac{a+b-2x}{b-a}\right|, 1 \right\} \bigvee_a^b (f), \quad (2.2.14)$$

for any $x \in [a, b]$.

Consider the function $f : [a, b] \to \{0, 1\}$ defined as

$$f(t) = \begin{cases} 0, & t \in (a, b); \\ 1, & t \in \{a, b\}. \end{cases}$$

For $x = a$, we have

$$\int_a^b f(t)dt = 0$$

and

$$\bigvee_a^b (f) = 2.$$

By using (2.2.14), we obtain,

$$1 \leq 2C$$

or

$$\frac{1}{2} \leq C$$

and thus it is proved that the constant $\frac{1}{2}$ is sharp. □

Remark 2.2.8 If we replace $x = \frac{a+b}{2}$ in (2.2.13), then we get the midpoint inequality (Hermite-Hadamard left bound) in the following corollary.

Corollary 2.2.11 *Let all the assumptions of Theorem 2.2.10 be valid. Then*

$$\left| f\left(\frac{a+b}{2}\right) - \frac{1}{b-a}\int_a^b f(t)dt \right| \leq \frac{1}{2}\bigvee_a^b (f), \qquad (2.2.15)$$

where $\frac{1}{2}$ is the sharp constant. Inequality (2.2.15) is Corollary 2 of [7] by Alomari and Corollary 2.6 of [46] by of Dragomir.

Remark 2.2.9 If we replace $x = a$ or $x = b$ in (2.2.13), then we get the trapezoidal inequality (Hermite-Hadamard right bound) in the following corollary.

Corollary 2.2.12 *Let all the assumptions of Theorem 2.2.10 be valid. Then*

$$\left| \frac{f(a)+f(b)}{2} - \frac{1}{b-a}\int_a^b f(t)dt \right| \leq \frac{1}{2}\bigvee_a^b (f), \qquad (2.2.16)$$

where the constant $\frac{1}{2}$ is sharp. Inequality (2.2.16) is Corollary 2 of [7] by Alomari and Corollary 2.4 of [46] by Dragomir.

Now, we will discuss certain applications to numerical quadrature of our results given in this section.

2.2.3 Applications to Numerical Integration

Let $I_n : a = \zeta_0 < \zeta_1 < \ldots < \zeta_n = b$ be a partition of the interval $[a, b]$ and $\Delta \zeta_k = \zeta_{k+1} - \zeta_k, k \in \{0, 1, 2, \cdots, n-1\}$. Then

$$\int_a^b f(t)dt = Q_n(f, I_n) + R_n(f, I_n), \qquad (2.2.17)$$

where $R_n(f, I_n)$ be the remainder and $Q_n(f, I_n)$ is defined as

$$Q_n(f, I_n) := \sum_{k=0}^{n-1} \left[f(\xi_k) - \frac{f(\zeta_{k+1}) - f(\zeta_k)}{\Delta \zeta_k} \left(\xi_k - \frac{\zeta_{k+1} + \zeta_k}{2} \right) \right] \Delta \zeta_k. \qquad (2.2.18)$$

Theorem 2.2.13 *Let all the assumptions of Theorem 2.2.2 be valid. Equation (2.2.17) holds in which $Q_n(f, I_n)$ is given in the form of (2.2.18) and $R_n(f, I_n)$ becomes*

$$|R_n(f, I_n)| \le \left(\frac{1}{r+1} \right)^{\frac{1}{r}}$$

$$\times \sum_{k=0}^{n-1} \left[\left(\frac{\zeta_k + \zeta_{k+1}}{2} - \xi_k \right)^{r+1} + \left(\xi_k - \frac{\zeta_k + \zeta_{k+1}}{2} \right)^{r+1} + 2 \left(\frac{\Delta \zeta_k}{2} \right)^{r+1} \right]^{\frac{1}{r}} \|f'\|_q, \qquad (2.2.19)$$

$\forall \xi_k \in [\zeta_k, \zeta_{k+1}]$.

Proof If inequality (2.2.4) is applied on $[\zeta_k, \zeta_{k+1}]$, we get

$$R_k(f, I_k) = \int_{\zeta_k}^{\zeta_{k+1}} f(t)dt - \left[f(\xi_k) - \frac{f(\zeta_{k+1}) - f(\zeta_k)}{\Delta \zeta_k} \left(\xi_k - \frac{\zeta_k + \zeta_{k+1}}{2} \right) \right] \Delta \zeta_k,$$

we sum up $R_k(f, I_k)$ from 0 to $n-1$ over k. This produces

$$R_n(f, I_n) = \int_a^b f(t)dt - \sum_{k=0}^{n-1} \left[f(\xi_k) - \frac{f(\zeta_{k+1}) - f(\zeta_k)}{\Delta \zeta_k} \left(\xi_k - \frac{\zeta_k + \zeta_{k+1}}{2} \right) \right] \Delta \zeta_k.$$

It follows from (2.2.4) that

$$|R_n(f, I_n)| = \left| \int_a^b f(t)dt - \sum_{k=0}^{n-1} \left[f(\xi_k) - \frac{f(\zeta_{k+1}) - f(\zeta_k)}{\Delta \zeta_k} \left(\xi_k - \frac{\zeta_k + \zeta_{k+1}}{2} \right) \right] \Delta \zeta_k \right|$$

2.2 Generalized Ostrowski Type Inequalities for Functions of L_p Spaces and...

$$\leq \left(\frac{1}{r+1}\right)^{\frac{1}{r}} \sum_{k=0}^{n-1}\left[\left(\frac{\zeta_k+\zeta_{k+1}}{2}-\xi_k\right)^{r+1}+\left(\xi_k-\frac{\zeta_k+\zeta_{k+1}}{2}\right)^{r+1}\right.$$

$$\left.+2\left(\frac{\Delta\zeta_k}{2}\right)^{r+1}\right]^{\frac{1}{r}} \|f'\|_q.$$

$\square$

Corollary 2.2.14 *Let* $r = 1$ *(and* $q \to \infty$*) in (2.2.19). Then (2.2.17) is valid in which* $Q_n(I_n, f)$ *is given in the form of (2.2.18) and* $R_n(I_n, f)$ *becomes*

$$|R_n(f, I_n)| \leq \sum_{k=0}^{n-1}\left[\frac{[\Delta\zeta_k]^2}{4}+\left(\xi_k-\frac{\zeta_k+\zeta_{k+1}}{2}\right)^2\right]\|f'\|_\infty, \quad (2.2.20)$$

$\forall\, \xi_k \in [\zeta_k, \zeta_{k+1}]$.

Theorem 2.2.15 *Let all the assumptions of Theorem 2.2.10 be valid. Then (2.2.17) is valid in which* $Q_n(f, I_n)$ *is given in the form of (2.2.18) and* $R_n(f, I_n)$ *becomes*

$$|R_n(f, I_n)| \leq \sum_{k=0}^{n-1} \frac{\Delta\zeta_k}{2} \max\left\{\left|\frac{\zeta_k+\zeta_{k+1}-2\xi_k}{\Delta\zeta_k}\right|, 1\right\} \bigvee_a^b(f), \quad (2.2.21)$$

$\forall\, \xi_k \in [\zeta_k, \zeta_{k+1}]$.

Proof If inequality (2.2.13) is applied on $[\zeta_k, \zeta_{k+1}]$, we get

$$R_k(f, I_k) = \int_{\zeta_k}^{\zeta_{k+1}} f(t)dt - \left[f(\xi_k) - \frac{f(\zeta_{k+1})-f(\zeta_k)}{\Delta\zeta_k}\left(\xi_k-\frac{\zeta_k+\zeta_{k+1}}{2}\right)\right]\Delta\zeta_k.$$

We sum up $R_k(f, I_k)$ from 0 to $n-1$ over k. This produces

$$R_n(f, I_n) = \int_a^b f(t)dt - \sum_{k=0}^{n-1}\left[f(\xi_k)-\frac{f(\zeta_{k+1})-f(\zeta_k)}{\Delta\zeta_k}\left(\xi_k-\frac{\zeta_k+\zeta_{k+1}}{2}\right)\right]\Delta\zeta_k.$$

It follows from (2.2.13) that

$$|R_n(f, I_n)| = \left|\int_a^b f(t)dt - \sum_{k=0}^{n-1}\left[f(\xi_k)+\frac{f(\zeta_{k+1})-f(\zeta_k)}{\Delta\zeta_k}\left(\xi_k-\frac{\zeta_k+\zeta_{k+1}}{2}\right)\right]\Delta\zeta_k\right|$$

$$\leq \sum_{k=0}^{n-1}\frac{\Delta\zeta_k}{2}\max\left\{\left|\frac{\zeta_k+\zeta_{k+1}-2\xi_k}{\Delta\zeta_k}\right|, 1\right\}\bigvee_a^b(f).$$

$\square$

If we choose, $\xi_k = \frac{\zeta_k + \zeta_{k+1}}{2}$ in (2.2.18), then $Q_n(f, I_n)$ could be defined as

$$Q_n(f, I_n) := \sum_{k=0}^{n-1} \left[f\left(\frac{\zeta_k + \zeta_{k+1}}{2} \right) \right] \Delta \zeta_k. \qquad (2.2.22)$$

Remark 2.2.10 If (2.2.17) is valid and $Q_n(f, I_n)$ is given in the form of (2.2.22), then in the following result the remainder $R_n(f, I_n)$ from (2.2.6), (2.2.10) and (2.2.15) becomes respectively.

Corollary 2.2.16 *Let all the assumptions of Theorem 2.2.2 and Theorem 2.2.10 be valid. Then*

$$|R_n(f, I_n)| \le \frac{1}{2} \left(\frac{1}{r+1} \right)^{\frac{1}{r}} \sum_{k=0}^{n-1} [\Delta \zeta_k]^{\frac{r+1}{r}} \|f'\|_q,$$

$$|R_n(f, I_n)| \le \frac{1}{4} \sum_{k=0}^{n-1} [\Delta \zeta_k]^2 \|f'\|_\infty$$

and

$$|R_n(f, I_n)| \le \frac{1}{2} \sum_{k=0}^{n-1} \Delta \zeta_k \bigvee_a^b (f).$$

If we choose, $\xi_k = \zeta_k$ or $\xi_k = \zeta_{k+1}$ in (2.2.18), then $Q_n(f, I_n)$ can be defined as

$$Q_n(f, I_n) := \sum_{k=0}^{n-1} \left[\frac{f(\zeta_k) + f(\zeta_{k+1})}{2} \right] \Delta \zeta_k. \qquad (2.2.23)$$

Remark 2.2.11 If (2.2.17) holds and $Q_n(f, I_n)$ is given by formula (2.2.23), then in the following result the remainder $R_n(f, I_n)$ from (2.2.7), (2.2.8), (2.2.9), (2.2.11) and (2.2.16) becomes respectively.

Corollary 2.2.17 *Let all the assumptions of Theorems 2.2.2 and Theorem 2.2.10 be valid. Then*

$$|R_n(f, I_n)| \le \frac{1}{(r+1)^{\frac{1}{r}}} \sum_{k=0}^{n-1} \left[\left(\frac{-\Delta \zeta_k}{2} \right)^{r+1} + 3 \left(\frac{\Delta \zeta_k}{2} \right)^{r+1} \right]^{\frac{1}{r}} \|f'\|_q,$$

$$|R_n(f, I_n)| \le \frac{1}{2} \left(\frac{2}{r+1} \right)^{\frac{1}{r}} \sum_{k=0}^{n-1} (\Delta \zeta_k)^{\frac{r+1}{r}} \|f'\|_q,$$

$$|R_n(f, I_n)| \leq \frac{1}{2}\left(\frac{1}{r+1}\right)^{\frac{1}{r}} \sum_{k=0}^{n-1} (\Delta \zeta_k)^{\frac{r+1}{r}} \|f'\|_q,$$

$$|R_n(f, I_n)| \leq \frac{1}{2} \sum_{k=0}^{n-1} [\Delta \zeta_k]^2 \|f'\|_\infty$$

and

$$|R_n(f, I_n)| \leq \frac{1}{2} \sum_{k=0}^{n-1} \Delta \zeta_k \bigvee_a^b (f).$$

2.3 Generalized Weighted Ostrowski Type Inequality with Parameter

This section introduces weighted Ostrowski type inequality for twice differentiable functions with bounded second derivatives and absolutely continuous first derivatives. It is worth mentioning that throughout this section $\alpha = a + \epsilon \frac{b-a}{2}$ and $\beta = b - \epsilon \frac{b-a}{2}$, where $\epsilon \in [0, 1]$.

In 1976, Milovanović and Pečarić proved a generalization of Ostrowski inequality for $n-$times differentiable functions from which only the case of twice differentiable functions [134, p. 470] is mentioned.

Theorem 2.3.1 *Let* $f : I \to \mathbb{R}$ *be a twice differentiable function such that* $f'' : (a, b) \to \mathbb{R}$ *is bounded. Then*

$$\left| \frac{1}{2}\left[f(x) + \frac{(x-a)f(a) + (b-x)f(b)}{b-a} \right] - \frac{1}{b-a} \int_a^b f(t)dt \right|$$
$$\leq \frac{\|f''\|_\infty}{4}(b-a)^2 \left[\frac{1}{12} + \frac{(x - \frac{a+b}{2})^2}{(b-a)^2} \right],$$

$\forall\, x \in [a, b]$.

In year 1999, Cerone et al. in [36] presented the following inequality.

Theorem 2.3.2 *Under the assumptions of Theorem* 2.3.1, *the following inequality holds*

$$\left| f(x) - \left(x - \frac{a+b}{2}\right) f'(x) - \frac{1}{b-a} \int_a^b f(t)dt \right|$$

$$\leq \left[\frac{(b-a)^2}{24} + \frac{1}{2}\left(x - \frac{a+b}{2}\right)^2 \right] \|f''\|_\infty \leq \frac{(b-a)^2}{6} \|f''\|_\infty,$$

$\forall\, x \in [a, b]$.

In the same year, Dragomir and Barnett in [47] stated the following result.

Theorem 2.3.3 *Under the assumptions of Theorem 2.3.1, the following inequality holds*

$$\left| f(x) - \frac{f(b) - f(a)}{b - a}\left(x - \frac{a+b}{2}\right) - \frac{1}{b-a}\int_a^b f(t)dt \right|$$

$$\leq \frac{(b-a)^2}{2} \left\{ \left[\left(\frac{x - \frac{a+b}{2}}{b - a}\right)^2 + \frac{1}{4} \right]^2 + \frac{1}{12} \right\} \|f''\|_\infty \leq \frac{(b-a)^2}{6} \|f''\|_\infty,$$

$\forall\, x \in [a, b]$.

Cerone et al. established Ostrowski type inequality for functions with bounded second derivatives in [36]. Dragomir and Barnett established a similar inequality in [47]. Dragomir and Sofo in [49] identified Ostrowski type inequality of the same kind in the sense of [36] or [47], as given in the following theorem.

Theorem 2.3.4 *Let $f : [a, b] \to \mathbb{R}$ be an absolutely continuous function on $[a, b]$ and $f'' \in L_\infty[a, b]$. Then the following inequality holds*

$$\left| \int_a^b f(t) - \frac{1}{2}\left[f(x) + \frac{f(a) + f(b)}{2} \right](b - a) + \frac{(b-a)}{2}\left(x - \frac{a+b}{2}\right) f'(x) \right|$$

$$\leq \|f''\|_\infty \left(\frac{1}{3}\left|x - \frac{a+b}{2}\right|^3 + \frac{(b-a)^3}{48} \right), \quad (2.3.1)$$

$\forall\, x \in [a, b]$.

In 2000, Dragomir et al. established Montgomery identity with parameter [53], stated as follows

Theorem 2.3.5 *If $f : [a, b] \to \mathbb{R}$ is differentiable on $[a, b]$ with f' integrable on $[a, b]$, where $\epsilon \in [0, 1]$, then generalized integral identity holds*

$$(1 - \epsilon) f(x) + \epsilon \frac{f(a) + f(b)}{2} - \frac{1}{b-a}\int_a^b f(t)dt = \frac{1}{b-a}\int_a^b K_\epsilon(x, t) f'(t)dt,$$

$$(2.3.2)$$

2.3 Generalized Weighted Ostrowski Type Inequality with Parameter

where $K_\epsilon(x, t)$ is defined as

$$K_\epsilon(x, t) = \begin{cases} t - \left(a + \epsilon \dfrac{b-a}{2}\right), & t \in [a, x]; \\ t - \left(b - \epsilon \dfrac{b-a}{2}\right), & t \in (x, b], \end{cases} \quad (2.3.3)$$

$\forall\, x \in [\alpha, \beta]$.

Zafar and Mir established following general form of integral inequality in [193] with the help of (2.3.2), stated as:

Theorem 2.3.6 *Let all the assumptions of Theorem 2.3.4 be valid. Then the following inequality holds*

$$\left| \frac{1}{b-a} \int_a^b f(t)dt - \frac{1}{2}\left[(1-\epsilon)f(x) + (1+\epsilon)\left(\frac{f(a)+f(b)}{2}\right) \right.\right.$$
$$\left.\left. -(1-\epsilon)\left(x - \frac{a+b}{2}\right)f'(x) - \epsilon\frac{b-a}{4}\left(f'(b) - f'(a)\right)\right]\right|$$
$$\leq \|f''\|_\infty \frac{1}{(b-a)}\left[\frac{1}{3}\left|x - \frac{a+b}{2}\right|^3 + \frac{(b-a)^3}{48}\Psi(\epsilon)\right], \quad (2.3.4)$$

$\forall\, x \in [\alpha, \beta]$,
where $\Psi(\epsilon) = (1-\epsilon)[2(1-\epsilon)^2 - 1] + 2\epsilon$, $\epsilon \in [0, 1]$.

Pečarić and Savić in [154] first time discussed the weighted version of Ostrowski inequality. Due to the importance of this inequality, in the last few decades, researchers are continuously in an effort for gaining sharp bounds of Ostrowski's inequality in terms of weight. In the following subsection, we introduce weights on the Ostrowski type inequality (2.3.4) which was proved by Zafar and Mir in [193]. The obtained inequality is then applied to generate certain composite quadrature formulae. The results of current section can be seen in [74].

2.3.1 Weighted Ostrowski Type Inequality with Parameter

We need the following lemma from [74] to prove our next main result.

Lemma 2.3.1 *Let* $f : I \to \mathbb{R}$ *be absolutely continuous function. Further let* $p : [a, b] \to [0, \infty)$ *be a probability density function, where* $\epsilon \in [0, 1]$. *Then we have the identity*

$$f(x) \int_\alpha^\beta p(t)dt = \int_a^b p(t)f(t)dt + f(a) \int_\alpha^a p(t)dt - f(b) \int_\beta^b p(t)dt$$

$$+ \int_a^b K_{p,\epsilon}(x, t) f'(t)dt,$$

where the kernel $K_{p,\epsilon}(x, t) : [a, b] \times [a, b] \to \mathbb{R}$ *is given by:*

$$K_{p,\epsilon}(x, t) = \begin{cases} \int_\alpha^t p(u)du, & t \in [a, x]; \\ \int_\beta^t p(u)du, & t \in (x, b], \end{cases} \quad (2.3.5)$$

$\forall \, x \in [\alpha, \beta]$.

Proof Using kernel (2.3.5), after some computation, we obtain

$$\int_a^x \left(\int_\alpha^t w(u)du \right) f'(t)dt = f(x) \int_\alpha^x p(t)dt - f(a) \int_\alpha^a p(t)dt - \int_a^x p(t)f(t)dt$$

(2.3.6)

and

$$\int_x^b \left(\int_\beta^t w(u)du \right) f'(t)dt = f(x) \int_x^\beta p(t)dt + f(b) \int_\beta^b p(t)dt - \int_x^b p(t)f(t)dt.$$

(2.3.7)

By adding (2.3.6) and (2.3.7), we get the required identity (2.3.5). □

Theorem 2.3.7 *Let* $f : I \to \mathbb{R}$ *be a function whose first derivative is absolutely continuous on* $[a, b]$ *and* $f'' \in L_\infty[a, b]$. *Also let* $p : [a, b] \to [0, \infty)$ *be a probability density function, where* $\epsilon \in [0, 1]$. *Then*

$$\left| \int_a^b f(t)p(t)dt - \frac{1}{2} \left[f(x) \int_\alpha^\beta p(t)dt + \frac{b-a}{2} (f(a)p(a) + f(b)p(b)) \right.$$

$$+ f(b) \int_\beta^b p(t)dt - f(a) \int_\alpha^a p(t)dt - \int_\alpha^\beta p(t)dt \left(x - \frac{a+b}{2} \right) f'(x)$$

2.3 Generalized Weighted Ostrowski Type Inequality with Parameter

$$-\int_a^b f(t)\left(t - \frac{a+b}{2}\right)p'(t)dt - \frac{b-a}{2}\left(f'(a)\int_\alpha^a p(t)dt + f'(b)\int_\beta^b p(t)dt\right)\Bigg]\Bigg|$$

$$\leq \|f''\|_\infty \Bigg[\left(\int_\alpha^x p(u)du + \int_\beta^x p(u)du\right)\left(\frac{(a+b)x}{4} - \frac{x^2}{4}\right)$$

$$+ \frac{1}{2}\int_x^{\frac{a+b}{2}} \left((a+b)t - t^2\right)p(t)dt\Bigg|$$

$$+ \left(\int_\alpha^{\frac{a+b}{2}} p(u)du\right)\frac{(a+b)^2}{8} - \left(\int_a^\alpha p(u)du + \int_\beta^b p(u)du\right)\frac{ab}{4}$$

$$-2\int_\alpha^\beta \left(\frac{(a+b)t}{4} - \frac{t^2}{4}\right)p(t)dt + \int_a^b \left(\frac{(a+b)t}{4} - \frac{t^2}{4}\right)p(t)dt\Bigg], \qquad (2.3.8)$$

$\forall\, x \in [\alpha, \beta]$.

Proof We have the following identity from Lemma 2.3.1,

$$\int_\alpha^\beta p(t)dt\, g(x) = \int_a^b p(t)g(t)dt + \int_\alpha^a p(t)dt\, g(a) - \int_\beta^b p(t)dt\, g(b)$$

$$+ \int_a^b K_{p,\epsilon}(x,t)g'(t)dt.$$

Let us consider

$$g(x) = \left(x - \frac{a+b}{2}\right)f'(x).$$

Equation (2.3.5) implies

$$\int_\alpha^\beta p(t)dt \left(x - \frac{a+b}{2}\right)f'(x) = \int_a^b p(t)\left(t - \frac{a+b}{2}\right)f'(t)dt - \frac{b-a}{2}$$

$$\left(\int_\alpha^a p(t)dt f'(a) + \int_\beta^b p(t)dt f'(b)\right) + \int_a^b K_{p,\epsilon}(x,t)\left[f'(t) + \left(t - \frac{a+b}{2}\right)f''(t)\right]dt.$$

$$(2.3.9)$$

Integrating by parts, we have

$$\int_a^b \left(t - \frac{a+b}{2}\right)p(t)f'(t)dt = \frac{b-a}{2}[f(a)p(a) + f(b)p(b)] - \int_a^b f(t)p(t)dt$$

$$- \int_a^b f(t)\left(t - \frac{a+b}{2}\right)p'(t)dt, \qquad (2.3.10)$$

also

$$\int_a^b K_{p,\epsilon}(x,t)f'(t)dt = \int_\beta^b p(t)dt f(b) - \int_\alpha^a p(t)dt f(a)$$
$$+ f(x)\int_\alpha^\beta p(t)dt - \int_a^b f(t)p(t)dt. \quad (2.3.11)$$

Now using equations (2.3.10) and (2.3.11) in (2.3.9), we get

$$\int_\alpha^\beta p(t)dt \left(x - \frac{a+b}{2}\right) f'(x)$$
$$= \frac{b-a}{2}(f(a)p(a) + f(b)p(b)) + \int_\beta^b p(t)dt f(b) - \int_\alpha^a p(t)dt f(a)$$
$$- \frac{b-a}{2}\left(\int_\alpha^a p(t)dt f'(a) + \int_\beta^b p(t)dt f'(b)\right)$$
$$+ \int_\alpha^\beta p(t)dt f(x) - 2\int_a^b f(t)p(t)dt - \int_a^b f(t)\left(t - \frac{a+b}{2}\right)p'(t)dt$$
$$+ \int_a^b K_{p,\epsilon}(x,t)\left(t - \frac{a+b}{2}\right)f''(t)dt,$$

or we can write

$$\int_a^b p(t)f(t)dt = \frac{b-a}{4}(f(a)p(a) + f(b)p(b)) + \frac{1}{2}\int_\beta^b p(t)dt f(b)$$
$$- \frac{1}{2}\int_\alpha^a p(t)dt f(a) - \frac{b-a}{4}\left(\int_\alpha^a p(t)dt f'(a) + \int_\beta^b p(t)dt f'(b)\right)$$
$$+ \frac{1}{2}\int_\alpha^\beta p(t)dt \left(f(x) - \left(x - \frac{a+b}{2}\right)f'(x)\right)$$
$$- \frac{1}{2}\int_a^b f(t)\left(t - \frac{a+b}{2}\right)p'(t)dt$$
$$+ \frac{1}{2}\int_a^b K_{p,\epsilon}(x,t)\left(t - \frac{a+b}{2}\right)f''(t)dt,$$

$\forall\, x \in [\alpha, \beta]$, this gives us

$$\left|\int_a^b f(t)p(t)dt - \frac{1}{2}\left[\int_\alpha^\beta p(t)dt f(x) + \frac{b-a}{2}(f(a)p(a) + f(b)p(b))\right.\right.$$
$$+ \int_\beta^b p(t)dt f(b) - \int_\alpha^a p(t)dt f(a) - \int_\alpha^\beta p(t)dt \left(x - \frac{a+b}{2}\right)f'(x)$$

2.3 Generalized Weighted Ostrowski Type Inequality with Parameter

$$-\int_a^b f(t)\left(t-\frac{a+b}{2}\right)p'(t)dt - \frac{b-a}{2}\left(\int_\alpha^a p(t)dt f'(a) + \int_\beta^b p(t)dt f'(b)\right)\Big]\Big|$$

$$= \left|\frac{1}{2}\int_a^b K_{p,\epsilon}(x,t)\left(t-\frac{a+b}{2}\right)f''(t)dt\right|$$

$$\leq \frac{1}{2}\int_a^b |K_{p,\epsilon}(x,t)|\left|t-\frac{a+b}{2}\right||f''(t)|\,dt. \tag{2.3.12}$$

It can be easily seen that

$$\int_a^b |K_{p,\epsilon}(x,t)|\left|t-\frac{a+b}{2}\right||f''(t)|\,dt \leq \|f''\|_\infty \int_a^b |K_{p,\epsilon}(x,t)|\left|t-\frac{a+b}{2}\right|dt, \tag{2.3.13}$$

where

$$\|f''\|_\infty = \sup_{t\in(a,b)} |f''(t)| < \infty.$$

Also,

$$I = \int_a^b |K_{p,\epsilon}(x,t)|\left|t-\frac{a+b}{2}\right|dt$$

or

$$I = \int_a^x \left|\int_\alpha^t p(u)du\right|\left|t-\frac{a+b}{2}\right|dt + \int_x^b \left|\int_\beta^t p(u)du\right|\left|t-\frac{a+b}{2}\right|dt. \tag{2.3.14}$$

Now, we have two cases:

(a) For $x \in \left[a, \frac{a+b}{2}\right]$, we obtain

$$I = \int_a^\alpha \left(\int_t^\alpha p(u)du\right)\left(\frac{a+b}{2}-t\right)dt + \int_\alpha^x \left(\int_\alpha^t p(u)du\right)\left(\frac{a+b}{2}-t\right)dt$$

$$+ \int_x^{\frac{a+b}{2}} \left(\int_t^\beta p(u)du\right)\left(\frac{a+b}{2}-t\right)dt + \int_{\frac{a+b}{2}}^\beta \left(\int_t^\beta p(u)du\right)\left(t-\frac{a+b}{2}\right)dt$$

$$+ \int_\beta^b \left(\int_\beta^t p(u)du\right)\left(t-\frac{a+b}{2}\right)dt.$$

We got after some computations

$$I = -\left(\int_\alpha^x p(u)du - \int_x^\beta p(u)du\right)\left(\frac{x^2}{2} - \frac{(a+b)x}{2}\right) + 2\int_{\frac{a+b}{2}}^\beta p(u)du \frac{(a+b)^2}{8}$$

$$-\left(\int_a^\alpha p(u)du + \int_\beta^b p(u)du\right)\frac{ab}{2} + \int_a^\alpha \left(\frac{(a+b)t}{2} - \frac{t^2}{2}\right)p(t)dt$$

$$+\int_\beta^b \left(\frac{(a+b)t}{2} - \frac{t^2}{2}\right)p(t)dt + \int_x^{\frac{a+b}{2}} \left(\frac{(a+b)t}{2} - \frac{t^2}{2}\right)p(t)dt$$

$$-\int_\alpha^x \left(\frac{(a+b)t}{2} - \frac{t^2}{2}\right)p(t)dt + \int_\beta^{\frac{a+b}{2}} \left(\frac{(a+b)t}{2} - \frac{t^2}{2}\right)p(t)dt$$

$$= -\left(\int_\alpha^x p(u)du - \int_x^\beta p(u)du\right)\left(\frac{x^2}{2} - \frac{(a+b)x}{2}\right) + 2\int_{\frac{a+b}{2}}^\beta p(u)du \frac{(a+b)^2}{8}$$

$$-\left(\int_a^\alpha p(u)du + \int_\beta^b p(u)du\right)\frac{ab}{2} + 2\int_x^{\frac{a+b}{2}} \left(\frac{(a+b)t}{2} - \frac{t^2}{2}\right)p(t)dt$$

$$-2\int_\alpha^\beta \left(\frac{(a+b)t}{2} - \frac{t^2}{2}\right)p(t)dt + \int_a^b \left(\frac{(a+b)t}{2} - \frac{t^2}{2}\right)p(t)dt$$

$$= -\left[\left(\int_\alpha^x p(u)du - \int_x^\beta p(u)du\right)\left(\frac{x^2}{2} - \frac{(a+b)x}{2}\right) - 2\int_x^{\frac{a+b}{2}}\left(\frac{(a+b)t}{2}\right.\right.$$

$$\left.\left.-\frac{t^2}{2}\right)p(t)dt\right] + 2\int_{\frac{a+b}{2}}^\beta p(u)du\frac{(a+b)^2}{8} - \left(\int_a^\alpha p(u)du + \int_\beta^b p(u)du\right)\frac{ab}{2}$$

$$-2\int_\alpha^\beta \left(\frac{(a+b)t}{2} - \frac{t^2}{2}\right)p(t)dt + \int_a^b \left(\frac{(a+b)t}{2} - \frac{t^2}{2}\right)p(t)dt. \qquad (2.3.15)$$

(b) Similarly,

$$I = \int_a^\alpha \left(\int_t^\alpha p(u)du\right)\left(\frac{a+b}{2} - t\right)dt + \int_\alpha^{\frac{a+b}{2}} \left(\int_\alpha^t p(u)du\right)\left(\frac{a+b}{2} - t\right)dt$$

$$+ \int_{\frac{a+b}{2}}^x \left(\int_\alpha^t p(u)du\right)\left(t - \frac{a+b}{2}\right)dt + \int_x^\beta \left(\int_t^\beta p(u)du\right)\left(t - \frac{a+b}{2}\right)dt$$

$$+ \int_\beta^b \left(\int_\beta^t p(u)du\right)\left(t - \frac{a+b}{2}\right)dt.$$

2.3 Generalized Weighted Ostrowski Type Inequality with Parameter

After some simplification, we get

$$I = \left(\int_\alpha^x p(u)du - \int_x^\beta p(u)du\right)\left(\frac{x^2}{2} - \frac{(a+b)x}{2}\right) + 2\int_\alpha^{\frac{a+b}{2}} p(u)du \frac{(a+b)^2}{8}$$

$$-\left(\int_a^\alpha p(u)du + \int_\beta^b p(u)du\right)\frac{ab}{2} + \int_a^\alpha \left(\frac{(a+b)t}{2} - \frac{t^2}{2}\right)p(t)dt$$

$$+\int_\beta^b \left(\frac{(a+b)t}{2} - \frac{t^2}{2}\right)p(t)dt - \int_x^{\frac{a+b}{2}}\left(\frac{(a+b)t}{2} - \frac{t^2}{2}\right)p(t)dt$$

$$-\int_\alpha^{\frac{a+b}{2}}\left(\frac{(a+b)t}{2} - \frac{t^2}{2}\right)p(t)dt - \int_x^\beta \left(\frac{(a+b)t}{2} - \frac{t^2}{2}\right)p(t)dt$$

$$= -\left(\int_\alpha^x p(u)du - \int_x^\beta p(u)du\right)\left(\frac{x^2}{2} - \frac{(a+b)x}{2}\right) + 2\int_{\frac{a+b}{2}}^\alpha p(u)du \frac{(a+b)^2}{8}$$

$$-\left(\int_a^\alpha p(u)du + \int_\beta^b p(u)du\right)\frac{ab}{2} - 2\int_x^{\frac{a+b}{2}}\left(\frac{(a+b)t}{2} - \frac{t^2}{2}\right)p(t)dt$$

$$-2\int_\alpha^\beta \left(\frac{(a+b)t}{2} - \frac{t^2}{2}\right)p(t)dt + \int_a^b \left(\frac{(a+b)t}{2} - \frac{t^2}{2}\right)p(t)dt$$

$$= \left[\left(\int_\alpha^x p(u)du - \int_x^\beta p(u)du\right)\left(\frac{x^2}{2} - \frac{(a+b)x}{2}\right) - 2\int_x^{\frac{a+b}{2}}\left(\frac{(a+b)t}{2}\right.\right.$$

$$\left.\left. -\frac{t^2}{2}\right)p(t)dt\right] + 2\int_{\frac{a+b}{2}}^\alpha p(u)du \frac{(a+b)^2}{8} - \left(\int_a^\alpha p(u)du + \int_\beta^b p(u)du\right)\frac{ab}{2}$$

$$-2\int_\alpha^\beta \left(\frac{(a+b)t}{2} - \frac{t^2}{2}\right)p(t)dt + \int_a^b \left(\frac{(a+b)t}{2} - \frac{t^2}{2}\right)p(t)dt, \quad (2.3.16)$$

$\forall x \in \left[\frac{a+b}{2}, b\right]$.

Using equations (2.3.13), (2.3.14), (2.3.15) and (2.3.16) we obtain

$$\left|\int_a^b f(t)p(t)dt - \frac{1}{2}\left[\int_\alpha^\beta p(t)dt f(x) + \frac{b-a}{2}(f(a)p(a) + f(b)p(b))\right.\right.$$

$$+ \int_\beta^b p(t)dt f(b) - \int_a^\alpha p(t)dt f(a) - \int_\alpha^\beta p(t)dt\left(x - \frac{a+b}{2}\right)f'(x)$$

$$\left.\left. -\int_a^b f(t)\left(t - \frac{a+b}{2}\right)p'(t)dt - \frac{b-a}{2}\left(\int_a^\alpha p(t)dt f'(a) + \int_\beta^b p(t)dt f'(b)\right)\right]\right|$$

$$\leq \frac{\|f''\|_\infty}{2}\left[\left(\int_\alpha^x p(u)du - \int_x^\beta p(u)du\right)\left(\frac{x^2}{2} - \frac{(a+b)x}{2}\right)\right.$$

$$-2\int_x^{\frac{a+b}{2}}\left(\frac{(a+b)t}{2}-\frac{t^2}{2}\right)p(t)dt\Bigg]$$

$$+2\int_\alpha^{\frac{a+b}{2}}p(u)du\frac{(a+b)^2}{8}-\left(\int_a^\alpha p(u)du+\int_\beta^b p(u)du\right)\frac{ab}{2}$$

$$-2\int_\alpha^\beta\left(\frac{(a+b)t}{2}-\frac{t^2}{2}\right)p(t)dt+\int_a^b\left(\frac{(a+b)t}{2}-\frac{t^2}{2}\right)p(t)dt$$

$$=\|f''\|_\infty\left[\left(\int_\alpha^x p(u)du-\int_x^\beta p(u)du\right)\left(\frac{(a+b)x}{4}-\frac{x^2}{4}\right)\right.$$

$$+\int_x^{\frac{a+b}{2}}\left(\frac{(a+b)t}{2}-\frac{t^2}{2}\right)p(t)dt\Bigg]$$

$$+\int_\alpha^{\frac{a+b}{2}}p(u)du\frac{(a+b)^2}{8}-\left(\int_a^\alpha p(u)du+\int_\beta^b p(u)du\right)\frac{ab}{4}$$

$$-2\int_\alpha^\beta\left(\frac{(a+b)t}{4}-\frac{t^2}{4}\right)p(t)dt+\int_a^b\left(\frac{(a+b)t}{4}-\frac{t^2}{4}\right)p(t)dt.$$

□

Special Case 1 If we simply put $p(t)\equiv\frac{1}{b-a}$ in (2.3.8), then we will get (2.3.4).

Remark 2.3.1 If we substitute $\epsilon=0$, then $\alpha=a$ and $\beta=b$ in (2.3.8), we get the following result.

Corollary 2.3.8 *Let all the assumptions of Theorem* 2.3.7 *be valid. Then*

$$\left|\int_a^b f(t)p(t)dt-\frac{1}{2}\left[f(x)+\frac{b-a}{2}[f(a)p(a)+f(b)p(b)]\right.\right.$$

$$\left.\left.-\int_a^b f(t)\left(t-\frac{a+b}{2}\right)p'(t)dt-\left(x-\frac{a+b}{2}\right)f'(x)\right]\right|$$

$$\leq\|f''\|_\infty\left[\left|\left(\int_a^x p(u)du+\int_b^x p(u)du\right)\left(\frac{(a+b)x}{4}-\frac{x^2}{4}\right)\right.\right.$$

$$+\int_x^{\frac{a+b}{2}}\left(\frac{(a+b)t}{2}-\frac{t^2}{2}\right)p(t)dt\Bigg|$$

$$+\left(\int_a^{\frac{a+b}{2}}p(u)du\right)\frac{(a+b)^2}{8}-\int_a^b\left(\frac{(a+b)t}{4}-\frac{t^2}{4}\right)p(t)dt\Bigg], \quad (2.3.17)$$

$\forall\,x\in[\alpha,\beta]$.

2.3 Generalized Weighted Ostrowski Type Inequality with Parameter

Special Case 2 If we simply put $p(t) \equiv \dfrac{1}{b-a}$ in (2.3.8), then we will get (2.3.1).

Corollary 2.3.9 *If we examine the estimates for the the midpoint* $x = \dfrac{a+b}{2}$ *and end points* $x = a, x = b$ *in (2.3.8), the midpoint provides us the best estimate*

$$\left| \int_a^b f(t)p(t)dt - \frac{1}{2}\left[\int_\alpha^\beta p(t)dt f\left(\frac{a+b}{2}\right) + \frac{b-a}{2}(f(a)p(a) + f(b)p(b)) \right. \right.$$
$$+ \int_\beta^b p(t)dt f(b) - \int_a^\alpha p(t)dt f(a) - \int_a^b f(t)\left(t - \frac{a+b}{2}\right) p'(t)dt$$
$$\left. -\frac{b-a}{2}\left(\int_\alpha^a p(t)dt f'(a) + \int_\beta^b p(t)dt f'(b) \right) \right] \Big|$$
$$\leq \|f''\|_\infty \left[\left| \left(\int_\alpha^{\frac{a+b}{2}} p(u)du + \int_\beta^{\frac{a+b}{2}} p(u)du \right) \left(\frac{(a+b)^2}{16}\right) \right| \right.$$
$$+ \left(\int_\alpha^{\frac{a+b}{2}} p(u)du \right) \frac{(a+b)^2}{8} - \left(\int_a^\alpha p(u)du + \int_\beta^b p(u)du \right) \frac{ab}{4}$$
$$\left. -2\int_\alpha^\beta \left(\frac{(a+b)t}{4} - \frac{t^2}{4} \right) p(t)dt + \int_a^b \left(\frac{(a+b)t}{4} - \frac{t^2}{4} \right) p(t)dt \right].$$

Remark 2.3.2 If we substitute $\epsilon = 0$, then $\alpha = a$ and $\beta = b$ in (2.3.8), we get the following result.

Corollary 2.3.10 *Let all the assumptions of Theorem 2.3.7 be valid. Then*

$$\left| \int_a^b f(t)p(t)dt - \frac{1}{2}\left[\int_a^b p(t)dt f\left(\frac{a+b}{2}\right) + \frac{b-a}{2}(f(a)p(a) \right. \right.$$
$$\left. + f(b)p(b)) - \int_a^b f(t)\left(t - \frac{a+b}{2}\right) p'(t)dt \right] \Big|$$
$$\leq \|f''\|_\infty \left[\left| \left(\int_a^{\frac{a+b}{2}} p(u)du + \int_b^{\frac{a+b}{2}} p(u)du \right) \left(\frac{(a+b)^2}{16}\right) \right| \right.$$
$$\left. + \left(\int_a^{\frac{a+b}{2}} p(u)du \right) \frac{(a+b)^2}{8} - \int_a^b \left(\frac{(a+b)t}{4} - \frac{t^2}{4} \right) p(t)dt \right]. \quad (2.3.18)$$

Remark 2.3.3 If we substitute $\epsilon = 1$, then $\alpha = \beta = \dfrac{a+b}{2}$ in (2.3.8), we get the following result.

Corollary 2.3.11 *Let all the assumptions of Theorem 2.3.7 be valid. Then*

$$\left| \int_a^b f(t)p(t)dt - \frac{1}{2}\left[\frac{b-a}{2}(f(a)p(a) + f(b)p(b)) \right. \right.$$

$$+ \int_{\frac{a+b}{2}}^b p(t)dt f(b) - \int_{\frac{a+b}{2}}^a p(t)dt f(a) - \int_a^b f(t)\left(t - \frac{a+b}{2}\right)p'(t)dt$$

$$\left. \left. - \frac{b-a}{2}\left(\int_{\frac{a+b}{2}}^a p(t)dt f'(a) + \int_{\frac{a+b}{2}}^b p(t)dt f'(b) \right) \right] \right|$$

$$\leq \|f''\|_\infty \left[-\frac{ab}{4} + \int_a^b \left(\frac{(a+b)t}{4} - \frac{t^2}{4} \right) p(t)dt \right]. \qquad (2.3.19)$$

Remark 2.3.4 If we substitute $\epsilon = \frac{3}{10}$, then $\alpha = \frac{17a+3b}{20}$ and $\beta = \frac{3a+17b}{20}$ in (2.3.8), we get the following result.

Corollary 2.3.12 *Let all the assumptions of Theorem 2.3.7 be valid. Then*

$$\left| \int_a^b f(t)p(t)dt - \frac{1}{2}\left[\int_{\frac{17a+3b}{20}}^{\frac{3a+17b}{20}} p(t)dt f(\frac{a+b}{2}) + \frac{b-a}{2}(f(a)p(a)\ f(b)p(b)) \right. \right.$$

$$+ \int_{\frac{3a+17b}{20}}^b p(t)dt f(b) - \int_{\frac{17a+3b}{20}}^a p(t)dt f(a) - \int_a^b f(t)\left(t - \frac{a+b}{2}\right)p'(t)dt$$

$$\left. \left. - \frac{b-a}{2}\left(\int_{\frac{17a+3b}{20}}^a p(t)dt f'(a) + \int_{\frac{3a+17b}{20}}^b p(t)dt f'(b) \right) \right] \right|$$

$$\leq \|f''\|_\infty \left[\left| \left(\int_{\frac{17a+3b}{20}}^{\frac{a+b}{2}} p(u)du + \int_{\frac{3a+17b}{20}}^{\frac{a+b}{2}} p(u)du \right) \left(\frac{(a+b)^2}{16} \right) \right| \right.$$

$$+ \left(\int_{\frac{17a+3b}{20}}^{\frac{a+b}{2}} p(u)du \right) \frac{(a+b)^2}{8} - \left(\int_a^{\frac{17a+3b}{20}} p(u)du + \int_{\frac{3a+17b}{20}}^b p(u)du \right) \frac{ab}{4}$$

$$\left. -2 \int_{\frac{17a+3b}{20}}^{\frac{3a+17b}{20}} \left(\frac{(a+b)t}{4} - \frac{t^2}{4} \right) p(t)dt + \int_a^b \left(\frac{(a+b)t}{4} - \frac{t^2}{4} \right) p(t)dt \right]. \qquad (2.3.20)$$

In the next subsection, we will discuss some applications of Composite Quadrature rules.

2.3.2 Applications to Numerical Integration

In order to get composite quadrature rule estimates with small errors, we can use inequality (2.3.8) to get better results.

Let $I_n : a = \zeta_0 < \zeta_1 < \ldots < \zeta_{n-1} < \zeta_n = b$ be a partition of interval $[a, b]$, $\Delta\zeta_k = \zeta_{k+1} - \zeta_k$, $\epsilon \in [0, 1]$, $\alpha_k \leq \xi_k \leq \beta_k$, where $\alpha_k = \zeta_k + h\dfrac{\Delta\zeta_k}{2}$, $\beta_k = \zeta_{k+1} - h\dfrac{\Delta\zeta_k}{2}$, $k \in \{0, \ldots, n-1\}$. Then

$$\int_a^b f(t)p(t)dt = Q_n(f, p, I_n) + R_n(f, p, I_n), \tag{2.3.21}$$

where

$$Q_n(f, p, I_n)$$
$$= \frac{1}{2}\sum_{k=0}^{n-1}\left[\int_{\alpha_k}^{\beta_k} p(t)dt\, f(\xi_k) + \frac{\Delta\zeta_k}{2}(f(\zeta_k)p(\zeta_k) + f(\zeta_{k+1})p(\zeta_{k+1}))\right.$$
$$+ \int_{\beta_k}^{\zeta_{k+1}} p(t)dt\, f(\zeta_{k+1}) - \int_{\alpha_k}^{\zeta_k} p(t)dt\, f(\zeta_k)$$
$$- \int_{\alpha_k}^{\beta_k} p(t)dt\left(\xi_k - \frac{\zeta_k + \zeta_{k+1}}{2}\right)f'(\xi_k) - \int_{\zeta_k}^{\zeta_{k+1}} f(t)\left(t - \frac{\zeta_k + \zeta_{k+1}}{2}\right)p'(t)dt$$
$$\left. - \frac{\Delta\zeta_k}{2}\left(\int_{\alpha_k}^{\zeta_k} p(t)dt\, f'(\zeta_k) + \int_{\beta_k}^{\zeta_{k+1}} p(t)dt\, f'(\zeta_{k+1})\right)\right]\Delta\zeta_k, \tag{2.3.22}$$

$\forall \xi_k \in [\alpha_k, \beta_k]$.

Theorem 2.3.13 *Let all the assumptions of Theorem 2.3.7 be valid. Equation (2.3.21) holds and $Q_n(f, p, I_n)$ is given in the form of (2.3.22). Then the remainder becomes*

$$|R_n(f, p, I_n)|$$
$$\leq \|f''\|_\infty\left[\left(\int_{\alpha_k}^{\xi_k} p(u)du + \int_{\beta_k}^{\xi_k} p(u)du\right)\left(\frac{(\zeta_k + \zeta_{k+1})\xi_k}{4} - \frac{\xi_k^2}{4}\right)\right.$$
$$+ \frac{1}{2}\int_{\xi_k}^{\frac{\zeta_k+\zeta_{k+1}}{2}}\left((\zeta_k + \zeta_{k+1})t - t^2\right)p(t)dt\bigg| + \left(\int_{\alpha_k}^{\frac{\zeta_k+\zeta_{k+1}}{2}} p(u)du\right)\frac{(\zeta_k + \zeta_{k+1})^2}{8}$$
$$- \left(\int_{\zeta_k}^{\alpha_k} p(u)du + \int_{\beta_k}^{\zeta_{k+1}} p(u)du\right)\frac{\zeta_k\zeta_{k+1}}{4} - 2\int_{\alpha_k}^{\beta_k}\left(\frac{(\zeta_k + \zeta_{k+1})t}{4} - \frac{t^2}{4}\right)p(t)dt$$

$$+ \int_{\zeta_k}^{\zeta_{k+1}} \left(\frac{(\zeta_k + \zeta_{k+1})t}{4} - \frac{t^2}{4} \right) p(t)dt \Bigg], \quad (2.3.23)$$

$\forall \, \xi_k \in [\alpha_k, \beta_k]$.

Proof Applying inequality (2.3.8) on $\xi_k \in [\alpha_k, \beta_k] = \left[\zeta_k + h\frac{\Delta \zeta_k}{2}, \zeta_{k+1} - h\frac{\Delta \zeta_k}{2} \right]$ and sum up from 0 to $n-1$ over k, also applying triangular inequality, we achieve (2.3.22) and (2.3.23). □

Special Case 1 If we fixed constant weight in (2.3.22) and (2.3.23) then we will get Theorem 2 of [193].

Corollary 2.3.14 *If we substitute* $\epsilon = 0$, *then* $\alpha_k = \zeta_k$ *and* $\beta_k = \zeta_{k+1}$ *in* (2.3.22) *and* (2.3.23), *we get*

$$Q_n(f, p, I_n)$$
$$= \frac{1}{2} \sum_{k=0}^{n-1} \left[\int_{\zeta_k}^{\zeta_{k+1}} p(t)dt f(\xi_k) + \frac{\Delta \zeta_k}{2} (f(\zeta_k)p(\zeta_k) + f(\zeta_{k+1})p(\zeta_{k+1})) \right.$$
$$- \int_{\zeta_k}^{\zeta_{k+1}} p(t)dt \left(\xi_k - \frac{\zeta_k + \zeta_{k+1}}{2} \right) f'(\xi_k)$$
$$\left. - \int_{\zeta_k}^{\zeta_{k+1}} f(t) \left(t - \frac{\zeta_k + \zeta_{k+1}}{2} \right) p'(t)dt \right] \Delta \zeta_k \quad (2.3.24)$$

and

$$|R_n(f, p, I_n)|$$
$$\leq \|f''\|_\infty \Bigg[\left| \left(\int_{\zeta_k}^{\xi_k} p(u)du + \int_{\zeta_{k+1}}^{\xi_k} p(u)du \right) \left(\frac{(\zeta_k + \zeta_{k+1})\xi_k}{4} - \frac{\xi_k^2}{4} \right) \right.$$
$$\left. + \frac{1}{2} \int_{\xi_k}^{\frac{\zeta_k+\zeta_{k+1}}{2}} \left((\zeta_k + \zeta_{k+1})t - t^2 \right) p(t)dt \right|$$
$$+ \left(\int_{\zeta_k}^{\frac{\zeta_k+\zeta_{k+1}}{2}} p(u)du \right) \frac{(\zeta_k + \zeta_{k+1})^2}{8} - \int_{\zeta_k}^{\zeta_{k+1}} \left(\frac{(\zeta_k + \zeta_{k+1})t}{4} - \frac{t^2}{4} \right) p(t)dt \Bigg].$$
$$(2.3.25)$$

2.3 Generalized Weighted Ostrowski Type Inequality with Parameter

Corollary 2.3.15 *For* $\xi_k = \dfrac{\zeta_k + \zeta_{k+1}}{2}$ *in (2.3.22) and (2.3.23) for* $k \in \{0, \ldots, n-1\}$, *then* $Q_n(f, p, I_n)$ *can be defined as*

$$Q_n(f, p, I_n)$$
$$= \frac{1}{2} \sum_{k=0}^{n-1} \left[\int_{\alpha_k}^{\beta_k} p(t)dt f\left(\frac{\zeta_k + \zeta_{k+1}}{2}\right) + \frac{\Delta \zeta_k}{2} (f(\zeta_k)p(\zeta_k) + f(\zeta_{k+1})p(\zeta_{k+1})) \right.$$
$$+ \int_{\beta_k}^{\zeta_{k+1}} p(t)dt f(\zeta_{k+1}) - \int_{\alpha_k}^{\zeta_k} p(t)dt f(\zeta_k) - \int_{\zeta_k}^{\zeta_{k+1}} f(t)\left(t - \frac{\zeta_k + \zeta_{k+1}}{2}\right) p'(t)dt$$
$$\left. - \frac{\Delta \zeta_k}{2} \left(\int_{\alpha_k}^{\zeta_k} p(t)dt f'(\zeta_k) + \int_{\beta_k}^{\zeta_{k+1}} p(t)dt f'(\zeta_{k+1}) \right) \right] \Delta \zeta_k \quad (2.3.26)$$

and

$$|R_n(f, p, I_n)|$$
$$\leq \|f''\|_\infty \left[\left| \left(\int_{\alpha_k}^{\frac{\zeta_k+\zeta_{k+1}}{2}} p(u)du + \int_{\beta_k}^{\frac{\zeta_k+\zeta_{k+1}}{2}} p(u)du \right) \left(\frac{(\zeta_k + \zeta_{k+1})^2}{16} \right) \right| \right.$$
$$+ \left(\int_{\alpha_k}^{\frac{\zeta_k+\zeta_{k+1}}{2}} p(u)du \right) \frac{(\zeta_k + \zeta_{k+1})^2}{8} - \left(\int_{\zeta_k}^{\alpha_k} p(u)du + \int_{\beta_k}^{\zeta_{k+1}} p(u)du \right) \frac{\zeta_k \zeta_{k+1}}{4}$$
$$\left. - 2\int_{\alpha_k}^{\beta_k} \left(\frac{(\zeta_k + \zeta_{k+1})t}{4} - \frac{t^2}{4} \right) p(t)dt + \int_{\zeta_k}^{\zeta_{k+1}} \left(\frac{(\zeta_k + \zeta_{k+1})t}{4} - \frac{t^2}{4} \right) p(t)dt \right]. \quad (2.3.27)$$

Corollary 2.3.16 *If we substitute* $\epsilon = 0$, *then* $\alpha_k = \zeta_k$ *and* $\beta_k = b$ *in (2.3.26) and (2.3.27),* $k \in \{0, \ldots, n-1\}$, *we get*

$$Q_n(f, p, I_n)$$
$$= \frac{1}{2} \sum_{k=0}^{n-1} \left[\int_{\zeta_k}^{\zeta_{k+1}} p(t)dt f\left(\frac{\zeta_k + \zeta_{k+1}}{2}\right) + \frac{\Delta \zeta_k}{2} (f(\zeta_k)p(\zeta_k) \right.$$
$$\left. + f(\zeta_{k+1})p(\zeta_{k+1})) - \int_{\zeta_k}^{\zeta_{k+1}} f(t)\left(t - \frac{\zeta_k + \zeta_{k+1}}{2}\right) p'(t)dt \right] \Delta \zeta_k \quad (2.3.28)$$

and

$$|R_n(f, p, I_n)|$$
$$\leq \|g''\|_\infty \left[\left| \left(\int_{\zeta_k}^{\frac{\zeta_k+\zeta_{k+1}}{2}} p(u)du + \int_{\zeta_{k+1}}^{\frac{\zeta_k+\zeta_{k+1}}{2}} p(u)du \right) \left(\frac{(\zeta_k + \zeta_{k+1})^2}{16} \right) \right| \right.$$
$$\left. + \left(\int_{\zeta_k}^{\frac{\zeta_k+\zeta_{k+1}}{2}} p(u)du \right) \frac{(\zeta_k + \zeta_{k+1})^2}{8} - \int_{\zeta_k}^{\zeta_{k+1}} \left(\frac{(\zeta_k + \zeta_{k+1})t}{4} - \frac{t^2}{4} \right) p(t)dt \right].$$
(2.3.29)

Corollary 2.3.17 *If we substitute* $\epsilon = 1$, *then* $\alpha_k = \beta_k = \dfrac{\zeta_k + \zeta_{k+1}}{2}$ *in* (2.3.26) *and* (2.3.27), $k \in \{0, \ldots, n-1\}$, *we get*

$$Q_n(f, p, I_n)$$
$$= \frac{1}{2} \sum_{k=0}^{n-1} \left[\frac{\Delta \zeta_k}{2} \left(f(\zeta_k) p(\zeta_k) + f(\zeta_{k+1}) p(b) \right) + \int_{\frac{\zeta_k+\zeta_{k+1}}{2}}^{\zeta_{k+1}} p(t)dt f(\zeta_{k+1}) \right.$$
$$- \int_{\frac{\zeta_k+\zeta_{k+1}}{2}}^{\zeta_k} p(t)dt f(\zeta_k) - \int_{\zeta_k}^{\zeta_{k+1}} f(t) \left(t - \frac{\zeta_k + \zeta_{k+1}}{2} \right) p'(t)dt$$
$$\left. - \frac{\Delta \zeta_k}{2} \left(\int_{\frac{\zeta_k+\zeta_{k+1}}{2}}^{\zeta_k} p(t)dt f'(\zeta_k) + \int_{\frac{\zeta_k+\zeta_{k+1}}{2}}^{\zeta_{k+1}} p(t)dt f'(\zeta_{k+1}) \right) \right] \Delta \zeta_k \quad (2.3.30)$$

and

$$|R_n(f, p, I_n)|$$
$$\leq \|f''\|_\infty \left[-\int_{\zeta_k}^{\zeta_{k+1}} p(u)du \frac{\zeta_k \zeta_{k+1}}{4} + \int_{\zeta_k}^{\zeta_{k+1}} \left(\frac{(\zeta_k + \zeta_{k+1})t}{4} - \frac{t^2}{4} \right) p(t)dt \right].$$
(2.3.31)

Corollary 2.3.18 *If we substitute* $\epsilon = \dfrac{3}{10}$, *then* $\alpha_k = \dfrac{17\zeta_k + 3\zeta_{k+1}}{20}$ *and* $\beta_k = \dfrac{3\zeta_k + 17\zeta_{k+1}}{20}$ *in* (2.3.26) *and* (2.3.27), $k \in \{0, \ldots, n-1\}$, *we get the inequality*

$$Q_n(f, p, I_n)$$
$$= \frac{1}{2} \sum_{k=0}^{n-1} \left[\int_{\frac{17\zeta_k+3\zeta_{k+1}}{20}}^{\frac{3\zeta_k+17\zeta_{k+1}}{20}} p(t)dt f\left(\frac{\zeta_k + \zeta_{k+1}}{2} \right) \right.$$

$$+\frac{\Delta\zeta_k}{2}\left(f(\zeta_k)p(\zeta_k)+f(\zeta_{k+1})p(\zeta_{k+1})\right)+\int_{\frac{3\zeta_k+17\zeta_{k+1}}{20}}^{\zeta_{k+1}}p(t)dtf(\zeta_{k+1})$$

$$-\int_{\frac{17\zeta_k+3\zeta_{k+1}}{20}}^{\zeta_k}p(t)dtf(\zeta_k)-\int_{\zeta_k}^{\zeta_{k+1}}f(t)\left(t-\frac{\zeta_k+\zeta_{k+1}}{2}\right)p'(t)dt$$

$$-\frac{\Delta\zeta_k}{2}\left(\int_{\frac{17\zeta_k+3\zeta_{k+1}}{20}}^{\zeta_k}p(t)dtf'(\zeta_k)+\int_{\frac{3\zeta_k+17\zeta_{k+1}}{20}}^{\zeta_{k+1}}p(t)dtf'(\zeta_{k+1})\right)\Bigg]\Bigg|\Delta\zeta_k$$

(2.3.32)

and

$$|R_n(f,p,I_n)|$$

$$\leq \|f''\|_\infty\Bigg[\Bigg|\left(\int_{\frac{17\zeta_k+3\zeta_{k+1}}{20}}^{\frac{\zeta_k+\zeta_{k+1}}{2}}p(u)du+\int_{\frac{3\zeta_k+17\zeta_{k+1}}{20}}^{\frac{\zeta_k+\zeta_{k+1}}{2}}p(u)du\right)\left(\frac{(\zeta_k+\zeta_{k+1})^2}{16}\right)\Bigg|$$

$$+\left(\int_{\frac{17\zeta_k+3\zeta_{k+1}}{20}}^{\frac{\zeta_k+\zeta_{k+1}}{2}}p(u)du\right)\frac{(\zeta_k+\zeta_{k+1})^2}{8}-\left(\int_{\zeta_k}^{\frac{17\zeta_k+3\zeta_{k+1}}{20}}p(u)du\right.$$

$$+\int_{\frac{3\zeta_k+17\zeta_{k+1}}{20}}^{\zeta_{k+1}}p(u)du\Bigg)\frac{\zeta_k\zeta_{k+1}}{4}-2\int_{\frac{17\zeta_k+3\zeta_{k+1}}{20}}^{\frac{3\zeta_k+17\zeta_{k+1}}{20}}\left(\frac{(\zeta_k+\zeta_{k+1})t}{4}-\frac{t^2}{4}\right)p(t)dt$$

$$+\int_{\zeta_k}^{\zeta_{k+1}}\left(\frac{(\zeta_k+\zeta_{k+1})t}{4}-\frac{t^2}{4}\right)p(t)dt\Bigg]. \qquad (2.3.33)$$

The next section deals with another important inequality that is Ostrowski-Grüss. We will discuss its generalization in terms of weight. We will also provide applications related to probability density function and numerical quadrature rules.

2.4 Generalized Weighted Ostrowski-Grüss Type Inequality with Parameter

This section is intended to generalize weighted Ostrowski-Grüss type inequality for differentiable functions and to deduce explicit error bounds for numerical quadrature formulae using weighted Peano kernel and korkine's identity. Its applications to the probability density function and numerical quadrature rules are also discussed. The current section's are from [55] and [78].

Here we need some results.

Grüss Inequality

Another classical inequality was introduced by Grüss, which depicts a relation between integral of product of two functions and product of integral of two functions which is known as Grüss inequality [65], in fact this inequality establishes a bounds on Čebyšev functional (see [135, p.297]). This type of inequality has great importance due to their applications in numerical integration, probability theory and integral operator theory etc. Grüss inequality is stated as:

Theorem 2.4.1 *Let* $f, g \in L[a,b]$ *where* $\gamma_1 \leq f(x) \leq \Gamma_1$ *and* $\gamma_2 \leq g(x) \leq \Gamma_2$ $\forall$ $x \in [a,b]$, *where* $\gamma_1, \Gamma_1, \gamma_2, \Gamma_2$ *are real constants. Then*

$$|T(f,g)| \leq \frac{1}{4}(\Gamma_1 - \gamma_1)(\Gamma_2 - \gamma_2), \quad (2.4.1)$$

where Čebyšev functional $T(f,g)$ *is defined as in* (1.1.57). *Here the constant* $\frac{1}{4}$ *is the best possible.*

Weighted Grüss Inequality

Dragomir established weighted Grüss inequality in [42] which may be stated as

Theorem 2.4.2 *Let all assumptions of Theorem* 2.4.1 *be valid. Then*

$$|T_p(f,g)| \leq \frac{1}{4}(\Gamma_1 - \gamma_1)(\Gamma_2 - \gamma_2), \quad (2.4.2)$$

where weighted functional is defined as

$$T_p(f,g) = \int_a^b p(x)dx \int_a^b f(x)g(x)dx - \left(\int_a^b f(x)p(x)dx\right)\left(\int_a^b g(x)p(x)dx\right) \quad (2.4.3)$$

and $p : [a,b] \to [0, \infty)$ *be a probability density function. Here the constant* $\frac{1}{4}$ *is the best possible.*

Remark 2.4.1 If we pur $f = g$ in above theorem, then we get another form of weighted Grüss inequality which is stated in [18].

$$0 \leq \int_a^b p(x)f^2(x)dx - \left(\int_a^b p(x)f(x)dx\right)^2 \leq \frac{1}{4}(\Gamma_1 - \gamma_1)^2. \quad (2.4.4)$$

Interested readers can see [135] for other related results.

2.4 Generalized Weighted Ostrowski-Grüss Type Inequality with Parameter

Ostrowski-Grüss Inequality

In 1997, following Ostrowski-Grüss inequality was presented by Dragomir and Wang in [50], that is a linkage between Ostrowski and Grüss inequality.

Theorem 2.4.3 *Let* $f : I \to \mathbb{R}$ *be a differentiable function in* I^0, $a, b \in I^0$ *and* $a < b$. *If for real constants* ϕ, Φ; $\phi \leq f'(x) \leq \Phi$, $\forall\, x \in [a, b]$, *then*

$$\left| f(x) - \frac{1}{b-a}\int_a^b f(t)dt - \frac{f(b) - f(a)}{b - a}\left(x - \frac{a+b}{2}\right)\right| \leq \frac{1}{4}(b-a)(\Phi - \phi).$$
(2.4.5)

In 2000, Ostrowski-Grüss inequality was improved by Matić et al. in [117].

Theorem 2.4.4 *Let* $f : I \to \mathbb{R}$ *be a differentiable function in* I^0 *and let* $a, b \in I^0$ *with* $a < b$. *If* $\phi \leq f'(x) \leq \Phi$, $\forall\, x \in [a, b]$ *for some constants* $\phi, \Phi \in \mathbb{R}$. *Then*

$$\left| f(x) - \frac{1}{b-a}\int_a^b f(t)dt - \frac{f(b) - f(a)}{b - a}\left(x - \frac{a+b}{2}\right)\right| \leq \frac{1}{4\sqrt{3}}(\Phi - \phi)(b-a).$$
(2.4.6)

In the same year in [19], Barnett et al. further improved that inequality (2.4.6), which may be stated as:

Theorem 2.4.5 *Let* $f : I \to \mathbb{R}$ *be an absolutely continuous function such that* $f' \in L_2[a, b]$, *if* $\phi \leq f'(x) \leq \Phi$, $\forall\, x \in [a, b]$ *for some constants* $\phi, \Phi \in \mathbb{R}$, *then*

$$\left| f(x) - \frac{1}{b-a}\int_a^b f(t)dt - \frac{f(b) - f(a)}{b - a}\left(x - \frac{a+b}{2}\right)\right|$$

$$\leq \frac{(b-a)}{2\sqrt{3}}\left[\frac{1}{b-a}\|f'\|_2^2 - \left(\frac{f(b) - f(a)}{b-a}\right)^2\right]^{\frac{1}{2}}$$

$$\leq \frac{1}{4\sqrt{3}}(\Phi - \phi)(b-a).$$
(2.4.7)

In 2010, Zafar and Mir in [194] using Montgomery identity with parameter (2.3.2) and generalized the inequality (2.4.7) in the following theorem.

Theorem 2.4.6 *Let all assumptions of Theorem 2.4.5 be valid. Then*

$$\left| (1-\epsilon)\left[f(x) - \frac{f(b) - f(a)}{b - a}\left(x - \frac{a+b}{2}\right)\right] + \epsilon\frac{f(a) + f(b)}{2} - \frac{1}{b-a}\int_a^b f(t)dt\right|$$

$$\leq \left[\frac{(b-a)^2}{12}(3\epsilon^2 - 3\epsilon + 1) + \epsilon(1-\epsilon)\left(x - \frac{a+b}{2}\right)^2\right]^{\frac{1}{2}}$$

$$\times \left[\frac{1}{b-a}\|f'\|_2^2 - \left(\frac{f(b)-f(a)}{b-a}\right)^2\right]^{\frac{1}{2}}$$

$$\leq \frac{1}{2}(\Phi - \phi)\left[\frac{(b-a)^2}{12}(3\epsilon^2 - 3\epsilon + 1) + \epsilon(1-\epsilon)\left(x - \frac{a+b}{2}\right)^2\right]^{\frac{1}{2}}, \quad (2.4.8)$$

$$\forall\, x \in \left[a + \epsilon\frac{b-a}{2}, b - \epsilon\frac{b-a}{2}\right], \quad \text{where} \quad \epsilon \in [0, 1].$$

Here we need a lemma, by using the following weighted Korkine's identity (2.4.9) from [18] and weighted Grüss inequality (2.4.4), in order to prove our results of this section.

Lemma 2.4.1 *Let $p, f, g : [a, b] \to \mathbb{R}$ be the measurable function for which the integrals involved in the following identity exist and finite. Then*

$$\int_a^b p(t)dt \int_a^b p(t)f(t)g(t)dt - \int_a^b p(t)f(t)dt \int_a^b p(t)g(t)dt$$

$$= \frac{1}{2}\int_a^b \int_a^b p(t)p(s)(f(t)-f(s))(g(t)-g(s))\,dtds. \quad (2.4.9)$$

In the next subsection, we generalize the inequality (2.4.7) and (2.4.8) for differentiable functions in terms of weight and parameter. The generalization of Ostrowski-Grüss inequality has established by introducing weighted Peano kernel. The parameter and weight could be adjusted to retrieve many established results. Throughout this section $\alpha = a + \epsilon\frac{b-a}{2}$ and $\beta = b - \epsilon\frac{b-a}{2}$, where $\epsilon \in [0, 1]$.

2.4.1 Weighted Ostrowski-Grüss Type Inequality with Parameter by Using Korkine's Identity

Theorem 2.4.7 *Let all the assumptions of Theorem 2.4.5 be valid. Further let $p : [a, b] \to (0, \infty)$ be probability density function. Then we get the following inequality*

$$\left| f(x)\int_\alpha^\beta p(u)du + f(a)\int_a^\alpha p(u)du + f(b)\int_\beta^b p(u)du - \int_a^b p(t)f(t)dt \right.$$

$$\left. - \left(x\int_\alpha^\beta p(u)du + a\int_a^\alpha p(u)du + a\int_\beta^b p(u)du - \int_a^b p(t)s\,dt\right)\left(\int_a^b p(t)f'(t)dt\right)\right|$$

2.4 Generalized Weighted Ostrowski-Grüss Type Inequality with Parameter

$$\leq \left[\int_a^b \frac{K_{p,\epsilon}^2(x,t)dt}{p(t)} - \left(\int_a^b K_{p,\epsilon}(x,t)dt \right)^2 \right]^{\frac{1}{2}}$$

$$\left[\int_a^b p(t)[f'(t)]^2 dt - \left(\int_a^b p(t)f'(t)dt \right)^2 \right]^{\frac{1}{2}}$$

$$\leq \frac{1}{2}(\Phi - \phi) H_{p,\epsilon}(x,t), \qquad (2.4.10)$$

where

$$H_{p,\epsilon}(x,t) = \int_a^b \frac{K_{p,\epsilon}^2(x,t)dt}{p(t)} - \left(\int_a^b K_{p,\epsilon}(x,t)dt \right)^2,$$

$\forall\, x \in [\alpha, \beta]$, where $\epsilon \in [0,1]$ and $K_{p,\epsilon}(x,t)$ is defined as in (2.3.5).

Proof From (2.4.9), we get the Korkine's identity in the form of

$$\int_a^b K_{p,\epsilon}(x,t) f'(t) dt - \int_a^b K_{p,\epsilon}(x,t) dt \int_a^b p(t) f'(t) dt$$

$$= \int_a^b \int_a^b p(t) p(s) \left(\frac{K_{p,\epsilon}(x,t)}{p(t)} - \frac{K_{p,\epsilon}(x,s)}{p(s)} \right) (f'(t) - f'(s))\, dt ds.$$

$$(2.4.11)$$

From [74], we have

$$\int_a^b K_{p,\epsilon}(x,t) f'(t) dt = f(x) \int_\alpha^\beta p(u) du + f(a) \int_a^\alpha p(u) du$$

$$+ f(b) \int_\beta^b p(u) du - \int_a^b p(t) f(t) dt \qquad (2.4.12)$$

and by using simple computation, we have

$$\int_a^b K_{p,\epsilon}(x,t) dt = x \int_\alpha^\beta p(u) du + a \int_a^\alpha p(u) du + b \int_\beta^b p(u) du - \int_a^b p(t) t\, dt.$$

$$(2.4.13)$$

By putting (2.4.12) and (2.4.13) in (2.4.11), we get

$$f(x)\int_\alpha^\beta p(u)du + f(a)\int_a^\alpha p(u)du + f(b)\int_\beta^b p(u)du - \int_a^b p(t)f(t)dt$$

$$-\left(x\int_\alpha^\beta p(u)du + a\int_a^\alpha p(u)du + b\int_\beta^b p(u)du - \int_a^b p(t)tdt\right)\left(\int_a^b p(t)f'(t)dt\right)$$

$$=\int_a^b\int_a^b p(t)p(s)\left(\frac{K_{p,\epsilon}(x,t)}{p(t)} - \frac{K_{p,\epsilon}(x,s)}{p(s)}\right)\left(f'(t) - f'(s)\right)dtds, \qquad (2.4.14)$$

$\forall\, x \in [\alpha, \beta]$.

Applying Cauchy-Schwartz inequality for double integrals, we get

$$\left|\frac{1}{2}\int_a^b\int_a^b p(t)p(s)\left(\frac{K_{p,\epsilon}(x,t)}{p(t)} - \frac{K_{p,\epsilon}(x,s)}{p(s)}\right)\left(f'(t) - f'(s)\right)dtds\right|$$

$$\leq \left(\frac{1}{2}\int_a^b\int_a^b p(t)p(s)\left(\frac{K_{p,\epsilon}(x,t)}{p(t)} - \frac{K_{p,\epsilon}(x,s)}{p(s)}\right)^2 dtds\right)^{\frac{1}{2}}$$

$$\times \left(\frac{1}{2}\int_a^b\int_a^b p(t)p(s)\left(f'(t) - f'(s)\right)^2 dtds\right)^{\frac{1}{2}}. \qquad (2.4.15)$$

By using (2.4.11), we get the following identities

$$\frac{1}{2}\int_a^b\int_a^b p(t)p(s)\left(\frac{K_{p,\epsilon}(x,t)}{p(t)} - \frac{K_{p,\epsilon}(x,s)}{p(s)}\right)^2 dtdt$$

$$= \int_a^b \frac{K_{p,\epsilon}^2(x,t)dt}{p(t)} - \left(\int_a^b K_{p,\epsilon}(x,t)dt\right)^2 \qquad (2.4.16)$$

and

$$\frac{1}{2}\int_a^b\int_a^b p(t)p(s)\left(f'(t) - f'(s)\right)^2 dtds = \int_a^b p(t)[f'(t)]^2 dt - \left(\int_a^b p(t)f'(t)dt\right)^2. \qquad (2.4.17)$$

Using weighted Grüss inequality (2.4.4), if $\phi \leq f'(t) \leq \Phi$ and $t \in (a,b)$, we get

$$0 \leq \int_a^b p(t)(f'(t))^2 dt - \left(\int_a^b p(t)f'(t)dt\right)^2 \leq \frac{1}{4}(\Phi - \phi)^2. \qquad (2.4.18)$$

2.4 Generalized Weighted Ostrowski-Grüss Type Inequality with Parameter

Using (2.4.14) – (2.4.18), we get

$$\left| f(x) \int_\alpha^\beta p(u)du + f(a) \int_a^\alpha p(u)du + f(b) \int_\beta^b p(u)du - \int_a^b p(t)f(t)dt \right.$$
$$\left. - \left(x \int_\alpha^\beta p(u)du + a \int_a^\alpha p(u)du + b \int_\beta^b p(u)du - \int_a^b p(t)t dt \right) \left(\int_a^b p(t)f'(t)dt \right) \right|$$

$$\leq \left[\int_a^b \frac{K_{p,\epsilon}^2(x,t)dt}{p(t)} - \left(\int_a^b K_{p,\epsilon}(x,t)dt \right)^2 \right]^{\frac{1}{2}}$$

$$\left[\int_a^b p(t)[f'(t)]^2 dt - \left(\int_a^b p(t)f'(t)dt \right)^2 \right]^{\frac{1}{2}}$$

$$\leq \frac{1}{2}(\Phi - \phi) \left[\int_a^b \frac{K_{p,\epsilon}^2(x,t)dt}{p(t)} - \left(\int_a^b K_{p,\epsilon}(x,t)dt \right)^2 \right]$$

$$= \frac{1}{2}(\Phi - \phi) H_{p,\epsilon}(x,t),$$

which proves our result (2.4.10). □

We can obtain some special results of (2.4.10) as follows:

Remark 2.4.2 If we put $p(t) \equiv \dfrac{1}{b-a}$ in (2.4.10), then we get the special result (2.4.8).

Remark 2.4.3 If we substitute $\epsilon = 0$ in (2.4.10), then $\alpha = a$ and $\beta = b$, then we get following result. (see also Theorem 1 of [78])

Corollary 2.4.8 *Let all the assumptions of Theorem 2.4.7 be valid. Then the following inequality holds*

$$\left| f(x) - \left(x - \int_a^b p(t)t dt \right) \left(\int_a^b p(t)f'(t)dt \right) - \int_a^b p(t)f(t)dt \right|$$

$$\leq \left[\int_a^b \frac{K_p^2(x,t)dt}{p(t)} - \left(\int_a^b K_p(x,t)dt \right)^2 \right]^{\frac{1}{2}}$$

$$\left[\int_a^b p(t)[f'(t)]^2 dt - \left(\int_a^b p(t)f'(t)dt \right)^2 \right]^{\frac{1}{2}}$$

$$\leq \frac{1}{2}(\Phi - \phi) H_p(x,t), \tag{2.4.19}$$

where $K_p(x,t)$ is defined in [78] as:

$$K_p(x,t) = \begin{cases} \int_a^s p(u)du, & t \in [a,x]; \\ \int_b^s p(u)du, & t \in (x,b]. \end{cases} \qquad (2.4.20)$$

Remark 2.4.4 If we put $p(t) \equiv \dfrac{1}{b-a}$ in (2.4.19), then we get the inequality (2.4.7) of [19].

Remark 2.4.5 If we substitute $\epsilon = 1$ in (2.4.10), then $\alpha = \beta = \dfrac{a+b}{2}$, then we get following result.

Corollary 2.4.9 *Let all the assumptions of Theorem 2.4.7 be valid. Then the following inequality holds*

$$\left| f(a) \int_a^{\frac{a+b}{2}} p(u)du + f(b) \int_{\frac{a+b}{2}}^b p(u)du - \int_a^b p(t)f(t)dt \right.$$

$$\left. - \left(a \int_a^{\frac{a+b}{2}} p(u)du + b \int_{\frac{a+b}{2}}^b p(u)du - \int_a^b p(t)t dt \right) \left(\int_a^b p(t)f'(t)dt \right) \right|$$

$$\leq \left[\int_a^b \frac{K_{p,1}^2(x,t)dt}{p(t)} - \left(\int_a^b K_{p,1}(x,t)dt \right)^2 \right]^{\frac{1}{2}}$$

$$\left[\int_a^b p(t)[f'(t)]^2 dt - \left(\int_a^b p(t)f'(t)dt \right)^2 \right]^{\frac{1}{2}}$$

$$\leq \frac{1}{2}(\Phi - \phi) H_{p,1}(x,t). \qquad (2.4.21)$$

Remark 2.4.6 If we put $p(t) \equiv \dfrac{b-a}{2}$ in (2.4.21), then we get following result.

Corollary 2.4.10 *Let all the assumptions of Theorem 2.4.7 be valid. Then the trapezoidal inequality (Hermite-Hadamard right bound) holds*

$$\left| \frac{f(a)+f(b)}{2} - \frac{1}{b-a} \int_a^b f(t)dt \right|$$

$$\leq \frac{b-a}{2\sqrt{3}} \left[\frac{1}{b-a} \|f'\|_2^2 - \left(\frac{f(b)-f(a)}{b-a} \right)^2 \right]^{\frac{1}{2}}$$

2.4 Generalized Weighted Ostrowski-Grüss Type Inequality with Parameter

$$\leq \frac{1}{4\sqrt{3}}(\Phi - \phi)(b - a). \tag{2.4.22}$$

The above inequality can also be found in [194].

Remark 2.4.7 If we put $x = \dfrac{a+b}{2}$ in (2.4.10), then we get following result.

Corollary 2.4.11 *Let all assumptions of Theorem 2.4.7 be valid. Then the following inequality holds*

$$\left| f\left(\frac{a+b}{2}\right) \int_\alpha^\beta p(u)du + f(a) \int_a^\alpha p(u)du + f(b) \int_\beta^b p(u)du - \int_a^b p(t)f(t)dt \right.$$

$$\left. - \left(\frac{a+b}{2} \int_\alpha^\beta p(u)du + a \int_a^\alpha p(u)du + b \int_\beta^b p(u)du - \int_a^b p(t)t\,dt\right)\left(\int_a^b p(t)f'(t)dt\right) \right|$$

$$\leq \left[\int_a^b \frac{K_{p,\epsilon}^2\left(\frac{a+b}{2},t\right)dt}{p(t)} - \left(\int_a^b K_{p,\epsilon}\left(\frac{a+b}{2},t\right)dt\right)^2 \right]^{\frac{1}{2}}$$

$$\left[\int_a^b p(t)[f'(t)]^2 dt - \left(\int_a^b p(t)f'(t)dt\right)^2 \right]^{\frac{1}{2}}$$

$$\leq \frac{1}{2}(\Phi - \phi) H_{p,\epsilon}\left(\frac{a+b}{2},t\right). \tag{2.4.23}$$

Remark 2.4.8 If we put $p(t) \equiv \dfrac{1}{b-a}$ in (2.4.23), then we get following result.

Corollary 2.4.12 *Let all the assumptions of Theorem 2.4.7 be valid. Then bound of average midpoint and trapezoidal inequality holds*

$$\left| (1-\epsilon)f\left(\frac{a+b}{2}\right) + \epsilon\frac{f(a)+f(b)}{2} - \frac{1}{b-a}\int_a^b f(t)dt \right|$$

$$\leq \frac{b-a}{2\sqrt{3}}\sqrt{3\epsilon^2 - 3\epsilon + 1}\left[\frac{1}{b-a}\|f'\|_2 - \left(\frac{f(b)-f(a)}{b-a}\right)^2\right]^{\frac{1}{2}}$$

$$\leq \frac{(\Phi-\phi)(b-a)}{4\sqrt{3}}\sqrt{3\epsilon^2 - 3\epsilon + 1}.$$

Remark 2.4.9 If we substitute $\epsilon = 1$ in (2.4.23), then we get following result.

Corollary 2.4.13 *Let all the assumptions of Theorem 2.4.7 be valid. Then the following inequality holds*

$$\left| f(a) \int_a^{\frac{a+b}{2}} p(u)du + f(b) \int_{\frac{a+b}{2}}^b p(u)du - \int_a^b p(t)f(t)dt \right.$$

$$\left. - \left(a \int_a^{\frac{a+b}{2}} p(u)du + b \int_{\frac{a+b}{2}}^b p(u)du - \int_a^b p(t)tdt \right) \left(\int_a^b p(t)f'(t)dt \right) \right|$$

$$\leq \left[\int_a^b \frac{K_{p,1}^2\left(\frac{a+b}{2},t\right)dt}{p(t)} - \left(\int_a^b K_{p,1}\left(\frac{a+b}{2},t\right)dt \right)^2 \right]^{\frac{1}{2}}$$

$$\left[\int_a^b p(t)[f'(t)]^2 dt - \left(\int_a^b p(t)f'(t)dt \right)^2 \right]^{\frac{1}{2}}$$

$$\leq \frac{1}{2}(\Phi - \phi)H_{p,1}\left(\frac{a+b}{2},t\right). \tag{2.4.24}$$

Remark 2.4.10 If we put $p(t) \equiv \dfrac{1}{b-a}$ in (2.4.24), then the trapezoidal inequality (Hermite-Hadamard right bound) holds as we achieved in (2.4.22).

Remark 2.4.11 If we substitute $\epsilon = 0$, then $\alpha = a$ and $\beta = b$ in (2.4.23), the we get following result. (see also Corollary 1 of [78])

Corollary 2.4.14 *Let all assumptions of Theorem 2.4.7 be valid. Then the weighted midpoint inequality holds*

$$\left| f\left(\frac{a+b}{2}\right) - \left(\frac{a+b}{2} - \int_a^b p(t)tdt\right) \left(\int_a^b p(t)f'(t)dt \right) - \int_a^b p(t)f(t)dt \right|$$

$$\leq \left[\int_a^b \frac{K_p^2\left(\frac{a+b}{2},t\right)dt}{p(t)} - \left(\int_a^b K_p\left(\frac{a+b}{2},t\right)dt \right)^2 \right]^{\frac{1}{2}}$$

$$\left[\int_a^b p(t)[f'(t)]^2 dt - \left(\int_a^b p(t)f'(t)dt \right)^2 \right]^{\frac{1}{2}}$$

$$\leq \frac{1}{2}(\Phi - \phi)H_p\left(\frac{a+b}{2},t\right). \tag{2.4.25}$$

Remark 2.4.12 If we put $p(t) \equiv \dfrac{1}{b-a}$ in (2.4.25), then we get following result.

2.4 Generalized Weighted Ostrowski-Grüss Type Inequality with Parameter

Corollary 2.4.15 *Let all the assumptions of Theorem 2.4.7 be valid. Then the midpoint inequality (Hermite-Hadamard right bound) holds*

$$\left| f\left(\frac{a+b}{2}\right) - \frac{1}{b-a} \int_a^b f(t)dt \right|$$
$$\leq \frac{(b-a)}{2\sqrt{3}} \left[\frac{1}{b-a} |f'|_2^2 - \left(\frac{f(b)-f(a)}{b-a}\right)^2 \right]^{\frac{1}{2}} \leq \frac{1}{4\sqrt{3}}(\Phi - \phi)(b-a).$$

The above inequality can also be found in [78] *and* [194].

Remark 2.4.13 *If we substitute* $\epsilon = \frac{1}{2}$ *in (2.4.23), then we get following result.*

Corollary 2.4.16 *Let all the assumptions of Theorem 2.4.7 be valid. Then the following inequality holds*

$$\left| f\left(\frac{a+b}{2}\right) \int_{\frac{3a+b}{4}}^{\frac{a+3b}{4}} p(u)du + f(a) \int_a^{\frac{3a+b}{4}} p(u)du \right.$$
$$+ f(b) \int_{\frac{a+3b}{4}}^b p(u)du - \int_a^b p(t)f(t)dt$$
$$- \left(\frac{a+b}{2} \int_{\frac{3a+b}{4}}^{\frac{a+3b}{4}} p(u)du + a \int_a^{\frac{3a+b}{4}} p(u)du + b \int_{\frac{a+3b}{4}}^b p(u)du \right.$$
$$\left. - \int_a^b p(t)t\,dt \right) \left. \left(\int_a^b p(t)f'(t)dt \right) \right|$$
$$\leq \left[\int_a^b \frac{K_{p,\frac{1}{2}}^2 \left(\frac{a+b}{2},t\right) dt}{p(t)} - \left(\int_a^b K_{p,\frac{1}{2}}\left(\frac{a+b}{2},t\right) dt \right)^2 \right]^{\frac{1}{2}}$$
$$\left[\int_a^b p(t)[f'(t)]^2 dt - \left(\int_a^b p(t)f'(t)dt \right)^2 \right]^{\frac{1}{2}}$$
$$\leq \frac{1}{2}(\Phi - \phi) H_{p,\frac{1}{2}}\left(\frac{a+b}{2},t\right). \tag{2.4.26}$$

Remark 2.4.14 *If we put* $p(t) \equiv \frac{1}{b-a}$ *in (2.4.26), then we get following result.*

Corollary 2.4.17 *Let all the assumptions of Theorem 2.4.7 be valid. Then the bound of average midpoint and trapezoidal inequality holds*

$$\left| \frac{1}{2} \left[\frac{f(a)+f(b)}{2} + f\left(\frac{a+b}{2}\right) \right] - \frac{1}{b-a} \int_a^b f(t)dt \right|$$

$$\leq \frac{(b-a)}{4\sqrt{3}} \left[\frac{1}{b-a} \|f'\|_2^2 - \left(\frac{f(b)-f(a)}{b-a} \right)^2 \right]^{\frac{1}{2}} \leq \frac{1}{8\sqrt{3}} (\Phi - \phi)(b-a).$$

The above inequality can also be found in [194].

Remark 2.4.15 If we substitute $\epsilon = \dfrac{1}{3}$ in (2.4.23), then we get following result.

Corollary 2.4.18 *Let all the assumptions of Theorem 2.4.7 be valid. Then the following inequality holds*

$$\left| f\left(\frac{a+b}{2}\right) \int_{\frac{5a+b}{6}}^{\frac{a+5b}{6}} p(u)du + f(a) \int_a^{\frac{5a+b}{6}} p(u)du \right.$$

$$+ f(b) \int_{\frac{a+5b}{6}}^b p(u)du - \int_a^b p(t)f(t)dt$$

$$- \left(\frac{a+b}{2} \int_{\frac{5a+b}{6}}^{\frac{a+5b}{6}} p(u)du + a \int_a^{\frac{5a+b}{6}} p(u)du + b \int_{\frac{a+5b}{6}}^b p(u)du \right.$$

$$\left. \left. - \int_a^b p(t)t \, dt \right) \left(\int_a^b p(t)f'(t)dt \right) \right|$$

$$\leq \left[\int_a^b \frac{K_{p,\frac{1}{3}}^2\left(\frac{a+b}{2}, t\right) dt}{p(t)} - \left(\int_a^b K_{p,\frac{1}{3}}\left(\frac{a+b}{2}, t\right) dt \right)^2 \right]^{\frac{1}{2}}$$

$$\left[\int_a^b p(t)[f'(t)]^2 dt - \left(\int_a^b p(t)f'(t)dt \right)^2 \right]^{\frac{1}{2}}$$

$$\leq \frac{1}{2} (\Phi - \phi) H_{p,\frac{1}{3}}\left(\frac{a+b}{2}, t\right). \tag{2.4.27}$$

Remark 2.4.16 If we put $p(t) \equiv \dfrac{1}{b-a}$ in (2.4.27), then we get following result.

Corollary 2.4.19 *Let all the assumptions of Theorem 2.4.7 be valid. Then the bound of* $\frac{1}{3}$ *Simpson's rule holds*

$$\left| \frac{1}{3}\left[\frac{f(a)+f(b)}{2} + 2f\left(\frac{a+b}{2}\right) \right] - \frac{1}{b-a}\int_a^b f(t)dt \right|$$

$$\leq \frac{(b-a)}{6}\left[\frac{1}{b-a}\|f'\|_2^2 - \left(\frac{f(b)-f(a)}{b-a}\right)^2 \right]^{\frac{1}{2}} \leq \frac{1}{12\sqrt{3}}(\Phi-\phi)(b-a).$$

The above inequality can also be found in [194].

Remark 2.4.17 *If we put* $x = a$ *or* $x = b$ *and* $p(t) \equiv \frac{1}{b-a}$ *in (2.4.10), then we get following result.*

Corollary 2.4.20 *Let all the assumptions of Theorem 2.4.7 be valid. Then the trapezoidal inequality (Hermite-Hadamard right bound) holds which is independent the value of* ϵ

$$\left| \frac{f(a)+f(b)}{2} - \frac{1}{b-a}\int_a^b f(t)dt \right|$$

$$\leq \frac{b-a}{2\sqrt{3}}\left[\frac{1}{b-a}\|f'\|_2 - \left(\frac{f(b)-f(a)}{b-a}\right)^2 \right]^{\frac{1}{2}} \leq \frac{(b-a)(\Phi-\phi)}{4\sqrt{3}}.$$

Now, we will discuss some applications for probability density function of our results in the next subsection.

2.4.2 Applications to Probability Theory

Suppose X be a continuous random variable with function of probability density $f : [a,b] \to \mathbb{R}_+$ and the cumulative distribution $F : [a,b] \to [0,1]$, i.e.,

$$F(x) = \int_a^x f(t)dt, \quad x \in [\alpha,\beta] \subset [a,b],$$

$$E(X) = \int_a^b tf(t)dt$$

and weighted expectation would be

$$E_p(X) = \int_a^b p(t)tf(t)dt.$$

Theorem 2.4.21 *Let all assumptions of Theorem 2.4.7 be valid. In addition we let probability density function belongs to $L_2[a,b]$ space. Then*

$$\left| F(x) \int_\alpha^\beta p(u)du + \int_\beta^b p(u)du - bp(b) + E_w(X) + \int_a^b p'(t)tF(t)dt \right.$$
$$- \left(x \int_\alpha^\beta p(u)du + a \int_a^\alpha p(u)du + b \int_\beta^b p(u)du - \int_a^b p(t)tdt \right)$$
$$\left. \left(p(b) - \int_a^b p'(t)F(t)dt \right) \right|$$
$$\leq \frac{1}{2}(\Phi - \phi) H_{p,\epsilon}(x,t), \qquad (2.4.28)$$

$\forall x \in [\alpha, \beta]$.

Proof Put $f = F$ in (2.4.10) and by using these two identities mention below, we get (2.4.28),

$$\int_a^b p(t)F(t)dt = bp(b) - E_p(X) - \int_a^b p'(t)tF(t)dt$$

and

$$\int_a^b p(t)F'(t)dt = p(b) - \int_a^b p'(t)F(t)dt.$$

$\square$

Corollary 2.4.22 *If we put $p(t) \equiv \dfrac{1}{b-a}$ in (2.4.28), then following inequality holds*

$$\left| (1-\epsilon) \left[F(x) - \frac{1}{b-a} \left(x - \frac{a+b}{2} \right) \right] + \frac{\epsilon}{2} - \frac{b - E(X)}{b-a} \right|$$
$$\leq \frac{1}{b-a} \left[\frac{1}{12}(3\epsilon^2 - 3\epsilon + 1) + \epsilon(1-\epsilon)\left(x - \frac{a+b}{2} \right)^2 \right]^{\frac{1}{2}} \left[(b-a)\|f\|_2^2 - 1 \right]^{\frac{1}{2}}$$
$$\leq \frac{\Phi - \phi}{2(b-a)} \left[\frac{1}{12}(3\epsilon^2 - 3\epsilon + 1) + \epsilon(1-\epsilon)\left(x - \frac{a+b}{2} \right)^2 \right]^{\frac{1}{2}}.$$

2.4 Generalized Weighted Ostrowski-Grüss Type Inequality with Parameter

Corollary 2.4.23 *If we substitute $\epsilon = 0$ in (2.4.28), then we get the following inequality*

$$\left| F(x) - p(b)b + E_p(X) - \left(x - \int_a^b p(t)tdt \right) \right.$$
$$\left. \left(p(b) - \int_a^b p'(t)F(t)dt \right) + \int_a^b p'(t)tF(t)dt \right|$$
$$\leq \frac{1}{2}(\Phi - \phi)H_p(x, t). \tag{2.4.29}$$

Corollary 2.4.24 *If we put $p(t) \equiv \dfrac{1}{b-a}$ in (2.4.29), then*

$$\left| F(x) - \frac{1}{b-a}\left(x - \frac{a+b}{2} \right) - \frac{b - E(X)}{b-a} \right|$$
$$\leq \frac{(b-a)}{2\sqrt{3}} \left[\frac{1}{b-a}\|f'\|_2^2 - \left(\frac{f(b) - f(a)}{b-a} \right)^2 \right]^{\frac{1}{2}} \leq \frac{1}{4\sqrt{3}}(\Phi - \phi)(b-a).$$

Next subsection covers some applications to numerical quadrature rules.

2.4.3 Applications to Numerical Integration

Let $I_n : a = \zeta_0 < \zeta_1 < \ldots < \zeta_n = b$ be a partition of the interval $[a, b]$ and $\Delta \zeta_k = \zeta_{k+1} - \zeta_k, k \in \{0, 1, 2, \ldots, n-1\}$. Then for constant weight following identity holds

$$\int_a^b p(t)f(t)dt = Q_n(f, p, I_n) + R_n(f, p, I_n), \tag{2.4.30}$$

where $Q_n(f, p, I_n)$ is defined as

$$Q_n(f, p, I_n)$$
$$= \sum_{k=0}^{n-1} \left[f(\xi) \int_{\alpha_k}^{\beta_k} p(u)du + f(\zeta_k) \int_{\zeta_k}^{\alpha_k} p(u)du + f(\zeta_{k+1}) \int_{\beta_k}^{\zeta_{k+1}} p(u)du \right.$$
$$- \left(\xi_k \int_{\alpha_k}^{\beta_k} p(u)du + \zeta_k \int_{\zeta_k}^{\alpha_k} p(u)du + \zeta_{k+1} \int_{\beta_k}^{\zeta_{k+1}} p(u)du - \int_{\zeta_k}^{\zeta_{k+1}} p(t)tdt \right)$$
$$\left. \times \left(\int_{\zeta_k}^{\zeta_{k+1}} p(t)f'(t)dt \right) \right]. \tag{2.4.31}$$

Theorem 2.4.25 *Let all assumptions of Theorem 2.4.7 be valid. Then identity (2.4.30) holds where $Q_n(I_n, f, p)$ is given in the form of (2.4.31) and the remainder $R_n(f, p, I_n)$ becomes*

$$|R_n(f, p, I_n)| \leq \sum_{k=0}^{n-1} \frac{(\Phi - \phi)}{2} H_{p,\epsilon}(\xi_k, t). \quad (2.4.32)$$

Proof If inequality (2.4.10) is applied on $[\zeta_k, \zeta_{k+1}]$, we get

$$R_k(f, p, I_k)$$
$$= \int_{\zeta_k}^{\zeta_{k+1}} p(t) f(t) dt - f(\xi_k) \int_{\alpha_k}^{\beta_k} p(u) du - f(\zeta_k) \int_{\zeta_k}^{\alpha_k} p(u) du$$
$$- f(\zeta_{k+1}) \int_{\beta_k}^{\zeta_{k+1}} p(u) du + \left(\xi_k \int_{\alpha_k}^{\beta_k} p(u) du + \zeta_k \int_{\zeta_k}^{\alpha_k} p(u) du \right.$$
$$+ \zeta_{k+1} \int_{\beta_k}^{\zeta_{k+1}} p(u) du - \int_{\zeta_k}^{\zeta_{k+1}} p(t) t\, dt \bigg) \times \left(\int_{\zeta_k}^{\zeta_{k+1}} p(t) f'(t) dt \right).$$

Summing $R_k(f, p, I_k)$ over k from 0 to $n - 1$. This yields

$$R_n(f, p, I_n)$$
$$= \int_a^b p(t) f(t) dt - \sum_{k=0}^{n-1} \bigg[f(\xi_k) \int_{\alpha_k}^{\beta_k} p(u) du + f(\zeta_k) \int_{\zeta_k}^{\alpha_k} p(u) du$$
$$+ f(\zeta_{k+1}) \int_{\beta_k}^{\zeta_{k+1}} p(u) du - \left(\xi_k \int_{\alpha_k}^{\beta_k} p(u) du + \zeta_k \int_{\zeta_k}^{\alpha_k} p(u) du \right.$$
$$+ \zeta_{k+1} \int_{\beta_k}^{\zeta_{k+1}} p(u) du - \int_{\zeta_k}^{\zeta_{k+1}} p(t) t\, dt \bigg) \times \left(\int_{\zeta_k}^{\zeta_{k+1}} p(t) f'(t) dt \right) \bigg].$$

Applying absolute property on the above identity, we get

$$|R_n(f, p, I_n)|$$
$$= \bigg| \int_a^b p(t) f(t) dt - \sum_{k=0}^{n-1} \bigg[f(\xi_k) \int_{\alpha_k}^{\beta_k} p(u) du + f(\zeta_k) \int_{\zeta_k}^{\alpha_k} p(u) du$$
$$+ f(\zeta_{k+1}) \int_{\beta_k}^{\zeta_{k+1}} p(u) du - \left(\xi_k \int_{\alpha_k}^{\beta_k} p(u) du + \zeta_k \int_{\zeta_k}^{\alpha_k} p(u) du \right.$$

2.4 Generalized Weighted Ostrowski-Grüss Type Inequality with Parameter

$$+\zeta_{k+1}\int_{\beta_k}^{\zeta_{k+1}} p(u)du - \int_{\zeta_k}^{\zeta_{k+1}} p(t)tdt\right) \times \left(\int_{\zeta_k}^{\zeta_{k+1}} p(t)f'(t)dt\right)\right]\right|$$

$$\leq \frac{1}{2}(\Phi - \phi)H_{p,\epsilon}(\xi_k, t).$$

□

Corollary 2.4.26 *If we fixed constant weight in (2.4.30) and (2.4.31), identity*

$$\int_a^b f(t)dt = Q_n(f, I_n) + R_n(f, I_n) \qquad (2.4.33)$$

holds, where $Q_n(f, I_n)$ is defined as

$$Q_n(f, I_n) = \sum_{k=0}^{n-1}\left[(1-\epsilon)\left[f(\xi_k) - \frac{f(\zeta_{k+1}) - f(\zeta_k)}{\Delta\zeta_k}\left(\xi_k - \frac{\zeta_k + \zeta_{k+1}}{2}\right)\right.\right.$$

$$\left.\left. +\epsilon\frac{f(\zeta_k) + f(\zeta_{k+1})}{2}\right]\Delta\zeta_k, \qquad (2.4.34)$$

then remainder $R_n(f, I_n)$ satisfies the estimates

$$|R_n(f, I_n)|$$

$$\leq \sum_{k=0}^{n-1}\frac{(\Phi-\phi)}{2}\left[\frac{(\Delta\zeta_k)^2}{12}(3\epsilon^2 - 3\epsilon + 1) + \epsilon(1-\epsilon)\left(\xi_k - \frac{\zeta_k + \zeta_{k+1}}{2}\right)^2\right]^{\frac{1}{2}}.$$

Corollary 2.4.27 *If we substitute $\epsilon = 0$ in (2.4.31), identity (2.4.30) holds, where $Q_n(f, p, I_n)$ is defined as*

$$Q_n(I_n, f, p) = \sum_{k=0}^{n-1}\left[f(\xi_k) - \left(\xi_k - \int_{\zeta_k}^{\zeta_{k+1}} p(t)tdt\right)\left(\int_{\zeta_k}^{\zeta_{k+1}} p(t)f'(t)dt\right)\right],$$

$$(2.4.35)$$

then the remainder $R_n(f, p, I_n)$ satisfies the estimates

$$|R_n(f, p, I_n)| \leq \sum_{k=0}^{n-1}\frac{(\Phi-\phi)}{2}H_p(\xi_k, t). \qquad (2.4.36)$$

Corollary 2.4.28 *If we fixed constant weight in (2.4.35), identity (2.4.33) holds where the $Q_n(f, I_n)$ is defined as*

$$Q_n(f, I_n) = \sum_{k=0}^{n-1} \left[f(\xi_k) - \frac{f(\zeta_{k+1}) - f(\zeta_k)}{\Delta \zeta_k} \left(\xi_k - \frac{\zeta_k + \zeta_{k+1}}{2} \right) \right] \Delta \zeta_k,$$

(2.4.37)

then remainder $R_n(f, I_n)$ satisfies the estimates

$$|R_n(f, I_n)| \leq \sum_{k=0}^{n-1} \frac{(\Phi - \phi) \Delta \zeta_k^2}{8\sqrt{3}}.$$

In the next section, we would like to present generalized Montgomery type identity using Riemann-Liouville fractional integtral. We will also discuss some generalization of fractional Ostrowski-Grüss type inequality.

2.5 Generalized Fractional Ostrwoski Type Inequality with Parameter

In the current section, we use Riemann-Liouville fractional integral to provide generalization of Ostrowski type integral inequality with bounded derivatives. We also get better bounds of the inequality under consideration.

Fractional Calculus represents a complex phenomenon in a more accurate and efficient way than classical calculus. It accepts integrals and derivatives of any positive order. This subject has gained importance and popularity during the last four decades because it has various applications in the field of science and engineering. The classical form of fractional calculus is given by Riemann and Liouville. Weyl, Fourier, Abel, Lacrois, Leibniz and Gunwald have also done significant contributions in this field. Fractional integrals play increasingly important role on some mathematical inequalities such as Ostrowski, Grüss, Ostrowski-Grüss and Hadamard inequality etc.

We quote from [14],

> The subject of fractional calculus (that is, calculus of integrals and derivatives of an arbitrary real or complex order) was planted over 300 years ago. Since that time the fractional calculus has drawn the attention of many researchers in. In recent years, the fractional calculus has played a significant role in many areas of science and engineering.

Over the last two decades, fractional integral inequalities have been one of the most important tools for the advancement of many branches of mathematics. Many authors have discussed certain generalizations of fractional integral inequalities. Recent generalizations can be found in the articles [53, 80, 169–172].

2.5 Generalized Fractional Ostrwoski Type Inequality with Parameter

For our next result we need here definition of Riemann-Liouville fractional integral from [64].

Definition 2.5.1 The Riemann-Liouville fractional integral operator of order $\gamma \geq 0$ is defined as

$$J_a^\gamma f(x) = \frac{1}{\Gamma(\gamma)} \int_a^x (x-t)^{\gamma-1} f(t) dt$$

$$J_a^0 f(x) = f(x).$$

where gamma function $\Gamma(\gamma)$ is defined as

$$\Gamma(\gamma) = \int_0^\infty x^{\gamma-1} e^{-x} dx.$$

In [11] by using Riemann-Liouville fractional integrals, the authors obtained fractional Montgomery identity.

Theorem 2.5.1 Let $f: I \to \mathbb{R}$ be differentiable function on I^0 with $a, b \in I$ $a < b$, $f' \in L[a, b]$ and for $\gamma \geq 1$, then Montgomery fractional identity holds

$$f(x) = \frac{\Gamma(\gamma)}{b-a}(b-x)^{1-\gamma} J_a^\gamma f(b) - J_a^{\gamma-1}(K^\gamma(x, b) f(b)) + J_a^\gamma (K^\gamma(x, b) f'(b)),$$

(2.5.1)

where $K^\gamma(x, t)$ is the fractional Peano kernel defined as

$$K^\gamma(x, t) = \begin{cases} \dfrac{t-a}{b-a}(b-x)^{1-\gamma} \Gamma(\gamma), & t \in [a, x]; \\[2mm] \dfrac{t-b}{b-a}(b-x)^{1-\gamma} \Gamma(\gamma), & t \in [x, b]. \end{cases} \quad (2.5.2)$$

In the following subsection, we propose some new results of fractional integral inequalities of Ostrowski type. We begin by establishing Montgomery identity with parameters using Riemann-Liouville fractional integrals and then use this useful identity to generate a lemma, which is further required in our main theorem.

2.5.1 Fractional Ostrowski Type Inequality Involving Parameter

We need to proof the following lemmas for our main result.

Lemma 2.5.1 *Let $f : I \to \mathbb{R}$ be an absolutely continuous function on I^0 with $a, b \in I$ $a < b$, $f' \in L[a,b]$ and for $\gamma \geq 1$, Then the following identity holds*

$$(1-\epsilon)f(x) = \frac{\Gamma(\gamma)}{b-a}(b-x)^{1-\gamma} J_a^\gamma f(b) - J_a^{\gamma-1}\left(K_\epsilon^\gamma(x,b)f(b)\right)$$
$$-\frac{\epsilon(b-x)^{1-\gamma}}{2(b-a)^{1-\gamma}} J_a^0 f(a) + J_a^\gamma \left(K_\epsilon^\gamma(x,b)f'(b)\right), \quad (2.5.3)$$

$\forall\, x \in \left[a+\epsilon\frac{b-a}{2}, b+\epsilon\frac{b-a}{2}\right]$, where $\epsilon \in [0,1]$, and $K_\epsilon^\gamma(x,t)$ is the fractional Peano kernel defined by

$$K_\epsilon^\gamma(x,t) = \begin{cases} \left[t-\left(a+\epsilon\frac{b-a}{2}\right)\right]\frac{(b-x)^{1-\gamma}}{b-a}\Gamma(\gamma), & t \in [a,x); \\ \left[t-\left(b-\epsilon\frac{b-a}{2}\right)\right]\frac{(b-x)^{1-\gamma}}{b-a}\Gamma(\gamma), & t \in [x,b]. \end{cases} \quad (2.5.4)$$

Proof Using Riemann-Liouville fractional integral operator twice, we get

$$J_a^\gamma(K_\epsilon^\gamma(x,b)f'(b))$$
$$= \frac{1}{\Gamma(\gamma)}\int_a^b (b-t)^{\gamma-1} K_\epsilon^\gamma(x,t) f'(t) dt$$
$$= \frac{(b-x)^{1-\gamma}}{b-a}\left[\int_a^x (b-t)^{\gamma-1}\left(t-\left(a+\epsilon\frac{b-a}{2}\right)\right) f'(t) dt\right.$$
$$\left.+\int_x^b (b-t)^{\gamma-1}\left(t-\left(b-\epsilon\frac{b-a}{2}\right)\right) f'(t) dt\right]$$
$$= (1-\epsilon)f(x) - \frac{(b-x)^{1-\gamma}}{(b-a)}\int_a^b (b-t)^{\gamma-1} f(t) dt$$
$$+\frac{\epsilon(b-x)^{1-\gamma}}{2(b-a)^{1-\gamma}} f(a) + \frac{\gamma-1}{\Gamma(\gamma)}\int_a^b (b-t)^{\gamma-2} K_\epsilon^\gamma(x,t) f(t) dt$$
$$= (1-\epsilon)f(x) - \frac{(b-x)^{1-\gamma}}{(b-a)}\Gamma(\gamma) J_a^\gamma f(b) + \frac{\epsilon(b-x)^{1-\gamma}}{2(b-a)^{1-\gamma}} J_a^0 f(a) + J_a^{\gamma-1}(K_\epsilon^\gamma(x,b)).$$

After rearranging the terms, we get (2.5.3). $\square$

Remark 2.5.1 If we substitute $\gamma = 1$ in (2.5.3), then we get Montgomery identity with parameter as stated in (2.3.2).

Remark 2.5.2 If we substitute $\epsilon = 0$ in (2.5.3), then we get the identity (2.5.1).

2.5 Generalized Fractional Ostrwoski Type Inequality with Parameter

Lemma 2.5.2 *Let all the assumptions of Lemma 2.5.1 be valid. Then the generalization of (2.5.3) holds*

$$(1-\epsilon)f(x)$$
$$= 2\Gamma(\gamma)\frac{(b-x)^{1-\gamma}}{b-a}J_a^\gamma f(b) - J_a^{\gamma-1}\left(K_\epsilon^\gamma(x,b)f(b)\right) - \frac{(b-x)^{2-\gamma}}{b-a}\Gamma(\gamma)J_a^{\gamma-1}f(b)$$
$$-\frac{(b-x)^{1-\gamma}}{(b-a)^{2-\gamma}}\left(x-a+\epsilon\frac{b-a}{2}\right)J_a^0 f(a) + 2J_a^\gamma\left(K_{*\epsilon}^\gamma(x,b)f'(b)\right), \qquad (2.5.5)$$

where $K_{*\epsilon}^\gamma(x,t)$ *is the fractional Peano kernel defined by*

$$K_{*\epsilon}^\gamma(x,t) = \begin{cases} \left[t-\left(\dfrac{a+x}{2}+\epsilon\dfrac{b-a}{4}\right)\right]\dfrac{(b-x)^{1-\gamma}}{b-a}\Gamma(\gamma), & \text{if } t \in [a,x); \\[2ex] \left[t-\left(\dfrac{b+x}{2}-\epsilon\dfrac{b-a}{4}\right)\right]\dfrac{(b-x)^{1-\gamma}}{b-a}\Gamma(\gamma), & \text{if } t \in [x,b]. \end{cases}$$

Proof Using Riemann-Liouville fractional integral operator on $K_{*\epsilon}^\gamma(x,t)$, we get

$$J_a^\gamma\left(K_{*\epsilon}^\gamma(x,b)f'(b)\right)$$
$$= \frac{1}{\Gamma(\gamma)}\int_a^b (b-t)^{\gamma-1}K_{*\epsilon}^\gamma(x,t)f'(t)dt$$
$$= \frac{(b-x)^{1-\gamma}}{b-a}\left[\int_a^x (b-t)^{\gamma-1}\left(t-\left(\frac{a+x}{2}+\epsilon\frac{b-a}{4}\right)\right)f'(t)dt \right.$$
$$\left. + \int_x^b (b-t)^{\gamma-1}\left(t-\left(\frac{b+x}{2}-\epsilon\frac{b-a}{4}\right)\right)f'(t)dt\right],$$

after some computations, we get

$$J_a^\gamma\left(K_{*\epsilon}^\gamma(x,b)f'(b)\right)$$
$$= \frac{1}{2}\left[J_a^\gamma\left(K_\epsilon^\gamma(x,b)f'(b)\right) + \frac{(b-x)^{1-\gamma}}{b-a}\int_a^b (b-t)^{\gamma-1}(t-x)f'(t)dt\right]. \qquad (2.5.6)$$

We also have

$$\int_a^b (b-t)^{\gamma-1}(t-x)f'(t)dt$$
$$= (x-a)(b-a)^{\gamma-1}J_a^0 f(a) + (b-x)\Gamma(\gamma)J_a^{\gamma-1}f(b) - \Gamma(\gamma)J_a^\gamma f(b). \qquad (2.5.7)$$

Now from Lemma 2.5.1, we have

$$J_a^\gamma \left(K_\epsilon^\gamma (x,b) f'(b) \right)$$
$$= (1-\epsilon) f(x) - \frac{\Gamma(\gamma)}{b-a} (b-x)^{1-\gamma} J_a^\gamma f(b) + J_a^{\gamma-1} \left(K_\epsilon^\gamma (x,b) f(b) \right)$$
$$+ \frac{\epsilon (b-x)^{1-\gamma}}{2(b-a)^{1-\gamma}} J_a^0 f(a). \tag{2.5.8}$$

Using (2.5.7) and (2.5.8) in (2.5.6), we get

$$J_a^\gamma (K_{*\epsilon}^\gamma (x,b) f'(b))$$
$$= \frac{1}{2}(1-\epsilon) f(x) - \Gamma(\gamma) \frac{(b-x)^{1-\gamma}}{2(b-a)} J_a^\gamma f(b) + \frac{1}{2} J_a^{\gamma-1} (K_\epsilon^\gamma (x,b) f(b))$$
$$+ \frac{(b-x)^{1-\gamma}}{2(b-a)^{2-\gamma}} J_a^0 f(a) \left(x - \left(a + \epsilon \frac{b-a}{2} \right) \right) + \frac{(b-a)^{1-\gamma}}{2(b-a)} (b-x) \Gamma(\gamma) J_a^{\gamma-1} f(b),$$

which yields the required result. $\square$

Remark 2.5.3 If we substitute $\gamma = 1$ in (2.5.5), we get the following corollary.

Corollary 2.5.2 *Let all the assumptions of Lemma 2.5.1 be valid. Then*

$$\frac{1}{2}(1-\epsilon) f(x) = \frac{1}{b-a} \int_a^b f(t) dt + \frac{1}{2(b-a)} \left[f(b) \left(x - \left(b + \epsilon \frac{b-a}{2} \right) \right) \right.$$
$$\left. - f(a) \left(x - \left(a - \epsilon \frac{b-a}{2} \right) \right) \right] + \frac{1}{b-a} \int_a^b K_{*\epsilon}(x,t) f'(t) dt, \tag{2.5.9}$$

where

$$K_{*\epsilon}(x,t) = \begin{cases} \left[t - \left(\frac{a+x}{2} + \epsilon \frac{b-a}{4} \right) \right], & t \in [a,x]; \\ \left[t - \left(\frac{b+x}{2} - \epsilon \frac{b-a}{4} \right) \right], & t \in [x,b]. \end{cases}$$

Remark 2.5.4 If we substitute $\epsilon = 0$ in (2.5.5), we get the following corollary.

2.5 Generalized Fractional Ostrwoski Type Inequality with Parameter

Corollary 2.5.3 *Let all the assumptions of Lemma 2.5.1 be valid. Then*

$$f(x) = 2\Gamma(\gamma)\frac{(b-x)^{1-\gamma}}{b-a}J_a^\gamma f(b) - J_a^{\gamma-1}\left(K^\gamma(x,b)f(b)\right)$$

$$-\frac{(b-x)^{2-\gamma}}{b-a}\Gamma(\gamma)J_a^{\gamma-1}f(b)$$

$$-\frac{(b-x)^{1-\gamma}}{(b-a)^{2-\gamma}}(x-a)J_a^0 f(a) + 2J_a^\gamma\left(K_*^\gamma(x,b)f'(b)\right), \quad (2.5.10)$$

where

$$K_*^\gamma(x,t) = \begin{cases} \left(t - \dfrac{a+x}{2}\right)\dfrac{(b-x)^{1-\gamma}}{b-a}\Gamma(\gamma), & t \in [a,x); \\ \left(t - \dfrac{b+x}{2}\right)\dfrac{(b-x)^{1-\gamma}}{b-a}\Gamma(\gamma), & t \in [x,b], \end{cases}$$

and $K^\gamma(x,t)$ is defined in (2.5.2).

Remark 2.5.5 If we substitute $\gamma = 1$ in (2.5.10), we get the following corollary which can be found in [183] by Tong and Guan.

Corollary 2.5.4 *Let all the assumptions of Lemma 2.5.1 be valid. Then*

$$\frac{1}{2}f(x) = \frac{1}{b-a}\int_a^b f(t)dt + \frac{(x-b)f(b) - (x-a)f(a)}{2(b-a)} + \frac{1}{b-a}\int_a^b K_*(x,t)f'(t)dt,$$

where

$$K_*(x,t) = \begin{cases} t - \dfrac{a+x}{2}, & t \in [a,x); \\ t - \dfrac{b+x}{2}, & t \in [x,b], \end{cases}$$

By using Lemma 2.5.2, we obtain generalized Ostrowski fractional integral inequality in the following theorem.

Theorem 2.5.5 *Let $f : I \to \mathbb{R}$ be an absolutely continuous function on I^o such that $a, b \in I$ and $a < b$. If $|f'(x)| \le M$ a. e. $\forall x \in (a,b)$ where M is positive real constant, then the following inequality holds*

$$\left| \frac{1}{2}(1-\epsilon)f(x) - \Gamma(\gamma)\frac{(b-x)^{1-\gamma}}{(b-a)}J_a^\gamma f(b) + \frac{1}{2}J_a^{\gamma-1}(K_\epsilon^\gamma(x,b)f(b)) \right.$$

$$\left. +\frac{(b-x)^{2-\gamma}}{2(b-a)}\Gamma(\gamma)J_a^{\gamma-1}f(b) + \frac{(b-x)^{1-\gamma}}{2(b-a)^{2-\gamma}}J_a^0 f(a)\left(x - a + \epsilon\frac{b-a}{2}\right) \right|$$

$$\leq \frac{M(b-x)^{1-\gamma}}{b-a} \left[\frac{(x-a)}{2\gamma} \left\{ (b-a)^\gamma - (b-x)^\gamma + (b-x)^{\gamma+1} + \frac{\epsilon}{2}(b-a)^{\gamma+1} \right\} \right.$$

$$+ \frac{1}{\gamma(\gamma+1)} \left\{ 2\left(b - \frac{a+x}{2} + \epsilon\frac{b-a}{4}\right)^{\gamma+1} - 2(b-x)^{\gamma+1} - (b-a)^{\gamma+1} \right.$$

$$\left.\left. + 2\left(\frac{b-x}{2} - \epsilon\frac{b-a}{4}\right)^{\gamma+1} \right\} \right], \tag{2.5.11}$$

for $\gamma \geq 1$, *where* $\epsilon \in [0,1]$.

Proof From Lemma 2.5.2, consider

$$I = \left| \frac{1}{2}(1-\epsilon)f(x) - \Gamma(\gamma)\frac{(b-x)^{1-\gamma}}{(b-a)} J_a^\gamma f(b) + \frac{1}{2}J_a^{\gamma-1}\left(K_\epsilon^\gamma(x,b)f(b)\right) \right.$$

$$\left. + \frac{(b-x)^{2-\gamma}}{2(b-a)}\Gamma(\gamma)J_a^{\gamma-1}f(b) + \frac{(b-x)^{1-\gamma}}{2(b-a)^{2-\gamma}}\left(x-a+\epsilon\frac{b-a}{2}\right)J_a^0 f(a) \right|$$

$$= \left| J_a^\gamma \left(K_{*\epsilon}^\gamma(x,b)f'(b)\right) \right|$$

$$= \left| \frac{1}{\Gamma(\gamma)} \int_a^b (b-t)^{\gamma-1} K_{*\epsilon}^\gamma(x,t) f'(t) dt \right|$$

$$\leq \frac{1}{\Gamma(\gamma)} \int_a^b (b-t)^{\gamma-1} \left|K_{*\epsilon}^\gamma(x,t)\right| \left|f'(t)\right| dt$$

$$\leq \frac{M}{\Gamma(\gamma)} \int_a^b (b-t)^{\gamma-1} \left|K_{*\epsilon}^\gamma(x,t)\right| dt$$

$$= M \frac{(b-x)^{1-\gamma}}{b-a} \left[\int_a^x (b-t)^{\gamma-1} \left|t - \left(\frac{a+x}{2} + \epsilon\frac{b-a}{4}\right)\right| dt \right.$$

$$\left. + \int_x^b (b-t)^{\gamma-1} \left|t - \left(\frac{b+x}{2} - \epsilon\frac{b-a}{4}\right)\right| dt \right]$$

$$= M \frac{(b-x)^{1-\gamma}}{b-a} [I_1 + I_2 + I_3 + I_4], \tag{2.5.12}$$

where

$$I_1 = \int_a^{\frac{a+x}{2}+\epsilon\frac{b-a}{4}} (b-t)^{\gamma-1} \left\{ \left(\frac{a+x}{2} + \epsilon\frac{b-a}{4}\right) - t \right\} dt$$

$$= \left(\frac{x-a}{2} + \epsilon\frac{b-a}{4}\right)\frac{(b-a)^\gamma}{\gamma} + \frac{1}{\gamma(\gamma+1)}\left(b - \frac{a+x}{2} - \epsilon\frac{b-a}{4}\right)^{\gamma+1}$$

$$- \frac{(b-a)^{\gamma+1}}{\gamma(\gamma+1)}, \tag{2.5.13}$$

2.5 Generalized Fractional Ostrwoski Type Inequality with Parameter

$$I_2 = \int_{\frac{a+x}{2}+\epsilon\frac{b-a}{4}}^{x} (b-t)^{\gamma-1} \left\{ t - \left(\frac{a+x}{2} + \epsilon \frac{b-a}{4} \right) \right\} dt$$

$$= \left(\frac{a-x}{2} + \epsilon \frac{b-a}{4} \right) \frac{(b-x)^{\gamma}}{\gamma} + \frac{1}{\gamma(\gamma+1)} \left(b - \frac{a+x}{2} - \epsilon \frac{b-a}{4} \right)^{\gamma+1}$$

$$- \frac{(b-x)^{\gamma+1}}{\gamma(\gamma+1)}, \tag{2.5.14}$$

$$I_3 = \int_{x}^{\frac{b+x}{2}-\epsilon\frac{b-a}{4}} (b-t)^{\gamma-1} \left\{ \left(\frac{b+x}{2} - \epsilon \frac{b-a}{4} \right) \right\} dt$$

$$= \left(\frac{b-x}{2} - \epsilon \frac{b-a}{4} \right) \frac{(b-x)^{\gamma}}{\gamma} + \frac{1}{\gamma(\gamma+1)} \left(\frac{b-x}{2} + \epsilon \frac{b-a}{4} \right)^{\gamma+1}$$

$$- \frac{(b-x)^{\gamma+1}}{\gamma(\gamma+1)} \tag{2.5.15}$$

and

$$I_4 = \int_{\frac{b+x}{2}-\epsilon\frac{b-a}{4}}^{b} (b-t)^{\gamma-1} \left\{ t - \left(\frac{b+x}{2} - \epsilon \frac{b-a}{4} \right) \right\} dt$$

$$= \frac{1}{\gamma(\gamma+1)} \left(\frac{b-x}{2} - \epsilon \frac{b-a}{4} \right)^{\gamma+1}. \tag{2.5.16}$$

Using (2.5.13)–(2.5.16) in (2.5.12), we get the bound

$$I \leq \frac{M(b-x)^{1-\gamma}}{b-a} \left[\frac{(x-a)}{2\gamma} \left\{ (b-a)^{\gamma} - (b-x)^{\gamma} + (b-x)^{\gamma+1} + \frac{\epsilon}{2}(b-a)^{\gamma+1} \right\} \right.$$

$$+ \frac{1}{\gamma(\gamma+1)} \left\{ 2 \left(b - \frac{a+x}{2} + \epsilon \frac{b-a}{4} \right)^{\gamma+1} - 2(b-x)^{\gamma+1} - (b-a)^{\gamma+1} \right.$$

$$\left. \left. + 2 \left(\frac{b-x}{2} - \epsilon \frac{b-a}{4} \right)^{\gamma+1} \right\} \right].$$

□

Remark 2.5.6 If we substitute $\gamma = 1$ in (2.5.11), we get the following corollary.

Corollary 2.5.6 *Let all the assumptions of Theorem 2.5.5 be valid. Then*

$$\left| \frac{1}{2}(1-\epsilon)f(x) - \frac{1}{2(b-a)} \left\{ \left(x - \left(b + \epsilon\frac{b-a}{2}\right)\right) f(b) \right. \right.$$
$$\left. \left. - \left(x - \left(a - \epsilon\frac{b-a}{2}\right)\right) f(a) \right\} - \frac{1}{b-a} \int_a^b f(t)dt \right|$$
$$\leq \frac{M}{4(b-a)} \left[(x-a)^2 + (b-x)^2 + \frac{\epsilon(b-a)^2}{4} \right], \qquad (2.5.17)$$

where $\epsilon \in [0, 1]$.

Remark 2.5.7 If we substitute $\epsilon = 0$ in (2.5.11), we get the following corollary.

Corollary 2.5.7 *Let all the assumptions of Theorem 2.5.5 be valid. Then*

$$\left| \frac{1}{2}f(x) - \Gamma(\gamma)\frac{(b-x)^{1-\gamma}}{(b-a)} J_a^\gamma f(b) + \frac{1}{2} J_a^{\gamma-1}(P_1(x,b)f(b)) \right.$$
$$\left. + \frac{(b-x)^{2-\gamma}}{2(b-a)}\Gamma(\gamma) J_a^{\gamma-1} f(b) + \frac{(b-x)^{1-\gamma}}{2(b-a)^{2-\gamma}} J_a^0 f(a)(x-a) \right|$$
$$\leq \frac{M(b-x)^{1-\gamma}}{b-a} \left[\frac{(x-a)}{2\gamma} \left\{ (b-a)^\gamma - (b-x)^\gamma + (b-x)^{\gamma+1} \right\} + \frac{1}{\gamma(\gamma+1)} \right.$$
$$\left. \times \left\{ 2\left(b - \frac{a+x}{2}\right)^{\gamma+1} - 2(b-x)^{\gamma+1} - (b-a)^{\gamma+1} + 2\left(\frac{b-x}{2}\right)^{\gamma+1} \right\} \right], \qquad (2.5.18)$$

for $\gamma \geq 1$.

Remark 2.5.8 If we substitute either $\epsilon = 0$ in (2.5.17) or $\gamma = 1$ in (2.5.18), we get the following corollary.

Corollary 2.5.8 *Let all the assumptions of Theorem 2.5.5 be valid. Then*

$$\left| \frac{1}{2}f(x) - \frac{1}{2(b-a)} \{(x-b)f(b) - (x-a)f(a)\} - \frac{1}{b-a} \int_a^b f(t)dt \right|$$
$$\leq \frac{M}{4(b-a)} \left[(x-a)^2 + (b-x)^2 \right]. \qquad (2.5.19)$$

Remark 2.5.9 If we put $x = \frac{a+b}{2}$ in (2.5.17), we get the following corollary.

2.5 Generalized Fractional Ostrwoski Type Inequality with Parameter

Corollary 2.5.9 *Let all the assumptions of Theorem 2.5.5 be valid. Then*

$$\left| \frac{1}{2}(1-\epsilon) f\left(\frac{a+b}{2}\right) + (1+\epsilon)\frac{f(a)+f(b)}{4} - \frac{1}{b-a}\int_a^b f(t)dt \right| \leq \frac{M}{16}(b-a)(2+\epsilon),$$

(2.5.20)

where $\epsilon \in [0, 1]$.

Remark 2.5.10 If we substitute $\epsilon = 0$ in (2.5.20), we get the bound for average midpoint and trapezoidal inequality in the following corollary.

Corollary 2.5.10 *Let all the assumptions of Theorem 2.5.5 be valid. Then*

$$\left| \frac{1}{2} f\left(\frac{a+b}{2}\right) + \frac{f(a)+f(b)}{4} - \frac{1}{b-a}\int_a^b f(t)dt \right| \leq \frac{M}{8}(b-a).$$

Remark 2.5.11 If we substitute $\epsilon = \frac{1}{2}$ in (2.5.20), we get the bound for perturbed trapezoidal inequality in the following corollary.

Corollary 2.5.11 *Let all the assumptions of Theorem 2.5.5 be valid. Then*

$$\left| \frac{1}{4} f\left(\frac{a+b}{2}\right) + \frac{3}{8}\left(f(a)+f(b)\right) - \frac{1}{b-a}\int_a^b f(t)dt \right| \leq \frac{5}{32}(b-a)M.$$

Remark 2.5.12 If we substitute $\epsilon = \frac{1}{3}$ in (2.5.20), we get the bound for perturbed trapezoidal inequality in the following corollary.

Corollary 2.5.12 *Let all the assumptions of Theorem 2.5.5 be valid. Then*

$$\left| \frac{1}{3} f\left(\frac{a+b}{2}\right) + \frac{1}{3}\left(f(a)+f(b)\right) - \frac{1}{b-a}\int_a^b f(t)dt \right| \leq \frac{7}{48}(b-a)M.$$

Remark 2.5.13 If we substitute $\epsilon = \frac{1}{4}$ in (2.5.20), we get the bound for perturbed trapezoidal inequality in the following corollary.

Corollary 2.5.13 *Let all the assumptions of Theorem 2.5.5 be valid. Then*

$$\left| \frac{3}{8} f\left(\frac{a+b}{2}\right) + \frac{5}{16}\left(f(a)+f(b)\right) - \frac{1}{b-a}\int_a^b f(t)dt \right| \leq \frac{9}{64}(b-a)M.$$

Remark 2.5.14 If we put $x = a$ or $x = b$ in (2.5.19), we get the bound for trapezoidal inequality (also Hermite-hadamard right bound) in the following corollary.

Corollary 2.5.14 *Let all the assumptions of Theorem 2.5.5 be valid. Then*

$$\left| \frac{f(a)+f(b)}{2} - \frac{1}{b-a}\int_a^b f(t)dt \right| \leq \frac{M}{4}(b-a).$$

In the upcoming section, we would like to generalize Montgomery identity and Ostrowski and Čebyšev type inequality for two variables with parameters.

2.6 Generalized Inequalities for Functions of L_p Spaces via Montgomery Identity with Parameters

Present section deals with the generalization of Montgomery identity for two independent variables using second order differentiable functions with parameters. Some new Ostrowski type inequalities for L_p spaces with better bounds are presented as well. Moreover we will modify Grüss type inequality by using the acquired Montgomery identity. At places we get better bounds of some new obtained inequalities.

Montgomery identity is one of the classical results that creates many important inequalities such as Ostrowski, Grüss and Ostrowski-Grüss. Its bivariate form has introduced some new generalization and advancement in different inequalities. These inequalities have many applications in various fields of mathematics such as numerical integration and probability theory. We can also obtain special means with the help of these inequalities. In the last 20 years rapid advancement in generalization and improvement of these types of inequalities has been observed for references see [24, 54, 66, 76, 77, 143, 145, 156]. This section deals with its bivariate form in order to generate our proposed results of Ostrowski and Grüss inequalities in terms of parameters. The idea behind the results based on parameters is to make further generalization of those results of Ostrowski and Grüss which are non parametric based, as parameters extends the region of inequality more wider and provides a family of solutions and the quality of inequality will improve conclusively. If we talk about L_p spaces, this is the first ever combination of L_p space, parameters and bivariate differentiable functions, which some how connects our result with lebesgue measure.

In the first subsection we obtain Montgomery identity with parameters of two independent variables, while in the second subsection, we establish Ostrowski type inequality for two variables in terms of parameters for L_p spaces. In the last subsection, we will achieve Grüss type inequalities with its Čebyšev functional.

2.6.1 Montgomery Identity for Functions of Two Variables involving Parameters

For classical form of Montgomery identity, see the book [134, p. 565], as given in the Theorem 1.1.1. In 2001, Dragomir et al. introduced this identity with parameters in [53] as given in the Theorem 2.3.5. In [16], (see also [52]) authors proved the double integral Montgomery identity for two independent variables stated as follows:

Theorem 2.6.1 *Let $f : I \times J = [a, b] \times [c, d] \to \mathbb{R}$ is differentiable such that $\dfrac{\partial^2 f(t, s)}{\partial t \partial s}$ is integrable on interior of $I \times J$. Then*

$$f(x, y) = \frac{1}{(b-a)} \int_a^b f(t, y) dt + \frac{1}{(d-c)} \int_c^d f(x, s) ds$$

$$- \frac{1}{(b-a)(d-c)} \int_a^b \int_c^d f(t, s) ds dt$$

$$+ \int_a^b \int_c^d K(x, t) Q(y, s) \frac{\partial^2 f(t, s)}{\partial t \partial s} ds dt, \qquad (2.6.1)$$

where $K(x, t)$ and $Q(y, s)$ are the Peano kernels defined as

$$K(x, t) = \begin{cases} \dfrac{t-a}{b-a}, & t \in [a, x]; \\ \dfrac{t-b}{b-a}, & t \in (x, b]. \end{cases} \qquad (2.6.2)$$

and

$$Q(y, s) = \begin{cases} \dfrac{s-c}{d-c}, & s \in [c, y]; \\ \dfrac{s-d}{d-c}, & s \in (y, d]. \end{cases} \qquad (2.6.3)$$

Now, we are going to establish new Montgomery identity for two independent variables involving parameters, which will provide generalization of existing Montgomery identities.

Here we state our first main result.

Theorem 2.6.2 *Let* $f : I \times J \to \mathbb{R}$ *be absolutely continuous such that* $\dfrac{\partial^2 f(t,s)}{\partial t \partial s}$ *is integrable on interior of* $I \times J$. *Then*

$$(1-\epsilon)(1-\kappa)f(x,y) = \frac{(1-\kappa)}{b-a}\int_a^b f(t,y)dt + \frac{(1-\epsilon)}{d-c}\int_c^d f(x,s)ds$$

$$-\frac{1}{(b-a)(d-c)}\int_a^b \int_c^d f(t,s)dsdt + \frac{1}{2}\psi_{\epsilon,\kappa}(f)$$

$$+\frac{1}{(b-a)(d-c)}\int_a^b \int_c^d K_\epsilon(x,t)Q_1(y,s)\frac{\partial^2 f(t,s)}{\partial t \partial s}dsdt,$$

(2.6.4)

where

$$\psi_{\epsilon,\kappa}(f) = \frac{\kappa}{(b-a)}\int_a^b (f(t,c)+f(t,d))\,dt - \kappa(1-\epsilon)(f(x,c)+f(x,d))$$

$$+\frac{\epsilon}{(d-c)}\int_c^d (f(a,s)+f(b,s))\,ds - \epsilon(1-\kappa)(f(a,y)+f(b,y))$$

$$-\frac{\epsilon\kappa}{2}(f(a,c)+f(a,d)+f(b,c)+f(b,d))\,,$$

(2.6.5)

also $K_\epsilon(x,t)$ *is defined in* (2.3.3), $Q_1(y,s)$ *is defined as*

$$Q_1(y,s) = \begin{cases} s - \left(c + \kappa\dfrac{d-c}{2}\right), & s \in [c,y]; \\[1em] s - \left(d - \kappa\dfrac{d-c}{2}\right), & s \in (y,d]. \end{cases}$$

(2.6.6)

where $\epsilon, \kappa \in [0,1]$.

Proof By using (2.3.3) and (2.6.6), we have

$$\int_a^b \int_c^d K_\epsilon(x,t)Q_1(y,s)\frac{\partial^2 f(t,s)}{\partial t \partial s}dsdt \qquad (2.6.7)$$

$$= \int_a^x \int_c^y \left(t - \left(a + \epsilon\frac{b-a}{2}\right)\right)\left(s - \left(c + \kappa\frac{d-c}{2}\right)\right)\frac{\partial^2 f(t,s)}{\partial t \partial s}dsdt$$

$$+ \int_a^x \int_y^d \left(t - \left(a + \epsilon\frac{b-a}{2}\right)\right)\left(s - \left(d - \kappa\frac{d-c}{2}\right)\right)\frac{\partial^2 f(t,s)}{\partial t \partial s}dsdt$$

2.6 Generalized Inequalities for Functions of L_p Spaces via Montgomery...

$$+ \int_x^b \int_c^y \left(t - \left(b - \epsilon \frac{b-a}{2}\right)\right)\left(s - \left(c + \kappa \frac{d-c}{2}\right)\right) \frac{\partial^2 f(t,s)}{\partial t \partial s} ds dt$$

$$+ \int_x^b \int_y^d \left(t - \left(b - \epsilon \frac{b-a}{2}\right)\right)\left(s - \left(d - \kappa \frac{d-c}{2}\right)\right) \frac{\partial^2 f(t,s)}{\partial t \partial s} ds dt$$

$$= I_1 + I_2 + I_3 + I_4. \tag{2.6.8}$$

After some calculations and simplifications, we have

$$I_1 = \left(x - \left(a + \epsilon \frac{b-a}{2}\right)\right)\left(y - \left(c + \kappa \frac{d-c}{2}\right)\right) f(x,y)$$

$$- \left(y - \left(c + \kappa \frac{d-c}{2}\right)\right) \int_a^x f(t,y) dt - \left(x - \left(a + \epsilon \frac{b-a}{2}\right)\right) \int_c^y f(x,s) ds$$

$$+ \epsilon \frac{b-a}{2} \left[\left(y - \left(c + \kappa \frac{d-c}{2}\right)\right) f(a,y) - \int_c^y f(a,s) ds\right]$$

$$+ \kappa \frac{d-c}{2} \left[\left(x - \left(a + \epsilon \frac{b-a}{2}\right)\right) f(x,c) - \int_a^x f(t,c) dt\right]$$

$$+ \kappa \epsilon \frac{(b-a)(d-c)}{4} f(a,c) + \int_a^x \int_c^y f(t,s) ds dt,$$

$$I_2 = - \left(x - \left(a + \epsilon \frac{b-a}{2}\right)\right)\left(y - \left(d - \kappa \frac{d-c}{2}\right)\right) f(x,y)$$

$$+ \left(y - \left(d - \kappa \frac{d-c}{2}\right)\right) \int_a^x f(t,y) dt - \left(x - \left(a + \epsilon \frac{b-a}{2}\right)\right) \int_y^d f(x,s) ds$$

$$+ \epsilon \frac{b-a}{2} \left[-\left(y - \left(d - \kappa \frac{d-c}{2}\right)\right) f(a,y) - \int_y^d f(a,s) ds\right]$$

$$+ \kappa \frac{d-c}{2} \left[\left(x - \left(a + \epsilon \frac{b-a}{2}\right)\right) f(x,d) - \int_a^x f(t,d) dt\right]$$

$$+ \kappa \epsilon \frac{(b-a)(d-c)}{4} f(a,d) + \int_a^x \int_y^d f(t,s) ds dt,$$

$$I_3 = - \left(x - \left(b - \epsilon \frac{b-a}{2}\right)\right)\left(y - \left(c + \kappa \frac{d-c}{2}\right)\right) f(x,y)$$

$$- \left(y - \left(c + \kappa \frac{d-c}{2}\right)\right) \int_x^b f(t,y) dt + \left(x - \left(b - \epsilon \frac{b-a}{2}\right)\right) \int_c^y f(x,s) ds$$

$$+ \epsilon \frac{b-a}{2} \left[\left(y - \left(c + \kappa \frac{d-c}{2}\right)\right) f(b,y) - \int_c^y f(b,s) ds\right]$$

$$+ \kappa \frac{d-c}{2} \left[-\left(x - \left(b - \epsilon \frac{b-a}{2} \right) \right) f(x,c) - \int_x^b f(t,c)ds \right]$$

$$+ \kappa \epsilon \frac{(b-a)(d-c)}{4} f(b,c) + \int_x^b \int_c^y f(t,s)dsdt,$$

and

$$I_4 = \left(x - \left(b - \epsilon \frac{b-a}{2} \right) \right) \left(y - \left(d - \kappa \frac{d-c}{2} \right) \right) f(x,y)$$

$$+ \left(y - \left(d - \kappa \frac{d-c}{2} \right) \right) \int_x^b f(t,y)dt + \left(x - \left(b - \epsilon \frac{b-a}{2} \right) \right) \int_y^d f(x,s)ds$$

$$+ \epsilon \frac{b-a}{2} \left[-\left(y - \left(d - \kappa \frac{d-c}{2} \right) \right) f(b,y) - \int_y^d f(b,s)ds + \right]$$

$$+ \kappa \frac{d-c}{2} \left[-\left(x - \left(b - \epsilon \frac{b-a}{2} \right) \right) f(x,d) - \int_x^b f(t,d)dt \right]$$

$$+ \kappa \epsilon \frac{(b-a)(d-c)}{4} f(b,d) + \int_x^b \int_y^d f(t,s)dsdt.$$

By substituting the values of I_1, I_2, I_3 and I_4 in (2.6.8), we get

$$\int_a^b \int_c^d K_\epsilon(x,t) Q_1(y,s) \frac{\partial^2 f(t,s)}{\partial t \partial s} dsdt$$

$$= (1-\epsilon)(1-\kappa)(b-a)(d-c) f(x,y) - (1-\kappa)(d-c) \int_a^b f(t,y)dt$$

$$- (1-\epsilon)(b-a) \int_c^d f(x,s)ds - \int_a^b \int_c^d f(t,s)dsdt$$

$$- \kappa \frac{d-c}{2} \int_a^b (f(t,c) + f(t,d)) dt + \frac{\kappa(1-\epsilon)(b-a)(d-c)}{2}$$

$$\times (f(x,c) + f(x,d)) - \epsilon \frac{b-a}{2} \int_c^d (f(a,s) + f(b,s)) ds$$

$$+ \frac{\epsilon(1-\kappa)(b-a)(d-c)}{2} (f(a,y) + f(b,y))$$

$$+ \frac{\epsilon\kappa(b-a)(d-c)}{4} (f(a,c) + f(a,d) + f(b,c) + f(b,d)),$$

which produce the required identity. $\square$

2.6 Generalized Inequalities for Functions of L_p Spaces via Montgomery...

Remark 2.6.1

1. If we substitute $\epsilon, \kappa = 0$ in (2.6.4), then it gives (2.6.1) of [16] as stated in Theorem 2.6.1.
2. If we substitute $f(t, s) = h(t)h(s)$ and $x = y$ in (2.6.4), then it gives (2.3.2) of [53] as stated in Theorem 2.3.5.

Remark 2.6.2 If we substitute $\epsilon = \kappa$, then we get a special type of Montgomery identity as established in [109].

$$(1-\epsilon)^2 f(x, y) = \frac{(1-\epsilon)}{b-a} \int_a^b f(t, y) dt + \frac{(1-\epsilon)}{d-c} \int_c^d f(x, s) ds$$

$$- \frac{1}{(b-a)(d-c)} \int_a^b \int_c^d f(t, s) ds dt + \frac{1}{2} \psi_\epsilon(f)$$

$$+ \frac{1}{(b-a)(d-c)} \int_a^b \int_c^d K_\epsilon(x, t) Q_2(y, s) \frac{\partial^2 f(t, s)}{\partial t \partial s} ds dt,$$

where

$$\psi_\epsilon(f) = \frac{\epsilon}{(b-a)} \int_a^b (f(t, c) + f(t, d)) dt + \frac{\epsilon}{(d-c)} \int_c^d (f(a, s)$$

$$+ f(b, s)) ds - \epsilon(1-\epsilon) \Big(f(x, c) + f(x, d) + f(a, y) + f(b, y) \Big)$$

$$- \frac{\epsilon^2}{2} \Big(f(a, c) + f(a, d) + f(b, c) + f(b, d) \Big),$$

also Peano kernels $K_\epsilon(x, t)$ is defined as in (2.3.3), $Q_2(y, s)$ is defined as

$$Q_2(y, s) = \begin{cases} s - \left(c + \epsilon \frac{d-c}{2}\right), & s \in [c, y]; \\ s - \left(d - \epsilon \frac{d-c}{2}\right), & s \in (y, d]. \end{cases}$$

where $\epsilon \in [0, 1]$.

Lets recall the concept of Hölder's inequality from Theorem 1.1.13 that is useful in our results of the coming subsections.

2.6.2 Generalized Ostrowski Type Inequality

Now we are going to present Ostrowski type inequality for L_p and L_∞ spaces by using the Montgomery identity (2.6.4) as we obtained in the previous subsection.

After the publication of research article [141] by Ostrowski, researchers are in an effort to generalize the Ostrowski inequality and trying to get the sharp bounds. Ostrowski inequality discussed previously in Theorem 2.0.1.

In 1997, Dragomir and Wang established the following of Ostrowski type inequality for differentiable functions in [51] where $f' \in L_q$ space.

Theorem 2.6.3 *Let $f : I \to \mathbb{R}$ be a differentiable function on I^o where $a < b$ such that $f' \in L_q[a, b]$ where $1 \leq q \leq \infty$ and $\frac{1}{q} + \frac{1}{r} = 1$. Then*

$$\left| f(x) - \frac{1}{b-a} \int_a^b f(t)dt \right| \leq \frac{1}{b-a} [B_1(x)]^{\frac{1}{r}} \|f'\|_q, \qquad (2.6.9)$$

where,

$$B_1(x) = \frac{(x-a)^{r+1} + (b-x)^{r+1}}{r+1}, \qquad (2.6.10)$$

$\forall \, x \in [a, b]$.

In 2001, Ostrowski type inequality for double integrals was introduced by Barnett and Dragomir in [16].

Theorem 2.6.4 *Let $f : I \times J \to \mathbb{R}$ is differentiable such that $\frac{\partial^2 f(t,s)}{\partial t \partial s}$ is integrable on interior of $I \times J$ and is bounded in L_∞ space. Then*

$$\left| f(x, y) - \frac{1}{(b-a)} \int_a^b f(t, y)dt - \frac{1}{(d-c)} \int_c^d f(x, s)ds \right.$$
$$\left. + \frac{1}{(b-a)(d-c)} \int_a^b \int_c^d f(t, s)dsdt \right|$$
$$\leq \frac{1}{4(b-a)(d-c)} \left[(x-a)^2 + (b-x)^2 \right] \left[(y-c)^2 + (d-y)^2 \right] \left\| \frac{\partial^2 f(t,s)}{\partial t \partial s} \right\|_\infty.$$

$$(2.6.11)$$

Furthermore in 2000, Dragomir et al. in [54] generalized the results of [16] for L_q space.

Theorem 2.6.5 *If $f : I \times J \to \mathbb{R}$ is differentiable such that $\frac{\partial^2 f(t,s)}{\partial t \partial s}$ is integrable on interior of $I \times J$ and is bounded in L_q space where $1 \leq q \leq \infty$ and $\frac{1}{q} + \frac{1}{r} = 1$. Then*

$$\left| f(x, y) - \frac{1}{(b-a)} \int_a^b f(t, y)dt - \frac{1}{(d-c)} \int_c^d f(x, s)ds \right.$$

2.6 Generalized Inequalities for Functions of L_p Spaces via Montgomery...

$$+ \frac{1}{(b-a)(d-c)} \int_a^b \int_c^d f(t,s) ds dt \Bigg|$$

$$\leq \frac{1}{(b-a)(d-c)} \left\| \frac{\partial^2 f(t,s)}{\partial t \partial s} \right\|_q [B_1(x)]^{\frac{1}{r}} [B_2(y)]^{\frac{1}{r}}, \qquad (2.6.12)$$

where $B_1(x)$ is defined as in (2.6.10) and

$$B_2(y) = \frac{(y-c)^{r+1} + (d-y)^{r+1}}{r+1}. \qquad (2.6.13)$$

In 2000, Dragomir et al. in [53] generalized the classical Ostrowski inequality [141] as stated in the following Theorem:

Theorem 2.6.6 *Let $f : I \to \mathbb{R}$ be a differentiable functions on I° such that $f \in L[a,b]$, where $a < b$ whose derivative f' is bounded on (a,b), i.e., $\|f'\|_\infty := \sup_{t \in (a,b)} |f'(t)| < \infty$. Then*

$$\left| (1-\epsilon)f(x) + \epsilon \frac{f(a)+f(b)}{2} - \frac{1}{b-a} \int_a^b f(t) dt \right|$$

$$\leq (b-a) \left[\frac{1}{4} \left\{ \epsilon^2 + (\epsilon-1)^2 \right\} + \frac{\left(x - \frac{a+b}{2}\right)^2}{(b-a)^2} \right] \|f'(x)\|_\infty, \qquad (2.6.14)$$

where $\epsilon \in [0,1]$.

In 2003, Yang established Ostrowski inequality for L_p spaces in [190] that is infact a generalization of *(2.6.14)*.

Theorem 2.6.7 *Let all assumptions of Theorem 2.6.3 be true. Then*

$$\left| (1-\epsilon)f(x) + \epsilon \frac{f(a)+f(b)}{2} - \frac{1}{b-a} \int_a^b f(t) dt \right|$$

$$\leq \frac{1}{(b-a)(r+1)^{\frac{1}{r}}} \left[\left(x - \left(a + \epsilon \frac{b-a}{2} \right) \right)^{r+1} \right.$$

$$\left. + \left(\left(b - \epsilon \frac{b-a}{2} \right) - x \right)^{r+1} + 2 \left(\epsilon \frac{b-a}{2} \right)^{r+1} \right]^{\frac{1}{r}} \|f'(x)\|_q, \qquad (2.6.15)$$

where $\epsilon \in [0,1]$.

Now we are going to present Ostrowski inequality with parameters of double integrals for L_p and L_∞ space with parameters.

Theorem 2.6.8 Let $f : I \times J \to \mathbb{R}$ is differentiable such that $\dfrac{\partial^2 f(t,s)}{\partial t \partial s}$ is integrable on interior of $I \times J$ and is bounded in L_q space where $1 \le q \le \infty$ and $\dfrac{1}{q} + \dfrac{1}{r} = 1$. Then

$$\left| (1-\epsilon)(1-\kappa) f(x,y) - \frac{(1-\kappa)}{(b-a)} \int_a^b f(t,y) dt - \frac{(1-\epsilon)}{(d-c)} \int_c^d f(x,s) ds \right.$$
$$\left. + \frac{1}{(b-a)(d-c)} \int_a^b \int_c^d f(t,s) ds dt - \frac{1}{2} \psi_{\epsilon,\kappa}(f) \right|$$
$$\le \frac{1}{(b-a)(d-c)(r+1)^{\frac{2}{r}}} \left\| \frac{\partial^2 f(t,s)}{\partial t \partial s} \right\|_q$$
$$\times \left[\left(x - \left(a + \epsilon \frac{b-a}{2} \right) \right)^{r+1} + \left(\left(b - \epsilon \frac{b-a}{2} \right) - x \right)^{r+1} + 2 \left(\epsilon \frac{b-a}{2} \right)^{r+1} \right]^{\frac{1}{r}}$$
$$\times \left[\left(y - \left(c + \kappa \frac{d-c}{2} \right) \right)^{r+1} + \left(\left(d - \kappa \frac{d-c}{2} \right) - y \right)^{r+1} + 2 \left(\kappa \frac{d-c}{2} \right)^{r+1} \right]^{\frac{1}{r}},$$
(2.6.16)

where $\epsilon, \kappa \in [0,1]$.

Proof From Theorem 2.6.2, we have

$$(1-\epsilon)(1-\kappa) f(x,y) - \frac{(1-\kappa)}{(b-a)} \int_a^b f(t,y) dt - \frac{1}{2} \psi_{\epsilon,\kappa}(f)$$
$$- \frac{(1-\epsilon)}{(d-c)} \int_c^d f(x,s) ds + \frac{1}{(b-a)(d-c)} \int_a^b \int_c^d f(t,s) ds dt$$
$$= \frac{1}{(b-a)(d-c)} \int_a^b \int_c^d K_\epsilon(x,t) Q_1(y,s) \frac{\partial^2 f(t,s)}{\partial t \partial s} ds dt. \quad (2.6.17)$$

Applying absolute on both sides of (2.6.17) and using Hölder's inequality, we get

$$\left| (1-\epsilon)(1-\kappa) f(x,y) - \frac{(1-\kappa)}{(b-a)} \int_a^b f(t,y) dt - \frac{1}{2} \psi_{\epsilon,\kappa}(f) \right.$$
$$\left. - \frac{(1-\epsilon)}{(d-c)} \int_c^d f(x,s) ds + \frac{1}{(b-a)(d-c)} \int_a^b \int_c^d f(t,s) ds dt \right|$$
$$= \left| \frac{1}{(b-a)(d-c)} \int_a^b \int_c^d K_\epsilon(x,t) Q_1(y,s) \frac{\partial^2 f(t,s)}{\partial t \partial s} ds dt \right|$$

2.6 Generalized Inequalities for Functions of L_p Spaces via Montgomery...

$$\leq \frac{1}{(b-a)(d-c)} \int_a^b \int_c^d |K_\epsilon(x,t) Q_1(y,s)| \left|\frac{\partial^2 f(t,s)}{\partial t \partial s}\right| ds dt$$

$$\leq \frac{1}{(b-a)(d-c)} \left(\int_a^b \int_c^d |K_\epsilon(x,t) Q_1(y,s)|^r ds dt\right)^{\frac{1}{r}} \left(\int_a^b \int_c^d \left|\frac{\partial^2 f(t,s)}{\partial t \partial s}\right|^q ds dt\right)^{\frac{1}{q}}$$

$$= \frac{1}{(b-a)(d-c)(r+1)^{\frac{2}{r}}} \left\|\frac{\partial^2 f(t,s)}{\partial t \partial s}\right\|_q$$

$$\times \left[\left(x - \left(a + \epsilon \frac{b-a}{2}\right)\right)^{r+1} + \left(\left(b - \epsilon \frac{b-a}{2}\right) - x\right)^{r+1} + 2\left(\epsilon \frac{b-a}{2}\right)^{r+1}\right]^{\frac{1}{r}}$$

$$\left[\left(y - \left(c + \kappa \frac{d-c}{2}\right)\right)^{r+1} + \left(\left(d - \kappa \frac{d-c}{2}\right) - y\right)^{r+1} + 2\left(\kappa \frac{d-c}{2}\right)^{r+1}\right]^{\frac{1}{r}}.$$

Corollary 2.6.9 *Let all the assumptions of Theorem 2.6.6 be valid. If we select $r = 1$ and $q \to \infty$ in (2.6.16), then we get following result.*

$$\left|(1-\epsilon)(1-\kappa)f(x,y) - \frac{(1-\kappa)}{(b-a)} \int_a^b f(t,y) dt - \frac{1}{2}\psi_{\epsilon,\kappa}(f)\right.$$

$$\left. - \frac{(1-\epsilon)}{(d-c)} \int_c^d f(x,s) ds + \frac{1}{(b-a)(d-c)} \int_a^b \int_c^d f(t,s) ds dt\right|$$

$$\leq \frac{1}{4(b-a)(d-c)} \left[(x-a)^2 + (b-x)^2 - \epsilon(1-\epsilon)(b-a)^2\right]$$

$$\times \left[(y-c)^2 + (d-y)^2 - \kappa(1-\kappa)(d-c)^2\right] \left\|\frac{\partial^2 f(t,s)}{\partial t \partial s}\right\|_\infty$$

$$= (b-a)(d-c) \left[\frac{1}{4}\{\epsilon^2 + (\epsilon-1)^2\} + \frac{\left(x - \frac{a+b}{2}\right)^2}{(b-a)^2}\right]$$

$$\times \left[\frac{1}{4}\{\kappa^2 + (\kappa-1)^2\} + \frac{\left(y - \frac{c+d}{2}\right)^2}{(d-c)^2}\right] \left\|\frac{\partial^2 f(t,s)}{\partial t \partial s}\right\|_\infty, \quad (2.6.18)$$

where $\epsilon, \kappa \in [0,1]$.

Remark 2.6.3 It is to be noted the constant $\frac{1}{4}$ is sharp in (2.6.18) in the first and second bracket in the sense that it cannot be replaced by any smaller values.

To be more specific, if we suppose the inequality (2.6.18) be valid for constants $C_1, C_2 > 0$, i.e.,

$$\left| (1-\epsilon)(1-\kappa) f(x,y) - \frac{(1-\kappa)}{(b-a)} \int_a^b f(t,y) dt - \frac{1}{2} \psi_{\epsilon,\kappa}(f) \right.$$

$$\left. - \frac{(1-\epsilon)}{(d-c)} \int_c^d f(x,s) ds + \frac{1}{(b-a)(d-c)} \int_a^b \int_c^d f(t,s) ds dt \right|$$

$$\leq (b-a)(d-c) \left[C_1 \{\epsilon^2 + (\epsilon-1)^2\} + \frac{(x - \frac{a+b}{2})^2}{(b-a)^2} \right]$$

$$\times \left[C_2 \{\kappa^2 + (\kappa-1)^2\} + \frac{(y - \frac{c+d}{2})^2}{(d-c)^2} \right] \left\| \frac{\partial^2 f(t,s)}{\partial t \partial s} \right\|_\infty.$$

Consider $f(s,t) = st$, $x = a$, $y = c$, and $\epsilon, \kappa = 0$ then above inequality reduces to

$$\frac{1}{4} \leq \left(C_1 + \frac{1}{4} \right) \left(C_2 + \frac{1}{4} \right)$$

$$\frac{1}{2} \times \frac{1}{2} \leq \left(C_1 + \frac{1}{4} \right) \left(C_2 + \frac{1}{4} \right),$$

which gives that $C_1 \geq \frac{1}{4}$ and $C_2 \geq \frac{1}{4}$. Hence we are true in our claim.

In the similar manner one can find out that the improved bounds will be obtained by choosing $\epsilon, \kappa = \frac{1}{2}$.

From (2.6.16) and (2.6.18) we can get many results of Ostrowski type inequality.

Remark 2.6.4

1. If we substitute $\epsilon = \kappa = 0$ in (2.6.16), the it gives (2.6.12) of [54] as stated in Theorem 2.6.5.
2. If we substitute $\epsilon = \kappa = 0$ in (2.6.18), then it gives (2.6.11) of [16] as stated in Theorem 2.6.4.
3. If we substitute $f(t,s) = h(t)h(s)$, here h be absolutely continuous function, also let $\|h'\| < \infty$ and $x = y$ in (2.6.16), then it gives (2.6.15) of [190] as stated in Theorem 2.6.7. Further if we choose $\epsilon = \kappa = 0$, then we get (2.6.9) of [51] as stated in Theorem 2.6.3.
4. If we substitute $f(t,s) = h(t)h(s)$, here h be absolutely continuous function, also let $\|h'\| < \infty$ and $x = y$ in (2.6.18), then it gives (2.6.14) of [53] as stated in Theorem 2.6.6. Further if we choose $\epsilon = \kappa = 0$, then we get (2.0.1) of [141] as stated in Theorem 2.0.1.

Corollary 2.6.10 *If we take* $\epsilon = \kappa = 0$, $x = \dfrac{a+b}{2}$ *and* $y = \dfrac{c+d}{2}$ *in* (2.6.16), *then we get*

$$\left| f\left(\frac{a+b}{2}, \frac{c+d}{2}\right) - \frac{1}{(b-a)} \int_a^b f\left(t, \frac{c+d}{2}\right) dt \right.$$
$$\left. - \frac{1}{(d-c)} \int_c^d f\left(\frac{a+b}{2}, s\right) ds + \frac{1}{(b-a)(d-c)} \int_a^b \int_c^d f(t,s) ds dt \right|$$
$$\leq \frac{1}{4} \left[\frac{(b-a)(d-c)}{(r+1)^2} \right]^{\frac{1}{r}} \left\| \frac{\partial^2 f(t,s)}{\partial t \partial s} \right\|_q.$$

The above inequality is Corollary 5 of [54].

Corollary 2.6.11 *If we take* $\epsilon = \kappa = 0$, $x = \dfrac{a+b}{2}$ *and* $y = \dfrac{c+d}{2}$ *in* (2.6.18), *then we get*

$$\left| f\left(\frac{a+b}{2}, \frac{c+d}{2}\right) - \frac{1}{(b-a)} \int_a^b f\left(t, \frac{c+d}{2}\right) dt \right.$$
$$\left. - \frac{1}{(d-c)} \int_c^d f\left(\frac{a+b}{2}, s\right) ds + \frac{1}{(b-a)(d-c)} \int_a^b \int_c^d f(t,s) ds dt \right|$$
$$\leq \frac{(b-a)(d-c)}{16} \left\| \frac{\partial^2 f(t,s)}{\partial t \partial s} \right\|_\infty.$$

The above inequality is Corollary 2.2 of [16].

Remark 2.6.5 It is easy to see that in all our results, we get better bounds for substituting $x = \dfrac{a+b}{2}$, $y = \dfrac{c+d}{2}$ and $\epsilon = \kappa = \dfrac{1}{2}$.

Remark 2.6.6 We can also get many interesting results by varying the values of q and r in our main result (2.6.16). The case $q = r = 2$ is of special interest.

2.6.3 Generalized Grüss Type Inequalities

Grüss type inequalities usually provide the estimation of Čebyšev bounded functional. In the current subsection, we would like to generalize parametric based Čebyšev type inequalities of [66]. Čebyšev introduced the following inequality in his article [30] for two absolutely continuous functions, in the literature this inequality is named as Grüss inequality which is obtained by classical Montgomery identity defined previously in the Theorem 1.1.1. This inequality gives the estimation of bounded functional for two absolutely continuous functions. Here is the inequality as given in the Theorem stated below:

Theorem 2.6.12 *Let $f, g : I \to \mathbb{R}$ be two absolutely continuous function such that $f', g' \in L_\infty$ spaces, for $x \in [a, b]$. Then we have*

$$|T(f, g)| \leq \frac{1}{12}(b - a)^2 \|f'\|_\infty \|g'\|_\infty, \qquad (2.6.19)$$

where Čebyšev functional $T(f, g)$ defined as in (1.1.57).

Pachpatte [143] obtained the another generalized of *(2.6.19)*, which states that:

Theorem 2.6.13 *Let $f, g : I \to \mathbb{R}$ be two absolutely continuous function such that $f', g' \in L_q[a, b]$ spaces where $1 \leq q \leq \infty$, $\frac{1}{q} + \frac{1}{r} = 1$ and $T(f, g)$ is a Čebyšev functional defined in (1.1.57). Then*

$$|T(f, g)| \leq \frac{1}{(b - a)^3} \|f'\|_q \|g'\|_q \int_a^b (B_1(x))^{\frac{2}{r}} dx \qquad (2.6.20)$$

and

$$|T(f, g)| \leq \frac{1}{2(b - a)^2} \int_a^b \left[|g(x)|\|f'\|_q + |f(x)|\|g'\|_q\right] (B_1(x))^{\frac{1}{r}} dx, \qquad (2.6.21)$$

where $B_1(x)$ is defined as in (2.6.10), $\forall\, x \in [a, b]$.

In 2011, Gauezane-Lakoud and Aissaoui in [66] extended this inequality for two independent variable as can be seen in the following Theorem:

Theorem 2.6.14 *Let $f, g : I \times J \to \mathbb{R}$ be differentiable functions such that their second order partial derivatives $\dfrac{\partial^2 f(t, s)}{\partial t \partial s}$ and $\dfrac{\partial^2 g(t, s)}{\partial t \partial s}$ are integrable on $I \times J$. Then*

$$|T_*(f, g)| \leq \frac{49}{3600}(b - a)^2(d - c)^2 \left\|\frac{\partial^2 f(t, s)}{\partial t \partial s}\right\|_\infty \left\|\frac{\partial^2 g(t, s)}{\partial t \partial s}\right\|_\infty \qquad (2.6.22)$$

and

$$|T_*(f, g)| \leq \frac{1}{8(b - a)^2(d - c)^2}$$
$$\int_a^b \int_c^d \left(|g(x, y)|\left\|\frac{\partial^2 f(t, s)}{\partial t \partial s}\right\|_\infty + |f(x, y)|\left\|\frac{\partial^2 g(t, s)}{\partial t \partial s}\right\|_\infty\right)$$
$$\times \{(x - a)^2 + (b - x)^2\}\{(y - c)^2 + (d - y)^2\} dy dx, \qquad (2.6.23)$$

where

$$T_*(f,g) = \frac{1}{(b-a)(d-c)} \int_a^b \int_c^d f(x,y)g(x,y)dydx$$
$$- \frac{1}{(b-a)^2(d-c)} \int_a^b \int_c^d g(x,y) \int_a^b f(t,y)dtdydx$$
$$- \frac{1}{(b-a)(d-c)^2} \int_a^b \int_c^d g(x,y) \int_c^d f(x,s)dsdydx$$
$$+ \frac{1}{(b-a)^2(d-c)^2} \int_a^b \int_c^d g(x,y)dydx \int_a^b \int_c^d f(t,s)dsdt.$$
(2.6.24)

In recent years, a number of research papers related to Grüss type inequality and its Čebyšev functional have been published, we may mention the works [24, 66, 144–146, 156]. Now we would like to generalize Grüss type inequalities of [66] for functions of L_q space and by introducing some parameters.

Theorem 2.6.15 *Let $f,g : I \times J \to \mathbb{R}$ be differentiable functions such that their second order partial derivatives $\dfrac{\partial^2 f(t,s)}{\partial t \partial s}$ and $\dfrac{\partial^2 g(t,s)}{\partial t \partial s}$ are integrable on $I^o \times J^o$ and are bounded in L_q spaces where $1 \leq q \leq \infty$ and $\dfrac{1}{q} + \dfrac{1}{r} = 1$. Then*

$$|T_1(f,g;\epsilon,\kappa)| \leq \frac{1}{(b-a)^3(d-c)^3(r+1)^{\frac{4}{r}}} \left\| \frac{\partial^2 f(t,s)}{\partial t \partial s} \right\|_q \left\| \frac{\partial^2 g(t,s)}{\partial t \partial s} \right\|_q$$
$$\int_a^b \int_c^d \left[\left(x - \left(a + \epsilon \frac{b-a}{2}\right)\right)^{r+1} + \left(\left(b - \epsilon \frac{b-a}{2}\right) - x\right)^{r+1} + 2\left(\epsilon \frac{b-a}{2}\right)^{r+1} \right]^{\frac{2}{r}}$$
$$\left[\left(y - \left(c + \kappa \frac{d-c}{2}\right)\right)^{r+1} + \left(\left(d - \kappa \frac{d-c}{2}\right) - y\right)^{r+1}\right.$$
$$\left. + 2\left(\kappa \frac{d-c}{2}\right)^{r+1} \right]^{\frac{2}{r}} dydx,$$
(2.6.25)

where

$$T_1(f,g;\epsilon,\kappa) = \frac{(1-\epsilon)^2(1-\kappa)^2}{(b-a)(d-c)} \int_a^b \int_c^d f(x,y)g(x,y)dydx$$
$$- \frac{(1-2\epsilon)(1-\kappa)^2}{(b-a)^2(d-c)} \int_a^b \int_c^d g(x,y) \int_a^b f(t,y)dtdydx$$

$$-\frac{(1-\epsilon)^2(1-2\kappa)}{(b-a)(d-c)^2}\int_a^b\int_c^d g(x,y)\int_c^d f(x,s)ds\,dy\,dx$$

$$+\frac{2(2\epsilon\kappa-\epsilon-\kappa)+1}{(b-a)^2(d-c)^2}\int_a^b\int_c^d g(x,y)dy\,dx\int_a^b\int_c^d f(t,s)ds\,dt$$

$$-\frac{1}{2}\int_a^b\int_c^d \left(F\psi_{\epsilon,\kappa}(g)+G\psi_{\epsilon,\kappa}(f)+\frac{1}{2}\psi_{\epsilon,\kappa}(f)\psi_{\epsilon,\kappa}(g)\right)dy\,dx,$$

(2.6.26)

where $\epsilon, \kappa \in [0,1]$.

Proof Let F, G, $\tilde{F}$ and $\tilde{G}$ be defined as follows

$$F = (1-\epsilon)(1-\kappa)f(x,y) - \frac{(1-\kappa)}{(b-a)}\int_a^b f(s,y)ds$$

$$-\frac{(1-\epsilon)}{(d-c)}\int_c^d f(x,t)dt + \frac{1}{(b-a)(d-c)}\int_a^b\int_c^d f(t,s)ds\,dt - \frac{1}{2}\psi_{\epsilon,\kappa}(f),$$

$$\tilde{F} = \frac{1}{(b-a)(d-c)}\int_a^b\int_c^d K_\epsilon(x,t)Q_1(y,s)\frac{\partial^2 f(t,s)}{\partial t\partial s}ds\,dt,$$

$$G = (1-\epsilon)(1-\kappa)g(x,y) - \frac{(1-\kappa)}{(b-a)}\int_a^b g(s,y)ds$$

$$-\frac{(1-\epsilon)}{(d-c)}\int_c^d g(x,t)dt + \frac{1}{(b-a)(d-c)}\int_a^b\int_c^d g(t,s)ds\,dt - \frac{1}{2}\psi_{\epsilon,\kappa}(g)$$

and

$$\tilde{G} = \frac{1}{(b-a)(d-c)}\int_a^b\int_c^d K_\epsilon(x,t)Q_1(y,s)\frac{\partial^2 g(t,s)}{\partial t\partial s}ds\,dt.$$

Then using the condition,

$$FG = \tilde{F}\tilde{G},$$

multiplying the resultant by $\dfrac{1}{(b-a)(d-c)}$ and integrate from a to b over x and integrate c to d over y, we get

$$T_1(f,g;\epsilon,\kappa) = \frac{(1-\epsilon)^2(1-\kappa)^2}{(b-a)(d-c)}\int_a^b\int_c^d f(x,y)g(x,y)dy\,dx$$

$$-\frac{(1-2\epsilon)(1-\kappa)^2}{(b-a)^2(d-c)}\int_a^b\int_c^d g(x,y)\int_a^b f(t,y)dt\,dy\,dx$$

2.6 Generalized Inequalities for Functions of L_p Spaces via Montgomery...

$$-\frac{(1-\epsilon)^2(1-2\kappa)}{(b-a)(d-c)^2}\int_a^b\int_c^d g(x,y)\int_c^d f(x,s)dsdydx$$

$$+\frac{2(2\epsilon\kappa-\epsilon-\kappa)+1}{(b-a)^2(d-c)^2}\int_a^b\int_c^d g(x,y)dydx\int_a^b\int_c^d f(t,s)dsdt$$

$$-\frac{1}{2}\int_a^b\int_c^d\left(F\psi_{\epsilon,\kappa}(g)+G\psi_{\epsilon,\kappa}(f)+\frac{1}{2}\psi_{\epsilon,\kappa}(f)\psi_{\epsilon,\kappa}(g)\right)dydx$$

$$=\frac{1}{(b-a)^3(d-c)^3}\int_a^b\int_c^d\left(\int_a^b\int_c^d K_\epsilon(x,t)Q_1(y,s)\frac{\partial^2 f(t,s)}{\partial t\partial s}dsdt\right)$$

$$\times\left(\int_a^b\int_c^d K_\epsilon(x,t)Q_1(y,s)\frac{\partial^2 g(t,s)}{\partial t\partial s}dsdt\right)dydx.$$

Applying absolute, we get

$$|T_1(f,g;\epsilon,\kappa)|\leq\frac{1}{(b-a)^3(d-c)^3}\int_a^b\int_c^d\left(\int_a^b\int_c^d\left|K_\epsilon(x,t)Q_1(y,s)\frac{\partial^2 f(t,s)}{\partial t\partial s}\right|dsdt\right)$$

$$\times\left(\int_a^b\int_c^d\left|K_\epsilon(x,t)Q_1(y,s)\frac{\partial^2 g(t,s)}{\partial t\partial s}\right|dsdt\right)dydx.$$

Using Hölder's inequality, we get

$$|T_1(f,g;\epsilon,\kappa)|$$

$$\leq\frac{1}{(b-a)^3(d-c)^3}\int_a^b\int_c^d\left(\int_a^b\int_c^d|K_\epsilon(x,t)Q_1(y,s)|^r dsdt\right)^{\frac{1}{r}}\left\|\frac{\partial^2 f(t,s)}{\partial t\partial s}\right\|_q$$

$$\times\left(\int_a^b\int_c^d|K_\epsilon(x,t)Q_1(y,s)|^r dsdt\right)^{\frac{1}{r}}\left\|\frac{\partial^2 g(t,s)}{\partial t\partial s}\right\|_q\right)dydx$$

$$=\frac{1}{(b-a)^3(d-c)^3}\left\|\frac{\partial^2 f(t,s)}{\partial t\partial s}\right\|_q\left\|\frac{\partial^2 g(t,s)}{\partial t\partial s}\right\|_q$$

$$\times\int_a^b\int_c^d\left(\int_a^b\int_c^d|K_\epsilon(x,t)Q_1(y,s)|^r dsdt\right)^{\frac{2}{r}}dydx$$

$$=\frac{1}{(b-a)^3(d-c)^3}\left\|\frac{\partial^2 f(t,s)}{\partial t\partial s}\right\|_q\left\|\frac{\partial^2 g(t,s)}{\partial t\partial s}\right\|_q\int_a^b\int_c^d H(x,y)^2 dydx$$

$$\leq\frac{1}{(b-a)^3(d-c)^3(r+1)^{\frac{4}{r}}}\left\|\frac{\partial^2 f(t,s)}{\partial t\partial s}\right\|_q\left\|\frac{\partial^2 g(t,s)}{\partial t\partial s}\right\|_q$$

$$\times\int_a^b\int_c^d\left[\left(x-\left(a+\epsilon\frac{b-a}{2}\right)\right)^{r+1}+\left(\left(b-\epsilon\frac{b-a}{2}\right)-x\right)^{r+1}+2\left(\epsilon\frac{b-a}{2}\right)^{r+1}\right]^{\frac{2}{r}}$$

$$\times \left[\left(y - \left(c + \kappa \frac{d-c}{2} \right) \right)^{r+1} + \left(\left(d - \kappa \frac{d-c}{2} \right) - y \right)^{r+1} + 2 \left(\kappa \frac{d-c}{2} \right)^{r+1} \right]^{\frac{2}{r}} dydx.$$

Remark 2.6.7 If we substitute $\epsilon = \kappa = 0$ in (2.6.25), we get

$$|T_*(f,g)| \leq \frac{1}{(b-a)^3(d-c)^3} \left\| \frac{\partial^2 f(t,s)}{\partial t \partial s} \right\|_q \left\| \frac{\partial^2 g(t,s)}{\partial t \partial s} \right\|_q$$

$$\times \int_a^b \int_c^d [B_1(x)]^{\frac{2}{r}} [B_2(y)]^{\frac{2}{r}} dydx, \tag{2.6.27}$$

where $T_*(f,g)$ is defined in (2.6.24). The above result is generalized case for L_p spaces of (2.6.22) of [66].

Corollary 2.6.16 *If we substitute $r = 1$ and $p \to \infty$ in (2.6.25), then we get*

$$|T_1(f,g;\epsilon,\kappa)| \leq \frac{(b-a)^2(d-c)^2}{144} \left\| \frac{\partial^2 f(t,s)}{\partial t \partial s} \right\|_\infty \left\| \frac{\partial^2 g(t,s)}{\partial t \partial s} \right\|_\infty$$

$$\times \left[\frac{7}{5} + \epsilon(1-\epsilon)\{3(1-\epsilon) - 4\} \right] \left[\frac{7}{5} + \kappa(1-\kappa)\{3(1-\kappa) - 4\} \right],$$

$$\tag{2.6.28}$$

where $T_1(f,g;\epsilon,\kappa)$ is defined as in (2.6.26).

Remark 2.6.8 If we substitute $\epsilon = \kappa = 0$ in (2.6.28), or $r = 1$, then $p \to \infty$ in (2.6.27), then we get (2.6.22) of [66] as stated in the Theorem 2.6.14.

Now we are going to present the second main result of Grüss inequality.

Theorem 2.6.17 *Let all assumptions of Theorems 2.6.15 be valid. Then*

$$|T_2(f,g,\epsilon,\kappa)|$$

$$\leq \frac{1}{2(b-a)^2(d-c)^2(r+1)^{\frac{2}{r}}}$$

$$\times \int_a^b \int_c^d \left(|g(x,y)| \left\| \frac{\partial^2 f(t,s)}{\partial t \partial s} \right\|_q + |f(x,y)| \left\| \frac{\partial^2 g(t,s)}{\partial t \partial s} \right\|_q \right)$$

$$\times \left[\left(x - \left(a + \epsilon \frac{b-a}{2} \right) \right)^{r+1} + \left(\left(b - \epsilon \frac{b-a}{2} \right) - x \right)^{r+1} + 2 \left(\epsilon \frac{b-a}{2} \right)^{r+1} \right]^{\frac{1}{r}}$$

$$\left[\left(y - \left(c + \kappa \frac{d-c}{2} \right) \right)^{r+1} + \left(\left(d - \kappa \frac{d-c}{2} \right) - y \right)^{r+1} + 2 \left(\kappa \frac{d-c}{2} \right)^{r+1} \right]^{\frac{1}{r}} dydx,$$

$$\tag{2.6.29}$$

2.6 Generalized Inequalities for Functions of L_p Spaces via Montgomery...

where

$$T_2(f, g; \epsilon, \kappa) = \frac{(1-\epsilon)(1-\kappa)}{(b-a)(d-c)} \int_a^b \int_c^d f(x,y) g(x,y) dy dx$$

$$-\frac{(1-\kappa)}{(b-a)^2(d-c)} \int_a^b \int_c^d g(x,y) \int_a^b f(t,y) dt dy dx$$

$$-\frac{(1-\epsilon)}{(b-a)(d-c)^2} \int_a^b \int_c^d g(x,y) \int_c^d f(x,s) ds dy dx$$

$$+\frac{1}{(b-a)^2(d-c)^2} \int_a^b \int_c^d g(x,y) dy dx \int_a^b \int_c^d f(t,s) ds dt$$

$$-\frac{1}{2(b-a)(d-c)} \int_a^b \int_c^d \Big(f(x,y)\psi_{\epsilon,\kappa}(g) + g(x,y)\psi_{\epsilon,\kappa}(f)\Big) dy dx,$$

(2.6.30)

where $\epsilon, \kappa \in [0, 1]$.

Proof where identity (2.6.11) to the function g, we get

$$(1-\epsilon)(1-\kappa)g(x,y) = \frac{(1-\kappa)}{b-a} \int_a^b g(t,y) dt + \frac{(1-\epsilon)}{d-c} \int_c^d g(x,s) ds$$

$$-\frac{1}{(b-a)(d-c)} \int_a^b \int_c^d g(t,s) ds dt + \frac{1}{2}\psi_{\epsilon,\kappa}(g)$$

$$+\frac{1}{(b-a)(d-c)} \int_a^b \int_c^d K_\epsilon(x,t) Q_1(y,s) \frac{\partial^2 g(t,s)}{\partial t \partial s} ds dt.$$

(2.6.31)

Multiplying (2.6.4) by $\frac{1}{(b-a)(d-c)} g(x,y)$, (2.6.31) by $\frac{1}{(b-a)(d-c)} f(x,y)$, summing the resultant identities, then integrate from a to b over x and integrate c to d over y, we obtain

$$\frac{(1-\epsilon)(1-\kappa)}{(b-a)(d-c)} \int_a^b \int_c^d f(x,y) g(x,y) dy dx$$

$$= \frac{(1-\kappa)}{(b-a)^2(d-c)} \int_a^b \int_c^d g(x,y) \int_a^b f(t,y) dt dy dx$$

$$+\frac{(1-\epsilon)}{(b-a)(d-c)^2} \int_a^b \int_c^d g(x,y) \int_c^d f(x,s) ds dy dx$$

$$-\frac{1}{(b-a)^2(d-c)^2} \int_a^b \int_c^d g(x,y) dy dx \int_a^b \int_c^d f(t,s) ds dt$$

$$+\frac{1}{2(b-a)(d-c)}\int_a^b\int_c^d \Big(f(x,y)\psi_{\epsilon,\kappa}(g)+g(x,y)\psi_{\epsilon,\kappa}(f)\Big)dydx,$$

$$+\frac{1}{2(b-a)^2(d-c)^2}\int_a^b\int_c^d \bigg(g(x,y)\int_a^b\int_c^d K_\epsilon(x,t)Q_1(y,s)\frac{\partial^2 f(t,s)}{\partial t\partial s}dtds$$

$$+f(x,y)\int_a^b\int_c^d K_\epsilon(x,t)Q_1(y,s)\frac{\partial^2 g(t,s)}{\partial t\partial s}dtds\bigg)dydx, \qquad (2.6.32)$$

from that we deduce,

$$T_2(f,g;\epsilon,\kappa)=\frac{1}{2(b-a)^2(d-c)^2}$$

$$\times\int_a^b\int_c^d \bigg(g(x,y)\int_a^b\int_c^d K_\epsilon(x,t)Q_1(y,s)\frac{\partial^2 f(t,s)}{\partial t\partial s}dtds$$

$$+f(x,y)\int_a^b\int_c^d K_\epsilon(x,t)Q_1(y,s)\frac{\partial^2 g(t,s)}{\partial t\partial s}dtds\bigg)dydx. \qquad (2.6.33)$$

Consequently taking absolute value on it and then applying Hölder's Inequality, we have

$$|T_2(f,g;\epsilon,\kappa)|\leq \frac{1}{2(b-a)^2(d-c)^2(r+1)^{\frac{2}{r}}}$$

$$\times\int_a^b\int_c^d \bigg(|g(x,y)|\left\|\frac{\partial^2 f(t,s)}{\partial t\partial s}\right\|_q+|f(x,y)|\left\|\frac{\partial^2 g(t,s)}{\partial t\partial s}\right\|_q\bigg)$$

$$\times\bigg[\Big(x-\Big(a+\epsilon\frac{b-a}{2}\Big)\Big)^{r+1}+\Big(\Big(b-\epsilon\frac{b-a}{2}\Big)-x\Big)^{r+1}$$

$$+2\Big(\epsilon\frac{b-a}{2}\Big)^{r+1}\bigg]^{\frac{1}{r}}\bigg[\Big(y-\Big(c+\kappa\frac{d-c}{2}\Big)\Big)^{r+1}$$

$$+\Big(\Big(d-\kappa\frac{d-c}{2}\Big)-y\Big)^{r+1}+2\Big(\kappa\frac{d-c}{2}\Big)^{r+1}\bigg]^{\frac{1}{r}}dydx. \qquad (2.6.34)$$

Remark 2.6.9 If we substitute $\epsilon=\kappa=0$ in (2.6.29), then we get

$$|T_*(f,g)|\leq \frac{1}{2(b-a)^2(d-c)^2}$$

$$\times\int_a^b\int_c^d \bigg(|g(x,y)|\left\|\frac{\partial^2 f(t,s)}{\partial t\partial s}\right\|_q+|f(x,y)|\left\|\frac{\partial^2 g(t,s)}{\partial t\partial s}\right\|_q\bigg)$$

$$\times[B_1(x)]^{\frac{1}{r}}[B_2(y)]^{\frac{1}{r}}dydx, \qquad (2.6.35)$$

2.6 Generalized Inequalities for Functions of L_p Spaces via Montgomery...

where $T_*(f, g)$ is defined as in (2.6.24).

Remark 2.6.10 If we substitute $r = 1$ and $q \to \infty$ in (2.6.29), then we get

$$|T_2(f, g, \epsilon, \kappa)| \leq \frac{1}{8(b-a)^2(d-c)^2}$$

$$\int_a^b \int_c^d \left(|g(x, y)| \left\| \frac{\partial^2 f(t, s)}{\partial t \partial s} \right\|_\infty + |f(x, y)| \left\| \frac{\partial^2 g(t, s)}{\partial t \partial s} \right\|_\infty \right)$$

$$\times \left[(x-a)^2 + (b-x)^2 - \epsilon(1-\epsilon)(b-a)^2 \right]$$

$$\times \left[(y-c)^2 + (d-y)^2 - \kappa(1-\kappa)(d-c)^2 \right] dy\,dx, \quad (2.6.36)$$

where $T_2(f, g; \epsilon, \kappa)$ is defined as in (2.6.30).

Remark 2.6.11 If we substitute $\epsilon = \kappa = 0$ in (2.6.36), then we get inequality (2.6.23) of [66] as stated in Theorem 2.6.14.

Remark 2.6.12 We can get many interesting inequalities by varying the values of ϵ and κ. It is to be noted that the better bound for (2.6.25) and (2.6.29) is derived from $\epsilon, \kappa = \frac{1}{2}$.

Chapter 3
Functions with Nondecreasing Increments

> *Mathematics has been called the science of tautology; that is to say, mathematicians have been accused of spending their time proving that things are equal to themselves. This statement (appropriately by a philosopher) is rather inaccurate on two counts. In the first place, mathematics, although the language of science, is not a science. Rather, it is a creative art. Secondly, the fundamental results of mathematics are often* inequalities *rather than* equalities.
>
> —Edwin F. Beckenbach and Richard Bellman

The main purpose of the present chapter is to establish the relationship of functions with nondecreasing increments (FWNDI) with other functions of high importance. Our special emphasis will be on the role of representation and connection among functions with nondecreasing increments and arithmetic integral mean, Wright convex functions, convex functions, ∇-convex functions, Jensen m-convex functions, m-convex functions, m-∇-convex functions, k-monotonic functions, absolutely monotonic functions, completely monotonic functions, Laplace transform and exponentially convex functions, by using the finite difference operator as different cases of $\Delta_h^m f$. We will also consider function with nondecreasing increments of order three and get a generalization of the Levinson's type inequality and Jensen-Mercer's type inequality by using Jensen-Boas inequality and also deduce some results.

Some contents of the current chapter have been published in 2020 see [87].

In the year 1964, Brunk briefly discussed the "functions with nondecreasing increments in his research article [25, pp. 784]. According to him, Ciesielski (1957–1958) was a man who floated and introduced the idea of functions with nondecreasing increments in [38]". But Brunk has considerable contribution to raising and exposing the importance of FWNDI in a broader way, he introduced an interesting class of multivariate real-valued functions namely functions with nondecreasing increments and established important results, examples and properties using that class of functions. For further details about the historical literature of functions with nondecreasing increments (see [87, 118]).

Book [135] having detailed discussion on properties of this distinct topic, i.e., functions with nondecreasing increments. In the past few decades, this topic has gained popularity in several branches of Mathematics, which has increased interest in the study of functions with nondecreasing increments by using finite difference operators. There are interesting topics in numerical methods, differential equations, physics, biology and engineering that play an important role where we use finite difference operators (see [106]). There are many applications of these operators in the areas such as; networking, probabilistic, fractal and random media, fractionally dependent components, applications to mechanics, controls theory, transport phenomena, fractional equations and chaos, and future ideas as documented in the book [174]. Finally, these operators are naturally connected to different inequalities; various general inequalities for functions with nondecreasing increments, for present contribution see [85].

Let us recall a few important definitions and significant results extracted from [97, 113, 126, 158, 188]. Throughout the context we will use I to be an interval in $\mathbb{R}$, **I** and [**a**, **b**] both intervals in $\mathbb{R}^k$.

We start some notations to recall the definition of function with nondecreasing increments as follows:

Let $\mathbb{R}^k$ represent k-dimensional vector lattice of elements $\mathbf{x} = (x_1, x_2, \ldots, x_k)$, x_i be real, with partial ordering " $\leq$ " on $\mathbb{R}^k$ is here stated as $(x_1, x_2, \ldots, x_k) \leq (y_1, y_2, \ldots, y_k) \iff x_1 \leq y_1, \ldots, x_k \leq y_k$, that is, $x_i \leq y_i$; $i \in \{1, 2, \ldots, k\}$. We denote

$$a\mathbf{x} + b\mathbf{y} = (ax_1 + by_1, \ldots, ax_k + by_k),$$

where **0** stands for k-tuple $(0, \ldots, 0)$ and $a, b \in \mathbb{R}$. While $\mathbf{a}, \mathbf{b} \in \mathbb{R}^k$, $\mathbf{b} \geq \mathbf{a}$, a set $\{\mathbf{x} \in \mathbb{R}^k : \mathbf{a} \leq \mathbf{x} \leq \mathbf{b}\}$ should be in the *interval* [**a**, **b**].

H. D. Brunk stated the important definition below:

Definition 3.0.1 A function $f : \mathbf{I} \to \mathbb{R}$ is said to have nondecreasing increments if following inequality

$$f(\mathbf{a} + \mathbf{h}) - f(\mathbf{a}) \leq f(\mathbf{b} + \mathbf{h}) - f(\mathbf{b}), \qquad (3.0.1)$$

holds, where $\mathbf{I} \subset \mathbb{R}^k$ and k is a fixed positive integer, $\mathbf{0} \leq \mathbf{h} \in \mathbb{R}^k$, $\mathbf{b} \geq \mathbf{a}$; $\mathbf{a}, \mathbf{b} + \mathbf{h} \in \mathbf{I}$.

Brunk also stated that inequality (3.0.1) does not imply continuity even if $k = 1$. Further, *Wright-convex* is the one dimension case of above definition of function with nondecreasing increments.

Now we would give some examples and properties of function with nondecreasing increments which were given by Brunk in paper [25]. Also see [96] for more discussion.

3 Functions with Nondecreasing Increments

Examples of FWNDI
  (i) The simplest example of a FWNDI is a constant function.
  (ii) Lines in the form $\mathbf{x} = \mathbf{a}t + \mathbf{b}$, where $(0, \ldots, 0) \leq \mathbf{a} \in \mathbb{R}^k, \mathbf{b} \in \mathbb{R}^k$ whose direction cosines are nonnegative, also belong to the family of functions with nondecreasing increments.
  (iii) An important continuous function $\vartheta : \mathbb{R}^2 \to \mathbb{R}$ stated as $\vartheta(x, y) = xy$ is a function with nondecreasing increments.
  (iv) A continuous function $\nu : [0, \infty)^k \to \mathbb{R}$ stated as $\nu(\mathbf{x}) = \prod_{i=1}^{k} x_i$ is another useful function with nondecreasing increments.
  (v) $F(\mathbf{x} + \mathbf{y}) = F(\mathbf{x}) + F(\mathbf{y})$ is the Cauchy functional equation which is an interesting and widely used example of such functions.

Properties of FWNDI
FWNDI possesses the following properties:

  (i) A FWNDI need not be continuous.
  (ii) If function $f : \mathbf{I} \to \mathbb{R}$ has 1st order partial derivatives $\forall \mathbf{x} \in \mathbf{I}$. Then f has nondecreasing increments iff every of those partial derivatives is nondecreasing in every arguments.
  (iii) If $f : \mathbf{I} \to \mathbb{R}$ has partial derivatives of 2nd order $\forall \mathbf{x} \in \mathbf{I}$. Then f has nondecreasing increments iff every of those partial derivatives is non-negative.
  (iv) A function $\upsilon : [0, 1] \to \mathbb{R}$ is convex, stated as $\upsilon(t) = f(t\mathbf{a}+\mathbf{b})$, if f FWNDI is continuous in $\mathbf{b} \leq \mathbf{x} \leq \mathbf{b} + \mathbf{a}; 0 \leq \mathbf{a} \in \mathbb{R}^k$.

In many fields of mathematics several types of differences are used such as finite difference, forward difference, backward difference, divided difference etc. From application point of view there are large number of implementation of these types of differences in the fields as numerical analysis, statistics, vector calculus and physics (see books [2, 68, 179]), we have chosen ∇- operator for our book due to its wide range of application.

Now we would like to define further generalized convex functions that can be seen in [95, 97] and [158].

Definition 3.0.2 A function $f : I \to \mathbb{R}$, is known as ∇-*convex of mth order* or m-∇-convex, if $\forall$ $(m + 1)$ different points $x_i, x_{i+1}, \ldots, x_{i+m}$ we have $\nabla_{(m)} f(x_i) = (-1)^m \Delta_{(m)} f(x_i) \geq 0$ where $\Delta_{(m)} f(x_i)$ represents m-th order divided difference of function f as defined in Definition 1.0.4.

We refer to the book [158], for further details of other results about the higher order convex functions.

Definition 3.0.3 ([158]) The finite difference of a function of order m on $I = [a, b] \in \mathbb{R}$, where m is non- negative integers, is defined by

$$\Delta_h^0 f(x) = f(x)$$
$$\Delta_h^m f(x) = \Delta_h^{m-1} f(x + h) - \Delta_h^{m-1} f(x),$$

where $h \neq 0$, $x + ih \in I$; $i \in \{0, 1, 2, \ldots, m\}$. Then it can easy to write the statement as

$$\Delta_h^m f(x) = \sum_{i=0}^{m} \binom{m}{i} f(x + ih)(-1)^{m-i}.$$

Further, we conclude that if function $f : I \to \mathbb{R}$ is satisfied $\Delta_h^m f(x) \geq 0$, then function is called Jensen m-convex for all $x \in I$ and $h > 0$.

We define here a special type of function which belongs to the class of function with nondecreasing increments and themselves connect/contain the class of other functions that are already provided in the starting section of this chapter, by using several cases of $\Delta_h^m f$.

Here we also recall the important Definitions 3.0.3 and 1.0.4 of finite difference and divided difference of function respectively and we will use these definitions to connect with other definitions that will be seen in the next section.

We present some remarks about the relationship among finite difference, divided difference and derivative of the function.

Remark 3.0.1 Some important remarks are following:

(i) Let us denote $[x_i, \ldots, x_{i+m}; f]$ by $\Delta_{(m)} f(x_i)$. The value $[x_i, \ldots, x_{i+m}; f]$ is independent of elements order $x_i, x_{i+1}, \ldots, x_{i+m}$.
(ii) We may extend Definition 1.0.4 by including case in which few elements or all elements coincide by supposing that $x_i \leq \cdots \leq x_{i+m}$ (see [158]) and letting

$$[x_i, \ldots, x_{i+m}; f] = \frac{f^{(m)}(x_i)}{m!},$$

provided that $f^{(m)}(x_i)$ exists.
(iii) The following identity is valid (see [164]):

$$\Delta_{\mathbf{h}}^m f(\mathbf{x}) = m! h^m \Delta_m f(\mathbf{x}),$$

provided that x_i's are equally spaced.

Using finite difference operator, we would like to state alternative form of functions with nondecreasing increments as;

$$\Delta_{\mathbf{h}_1} \Delta_{\mathbf{h}_2} f(\mathbf{a}) \geq 0, \tag{3.0.2}$$

Since $\Delta_{\mathbf{h}_1}(\Delta_{\mathbf{h}_2} f(\mathbf{a})) = \Delta_{\mathbf{h}_1}(f(\mathbf{a} + \mathbf{h}_2) - f(\mathbf{a}))$
$= f(\mathbf{a} + \mathbf{h}_2 + \mathbf{h}_1) - f(\mathbf{a} + \mathbf{h}_2) - f(\mathbf{a} + \mathbf{h}_1) + f(\mathbf{a}) \geq 0.$

Setting $\mathbf{h} = \mathbf{h_1}$, $\mathbf{b} = \mathbf{a} + \mathbf{h_2}$ in above statement, then

$$f(\mathbf{b} + \mathbf{h}) - f(\mathbf{b}) - f(\mathbf{a} + \mathbf{h}) + f(\mathbf{a}) \geq 0.$$

By taking $\mathbf{h_1} = \mathbf{h_2} = \mathbf{h}$, then we can obtain special case of (3.0.2).

$$\Delta_\mathbf{h} f(\mathbf{b}) - \Delta_\mathbf{h} f(\mathbf{a}) \geq 0$$
$$\Delta_\mathbf{h}(f(\mathbf{b}) - f(\mathbf{a})) \geq 0$$
$$\Delta_\mathbf{h}^2 f(\mathbf{a}) \geq 0, \text{ where } \mathbf{a} \leq \mathbf{b}.$$

We know that $\Delta_{\mathbf{h_1}} f(\mathbf{x}) = f(\mathbf{x} + \mathbf{h_1}) - f(\mathbf{x})$ and further,

$$\Delta_{\mathbf{h_1}} \Delta_{\mathbf{h_2}} \cdots \Delta_{\mathbf{h_m}} f(\mathbf{x}) = \Delta_{\mathbf{h_1}}(\Delta_{\mathbf{h_2}} \cdots \Delta_{\mathbf{h_m}} f(\mathbf{x})) \text{ for } m \geq 2,$$

where $\mathbf{x}, \mathbf{x} + \mathbf{h_1} + \cdots + \mathbf{h_m} \in \mathbf{I}, 0 \leq \mathbf{h_i} \in \mathbb{R}^k$ for $i \in \{1, 2, \ldots, m\}$.
Similarly we can extend this definition for mth order as:

Definition 3.0.4 A function $f: \mathbf{I} \to \mathbb{R}$ is known as function with nondecreasing increments of mth order if

$$\Delta_{\mathbf{h_1}} \Delta_{\mathbf{h_2}} \cdots \Delta_{\mathbf{h_m}} f(\mathbf{x}) \geq 0 \qquad (3.0.3)$$

holds, where $\mathbf{x}, \mathbf{x} + \mathbf{h_1} + \cdots + \mathbf{h_m} \in \mathbf{I}, 0 \leq \mathbf{h_i} \in \mathbb{R}^k$ for $i \in \{1, 2, \ldots, m\}$. Then the special case is given by

$$\Delta_\mathbf{h}^m f(\mathbf{x}) \geq 0, \qquad (3.0.4)$$

where f is called FWNDI of order m with equally spaced $\mathbf{h}$.

3.1 Functions with Nondecreasing Increments in Real Life

We can see the applications of functions with nondecreasing increments in the ultramodular function which is the special case of FWNDI but the only difference is that the range of ultramodular function is [0, 1] while the range of FWNDI is $\mathbb{R}$ and ultramodular function is stated as:

Ultramodular Function
A function $f: \mathbf{I} \to [0, 1]$ is called ultramodular [168], if for all $\mathbf{x}, \mathbf{y} \in \mathbf{I} \in \mathbb{R}^k$ with $\mathbf{x} \leq \mathbf{y}$ and for all $\mathbf{h} \in \mathbb{R}^k$ with $\mathbf{h} \geq \mathbf{0}$ and $\mathbf{x} + \mathbf{h}, \mathbf{y} + \mathbf{h} \in \mathbf{I}$.

$$f(\mathbf{x} + \mathbf{h}) - f(\mathbf{x}) \leq f(\mathbf{y} + \mathbf{h}) - f(\mathbf{y}).$$

Statistics
In Statistics, ultramodular functions play an important role in modelling stochastic orders and positive dependence among random vectors [177].

Economics
In Economics, ultramodular functions $f : \mathbf{I} \to [0, 1]$ are often said to have nondecreasing increments [168].

Copulas
Class of absolutely monotonic functions is a subclass of completely monotonic function and this is subclass of FWNDI if differentiability exists. Absolutely monotonic function is used in copulas [136] and there are many applications of copulas in various branches of mathematics, economics, engineering and medicine which have been highlighted after the following definition.

Copula is used in probability theory and specially it is general tool to construct multivariate distributions and to investigate dependence structure between random variables [84].

A function $C : [0, 1] \times [0, 1] \to [0, 1]$ is a bivariate copula if $C(0, u) = C(u, 0) = 0$, $C(1, u) = C(u, 1) = u$ and $C(u_2, v_2) - C(u_2, v_1) - C(u_1, v_2) + C(u_1, v_1) \geq 0$ for all $0 \leq u_1 \leq u_2 \leq 1$ and $0 \leq v_1 \leq v_2 \leq 1$.

Quantitative Finance
In quantitative finance, copulas are applied to risk management, to portfolio management and optimization [112].

Civil Engineering
Recently, copula functions have been successfully applied to the database formulation for the reliability analysis of highway bridges, and to various multivariate simulation studies in civil [182], reliability of wind and earthquake engineering [191], mechanical and offshore engineering [195].

Reliability Engineering
Copulas are being used for reliability analysis of complex systems of machine components with competing failure modes [161].

Warranty Data Analysis
Copulas are being used for warranty data analysis in which the tail dependence is analysed [189].

Medicine
Copula functions have been successfully applied to the analysis of neuronal dependencies [57] and spike counts in neuroscience [140].

Solar Irradiance Variability
Copulas have been used to estimate the solar irradiance variability in spatial networks and temporally for single locations [137].

Hydrology Research
Copulas are used for research of hydrology for more information see in [107].

Climate and Weather Research

Copulas are used in climate and weather research for more detailed we can see in [173].

The current chapter has an aim to collect the established facts about the FWNDI, together with some other important functions and notions, that can help out for finding whether a given function is FWNDI or not. In the second section of current chapter, we would like to establish the connection among functions with nondecreasing increments and many other functions by using finite difference operator as different cases of $\Delta_h^m f$ with detailed examples. In the third section, we would like to obtain the generalization of the Levinson's type inequality and Jensen-Mercer's type inequality by using Jensen-Boas inequality for FWNDI of order 3 and also deduce some results.

3.2 Relationship Among Functions with Nondecreasing Increments and Many Others

We wouldfs like to use the Definition 3.0.4 to establish relationship among functions with nondecreasing increments and many other functions, the detailed list of other functions is already mentioned in the initial section. In current section we would like to recall some important definitions which are extracted from the articles [3, 22, 41, 59, 62, 71, 85, 97, 101, 113, 119, 121, 122, 124, 126, 128, 135, 147, 157, 158, 188] and these will relate to functions with nondecreasing increments, by using the finite difference operator as different cases of $\Delta_h^m f \geq 0$. In this connection we will use the relationships of finite difference, derivative and some other differences (see [41, 62, 158, 164]).

Arithmetic Integral Mean vs FWNDI

Let us have A is mean of arithmetic integral of function f on $[0, a]$ (see [85, 105]).

Definition 3.2.1 A function A (nondecreasing function) is said to be a Arithmetic integral mean in the interval $[0, a]$, such that

$$A(t) = \frac{1}{t} \int_0^t f(x)dx,$$

provided that $f : [0, a] \to \mathbb{R}$ is a non-negative and nondecreasing, where $a > 0$.

Now, we will recall extension of previously mentioned result to FWNDI of higher order, extracted from [85].

Theorem 3.2.1 *Let* $f : [\mathbf{a}, \mathbf{b}] \to \mathbb{R}$ *be continuous and with nondecreasing increments of mth order. Then A is stated as*

$$A(\mathbf{t}) = \left(\prod_{i=1}^k (t_i - a_i)\right)^{-1} \int_{a_1}^{t_1} \cdots \int_{a_k}^{t_k} f(\mathbf{v})d\mathbf{v},$$

is a FWNDI of order m in the interval $[\mathbf{a}, \mathbf{b}]$, where $\mathbf{v} = (v_1, \ldots, v_k)$ and $\mathbf{dv} = dv_1 \cdots dv_k$.

The Alternative form of Arithmetic integral mean of order m in the interval $[\mathbf{a}, \mathbf{b}]$, using Eq. (3.0.3) it is defined by

$$\Delta_{\mathbf{h}_1} \Delta_{\mathbf{h}_2} \cdots \Delta_{\mathbf{h}_m} A(\mathbf{t}) \geq 0, \qquad \mathbf{h_i} \geq \mathbf{0}.$$

Remark 3.2.1 By taking $\mathbf{h_1} = \mathbf{h_2} = \cdots = \mathbf{h_m}$ in above inequality, then obtain special case as $\Delta_{\mathbf{h}}^m A(\mathbf{t}) \geq 0$.

Example 3.2.1 Let $f : [0, a] \to \mathbb{R}_+$ be a function, stated as

$$f(x) = e^{x-c^2}.$$

Since $\Delta_h^m f(x) \geq 0$, then from the above remark also holds $\Delta_h^m A(t) \geq 0$ when h is very small, therefore we can say that A is Arithmetic integral mean of mth order in the interval $[0, a]$ for every $a > 0, c \in \mathbb{R}$.

Now first of all we would like to start from 1st dimensional case of function with nondecreasing increments which is Wright–convex function and we will give the equivalent form.

Wright–Convex Function vs FWNDI

Definition 3.2.2 ([158]) A function $f : [a, b] \to \mathbb{R}$ is known as Wright–convex, if following inequality is valid $\forall \ y \geq x; z \geq 0; x, y + z$ belong to $[a, b]$.

$$f(y + z) - f(y) \geq f(x + z) - f(x)$$

It can also be written as $\Delta_z^2 f(x) \geq 0 \qquad x \leq y; x, y + z \in [a, b]$.
This is computed same as function with nondecreasing increments, i.e.,

$$\Delta_h^2 f(x) \geq 0. \tag{3.2.1}$$

If there exists f'', then f is Wright–convex function iff

$$h^2 f''(x) \geq 0. \tag{3.2.2}$$

And the equivalent form of (3.2.2) on $[a, b] \in I \subset \mathbb{R}$, using Eq. (3.0.4) it is also defined by same inequality (3.2.1), when h is very small.

Example 3.2.2 Let $f : [a, b] \to \mathbb{R}$ be a function, stated as

$$f(x) = x(ax - b).$$

Since $h^2 f''(x) \geq 0$, i.e., $\Delta_h^2 f(x) \geq 0$, therefore f is Wright–convex in the interval $[a, b] \subset \mathbb{R}$ for every $a \geq 0, b \in \mathbb{R}$.

3.2 Relationship Among Functions with Nondecreasing Increments and Many...

Remark 3.2.2 Wright-convex function is a special case of FWNDI for $k = 1$.

Now we would like to state generalized convex function may be seen in [97, 158].

m-Convex Function vs FWNDI

Definition 3.2.3 A function $f : I \to \mathbb{R}$ is known as m-convex, if the inequality $\Delta_{(m)} f(x) \geq 0$ holds $\forall$ $(m + 1)$ different points $x_0, x_1, \ldots, x_m \in I$.

Special case of convex function of order m, using Eq. (3.0.4) and Remark 3.0.1 it is defined by

$$\frac{\Delta_h^m f(x)}{m! h^m} \geq 0, \qquad x \in I, h > 0. \tag{3.2.3}$$

If there exists $f^{(m)}$, then function is m-convex or mth order convex iff

$$\frac{f^{(m)}(x)}{m!} \geq 0. \tag{3.2.4}$$

And the equivalent form of (3.2.4) on $I \subset \mathbb{R}$, using Eq. (3.0.4) it is also defined by same inequality (3.2.3), when h is very small.

Example 3.2.3 Let $f : I \to \mathbb{R}$ be a function, stated as

$$f(x) = \frac{x^m}{m!}.$$

Since $\dfrac{f^{(m)}(x)}{m!} \geq 0$, then $\dfrac{\Delta_h^m f(x)}{m! h^m} \geq 0$, therefore f is m-convex in the interval I for $q > 0, m \in \{0, 1, 2, \cdots\}$.

m-∇-Convex Function vs FWNDI

Definition 3.2.4 A function $f : I \to \mathbb{R}$ is known as m-∇-convex, if $\forall$ $(m + 1)$ different points $x_0, x_1, \ldots, x_m \in I$ we have $\nabla_{(m)} f(x) = (-1)^m \Delta_{(m)} f(x) \geq 0$.

Special case of ∇-convex of order m, using Eq. (3.0.4) and Remark 3.0.1 it is defined by

$$\frac{(-1)^m \Delta_h^m f(x)}{m! h^m} \geq 0, \qquad x \in I, h > 0. \tag{3.2.5}$$

If there exists $f^{(m)}$, then function is m-∇-convex iff

$$\frac{(-1)^m f^{(m)}(x)}{m!} \geq 0. \tag{3.2.6}$$

And the equivalent form of (3.2.6) on $I \subset \mathbb{R}$, using Eq. (3.0.4) it is also defined by same inequality (3.2.5), when h is very small.

Example 3.2.4 Let $f : I \to \mathbb{R}$ be a function, stated as

$$f(x) = \frac{e^{-rx}}{r^m}.$$

Since $\dfrac{(-1)^m f^{(m)}(x)}{m!} \geq 0$, i.e., $\dfrac{(-1)^m \Delta_h^m f(x)}{m! h^m} \geq 0$, therefore f is m-∇-convex in the interval $I \subset \mathbb{R}_*$, where $r \in \mathbb{R}\setminus\{0\}$, $m \in \{0, 1, \ldots\}$.

Remark 3.2.3 Convex function and ∇-convex function are the special cases of m-convex function and m-∇-convex respectively, if we put $m = 2$.

Jensen m-Convex Function vs FWNDI

Definition 3.2.5 A function $f : I \to \mathbb{R}$ is known as Jensen m-convex or J-convex of order m if holds

$$\Delta_h^m f(x) \geq 0, \qquad \forall\, h > 0 \text{ and } x \in I. \tag{3.2.7}$$

This is done by finite difference operator and if there exists $f^{(m)}$, then f is J–convex of order m iff

$$h^m f^{(m)}(x) \geq 0. \tag{3.2.8}$$

And the equivalent form of (3.2.8) on $I \subset \mathbb{R}$, using Eq. (3.0.4) is also defined by same inequality (3.2.7), when h is very small.

Example 3.2.5 Let a function $f : I \to \mathbb{R}$ which is stated as

$$f(x) = \frac{e^{qx}}{q^m}.$$

Since $h^m f^{(m)}(x) \geq 0$, then $\Delta_h^m f(x) \geq 0$, therefore f is J-convex of mth order in the interval $I \subset \mathbb{R}_*$ for $q \in \mathbb{R}\setminus\{0\}$, $m \in \{0, 1, 2, \ldots\}$.

Remark 3.2.4 J-convex function is the special case of J-convex function of mth order, if we put $m = 2$.

Now we recall definitions of convex function and ∇-convex function and their connection with FWNDI using Eq. (3.0.4).

Convex Function vs FWNDI

Definition 3.2.6 ([158]) A continuous function $f : I \to \mathbb{R}$ is known as convex, if there exists non-negative second order divided difference, such that

$$\Delta_2 f(x) \geq 0, \qquad x \in I.$$

Special case of convex function on $I \subset \mathbb{R}$, using Eq. (3.0.4) and Remark 3.0.1 it is defined by

$$\frac{\Delta_h^2 f(x)}{2!h^2} \geq 0, \qquad x \in I, h > 0. \tag{3.2.9}$$

If there exists f'', then function is convex iff

$$\frac{f''(x)}{2!} \geq 0. \tag{3.2.10}$$

And the equivalent form of (3.2.10) on $I \subset \mathbb{R}$, using Eq. (3.0.4) it is also defined by same inequality (3.2.9), when h is very small.

Example 3.2.6 Let $f : I \to \mathbb{R}_*$ be a function, stated as

$$f(x) = mx^2 + m^2.$$

Since $\dfrac{f''(x)}{2!} \geq 0$, i.e., $\dfrac{\Delta_h^2 f(x)}{2!h^2} \geq 0$, therefore f is convex in the interval I for $m \in \{0, 1, 2, \cdots\}$.

Remark 3.2.5 If sets, "C" convex function, "W" Wright-convex function and "J" Jensen convex function then $C \subset W \subset J$. Moreover, each inclusion is proper (see [135, 158]).

∇-Convex Function vs FWNDI

Definition 3.2.7 [158] A continuous function $f : I \to \mathbb{R}$ is known as ∇-convex, if there is existence of $\nabla_2 f(x) = (-1)^2 \Delta_2 f(x)$ and satisfy

$$\nabla_2 f(x) \geq 0, \qquad x \in I.$$

Special case of ∇-convex function on $I \subset \mathbb{R}$, using Eq. (3.0.4) and Remark 3.0.1 it is defined by

$$\frac{(-1)^2 \Delta_h^2 f(x)}{2!h^2} \geq 0, \qquad x \in I, h > 0. \tag{3.2.11}$$

If there exists f'', then f is ∇-convex iff

$$\frac{(-1)^2 f''(x)}{2!} \geq 0. \tag{3.2.12}$$

And the equivalent form of (3.2.12) on $I \subset \mathbb{R}$, using Eq. (3.0.4) it is also defined by same inequality (3.2.11), when h is very small.

Example 3.2.7 Let $f : \mathbb{R}_+ \to \mathbb{R}$ be a function, stated as

$$f(x) = -\ln x.$$

Since $\dfrac{(-1)^2 f''(x)}{2!} \geq 0$, i.e., $\dfrac{(-1)^2 \Delta_h^2 f(x)}{2! h^2} \geq 0$, therefore f is ∇-convex in the interval $\mathbb{R}_+$.

Completely Monotonic Function vs FWNDI

Definition 3.2.8 ([59, 113, 128]) A continuous function f is known as completely monotonic, if there is existence of derivatives of all orders on $I \subset \mathbb{R}$, and satisfy

$$(-1)^i f^{(i)}(x) \geq 0, \qquad i \in \{0, 1, \ldots\}; \quad x \in I.$$

Equivalent form of completely monotonic function on $I \subset \mathbb{R}$, using Eq. (3.0.4) it is defined by

$$\frac{(-1)^i \Delta_h^i f(x)}{h^i} \geq 0, \qquad i \in \{0, 1, \ldots\}; \quad x \in I, h > 0,$$

when h is very small.

Example 3.2.8 Some examples of completely monotonic functions are following:

(i) $f(x) = \dfrac{\alpha}{x^{1-\alpha}}$, $\quad 0 \leq \alpha \leq 1, x > 0$.
(ii) $f(x) = -\ln(1 - 1/x)$, for every $x \in \mathbb{R}_+$.
(iii) $f(x) = e^{1/x}$, for every $x \in \mathbb{R}_+$.

Completely monotonic function is generalized form of absolutely monotonic function and k-monotonic function, now we will give connection of absolutely monotonic function and k-monotonic function with FWNDI using Eq. (3.0.4).

Absolutely Monotonic Function vs FWNDI

Definition 3.2.9 ([59, 113]) A continuous function f is known as absolutely monotonic, if there is existence of derivatives of all orders on $I \subset \mathbb{R}$, and satisfy

$$f^{(i)}(x) \geq 0, \qquad i \in \{0, 1, \ldots\}; \quad x \in I.$$

Equivalent form of absolutely monotonic function on $I \subset \mathbb{R}$, using Eq. (3.0.4) it is defined by

$$\frac{\Delta_h^i f(x)}{h^i} \geq 0, \qquad i \in \{0, 1, 2, \ldots\}; \quad x \in I, h > 0,$$

when h is very small.

3.2 Relationship Among Functions with Nondecreasing Increments and Many...

Example 3.2.9 Let a function $f : [-1, 1] \to \mathbb{R}$ which is stated as

$$f(x) = \arcsin x.$$

Since $f^{(i)}(x) \geq 0$, then $\dfrac{\Delta_h^i f(x)}{h^i} \geq 0$, therefore f is absolutely monotonic in the interval $[0, 1)$.

k-Monotonic Function vs FWNDI

Definition 3.2.10 ([59, 113]) A function is known as k–monotonic on $I \subset \mathbb{R}$, if all its derivatives $f^{(i)}(x)$ exist and satisfy

$$(-1)^i f^{(i)}(x) \geq 0, \quad i \in \{0, 1, \ldots, k\}, \text{ where } k \text{ is fixed}; \quad x \in I.$$

Equivalent form of k–monotonic function on $I \subset \mathbb{R}$, using Eq. (3.0.4) it is defined by

$$\frac{(-1)^i \Delta_h^i f(x)}{h^i} \geq 0, \quad i \in \{0, 1, \ldots, k\}; \quad x \in I, h > 0,$$

when h is very small.

Example 3.2.10 Let a function $f : I \to \mathbb{R}$ which is stated as

$$f(x) = x^{-r}.$$

Since $(-1)^i f^{(i)}(x) \geq 0$, then $\dfrac{(-1)^i \Delta_h^i f(x)}{h^i} \geq 0$, therefore the function f is k-monotonic in the interval $I \subset \mathbb{R}_*$ for $r \geq 0$.

Laplace Transform of f vs FWNDI

Definition 3.2.11 ([188]) Let f be function which satisfies $|f(x)| \leq Me^{ax}$ and piecewise continuous, where a and M are real constants. Then the Laplace transform of $f(x)$, stated as

$$F(s) = L\{f(x)\} = \int_0^\infty f(x)e^{-sx}dx, \quad s > a.$$

Similarly, the Laplace transformation of Borel measure $\varphi(t)$ on $\mathbb{R}_*$ is stated as

$$L\{\varphi(t)\} = \int_0^\infty e^{-xt}d\varphi(t).$$

At origin the Laplace transformation is continuous iff φ is finite.

Bernstein [22] proved that f in the interval $\mathbb{R}_*$ is completely monotonic, iff there is existence of increasing function $\varphi(t)$ on $\mathbb{R}_*$.

$$f(x) = \int_0^\infty e^{-xt} d\varphi(t).$$

Remark 3.2.6 Equivalent form of above statement can also be present as a function f on $\mathbb{R}_*$ satisfied the condition $\dfrac{(-1)^i \Delta_h^i f(x)}{h^i} \geq 0$ *(completely monotonic)* where h is very small and $x \in I \subset \mathbb{R}_*, i \in \{0, 1, \ldots\}$ iff there exists

$$f(x) = \int_0^\infty e^{-xt} d\varphi(t),$$

here $\varphi(t)$ is an increasing function in the interval $\mathbb{R}_*$.

Exponentially Convex Function vs FWNDI

Definition 3.2.12 ([97, 158]) A continuous function $\omega : I \to \mathbb{R}$ on open interval I is called exponentially convex, if

$$\sum_{i,j=1}^m \rho_i \rho_j \, \omega \left(x_i + x_j\right) \geq 0,$$

$\forall \, m \in \mathbb{N}$ and $\forall \, \rho_i, \rho_j \in \mathbb{R}$; such that $x_i + x_j \in I$ and $i, j \in \{1, \ldots, m\}$.

For detailed discussion on exponentially convex function we refer the readers to Chap. 1 (See Sect. 1.2.4).

Theorem 3.2.2 *The function $\omega : I \to \mathbb{R}$ in the interval I is exponentially convex iff*

$$\omega(x) = \int_{-\infty}^\infty e^{tx} d\varphi(t), \qquad x \in I,$$

for some $\varphi(t) : \mathbb{R} \to \mathbb{R}$ (nondecreasing function).

Proof See [3], p. 211. □

Little less obvious examples can be deduced by applying above integral representation and some results from Laplace transform are following.

Example 3.2.11 The following are examples of Exponentially convex functions as well as Laplace transform of $f(x)$ through above theorem on $\mathbb{R}_+$ include:

(i) $f(x) = x^{-\alpha}$, for every $\alpha > 0$.
(ii) $f(x) = e^{-\alpha \sqrt{x}}$, for every $\alpha > 0$.
(iii) $f(x) = e^{-x\sqrt{s}}$, for every $s > 0$.
(iv) $f(x) = e^{-tx}$, for every $t > 0$.

Remark 3.2.7 The above functions are also satisfying the condition $\frac{(-1)^i \Delta_h^i f(x)}{h^i} \geq 0$ (completely monotonic) by using Eq. (3.0.4) in these examples, when h is very small.

3.3 Functions with Nondecreasing Increments of Order 3

Consider a common scenario that a car is stopped at a traffic signal, as light of signal blinks green if the accelerator of the car is pushed then the engine takes pickup, that in turns causes the *jerk* when the car is accelerating. *Jerk* is an example a function with nondecreasing increments of order three.

Before we give the generalization of Levinson's type inequality and Jensen-Mercer's type inequality by using Jensen-Boas inequality for function with nondecreasing increments of order three, we must present following theorem about the inequality of Jensen-Steffensen type for a FWNDI which is collected from [148]. Here in current section $[c, d]$ is an interval in $\mathbb{R}$, $\mathbf{I}$ in $\mathbb{R}^k$ and $\mathbf{Y}(t)$ is a vector of functions.

Theorem 3.3.1 *Let* $\mathbf{Y} : [c, d] \to \mathbf{I}$ *be nondecreasing continuous map and let* $G \in BV[c, d]$ *such that*

$$G(c) \leq G(y) \leq G(d), \quad G(c) < G(d).$$

If $f : \mathbf{I} \to \mathbb{R}$ *is a continuous FWNDI, then*

$$f\left(\frac{\int_c^d \mathbf{Y}(t)\, dG(t)}{\int_c^d dG(t)}\right) \leq \frac{\int_c^d f(\mathbf{Y}(t))\, dG(t)}{\int_c^d dG(t)}$$

holds, where $\int_c^d \mathbf{Y}\, dG = \left(\int_c^d Y_1\, dG, \ldots, \int_c^d Y_k\, dG\right)$.

We also give generalized form of above theorem using Jensen-Boas inequality for function with nondecreasing increments, extracted from [94] as below.

Theorem 3.3.2 *Let* $\mathbf{Y} : [c, d] \to \mathbf{I}$ *be a monotonic (either non-increasing or nondecreasing) and continuous map in every of j intervals* (d_{i-1}, d_i). *Let* $G : [c, d] \to \mathbb{R}$ *be continuous or of BV satisfying*

$$G(c) \leq G(c_1) \leq G(d_1) \leq G(c_2) \leq \cdots \leq G(d_{j-1}) \leq G(c_j) \leq G(d) \quad (3.3.1)$$

for all $c_i \in (d_{i-1}, d_i)$ $(d_0 = c, d_j = d)$, and $G(d) > G(c)$. If f is continuous having nondecreasing increments in every of j intervals (d_{i-1}, d_i), then

$$f\left(\frac{\int_c^d \mathbf{Y}(t) dG(t)}{\int_c^d dG(t)}\right) \leq \frac{\int_c^d f(\mathbf{Y}(t)) dG(t)}{\int_c^d dG(t)}. \tag{3.3.2}$$

Remark 3.3.1

(i) If $\dfrac{\int_c^d (\mathbf{Y}(t)) dG(t)}{\int_c^d dG(t)} \in \mathbf{I}$ and $\forall\ y \in c_i \in (d_{i-1}, d_i)$ $(d_0 = c, d_j = d)$ we have either $G(y) \geq G(d)$ or $G(y) \leq G(c)$, then the inequality in (3.3.2) holds for reverse direction.

(ii) By putting $j = 1$, Theorem 3.3.2 gives a special case as Theorem 3.3.1.

Now we are able to give our main theorems which are generalizations of the Levinson's inequality and Jensen-Mercer's type inequality using Jensen-Boas inequality in the following next sub-sections.

3.3.1 On Levinson Type Inequalities

Inequality of Levinson is the generalization of inequality of Ky Fan, *i.e.*, in other word we say inequality of Ky. Fan is the special case of inequality of Levinson (see [21, 69]) and for generalizations of Levinson's inequality (see [15]).

The next theorem is the generalization of Levinson's type inequality using Jensen-Boas inequality.

Theorem 3.3.3 *Let* $G \in BV[c, d]$ *such that* (3.3.1) *valid and* $\mathbf{Y} : [c, d] \to [\mathbf{0}, \mathbf{q}]$, ($\mathbf{q} > \mathbf{0}$) *be a monotonic (either non-increasing or nondecreasing) and continuous map in every of j intervals (d_{i-1}, d_i). If f is a continuous FWNDI of third order in the interval* $\mathbf{J} = [\mathbf{0}, 2\mathbf{q}] \subset \mathbb{R}^k$, *then the inequality*

$$\frac{\int_c^d f(\mathbf{Y}(t)) dG(t)}{\int_c^d dG(t)} - f\left(\frac{\int_c^d \mathbf{Y}(t) dG(t)}{\int_c^d dG(t)}\right)$$

$$\leq \frac{\int_c^d f(2\mathbf{q} - \mathbf{Y}(t)) dG(t)}{\int_c^d dG(t)} - f\left(\frac{\int_c^d (2\mathbf{q} - \mathbf{Y}(t)) dG(t)}{\int_c^d dG(t)}\right) \tag{3.3.3}$$

holds.

Proof If f is a FWNDI of third order in the interval $\mathbf{J}$, then inequality

$$\Delta_{\mathbf{h}_1} \Delta_{\mathbf{h}_2} \Delta_{\mathbf{h}_3} f(\mathbf{y}) \geq 0 \quad \text{for} \quad \mathbf{y}, \mathbf{y} + \mathbf{h}_1 + \mathbf{h}_2 + \mathbf{h}_3 \in \mathbf{J}, \quad \mathbf{h}_1, \mathbf{h}_2, \mathbf{h}_3 \in \mathbb{R}_*^k$$

3.3 Functions with Nondecreasing Increments of Order 3

holds,
i.e.,

$$\Delta_{\mathbf{h}_1} \Delta_{\mathbf{h}_2} (f(\mathbf{y} + \mathbf{h}_3) - f(\mathbf{y})) \geq 0. \tag{3.3.4}$$

If $\mathbf{y} \in \mathbf{J}$ and $\mathbf{h}_3 = 2\mathbf{q} - 2\mathbf{y}$, we have

$$\Delta_{\mathbf{h}_1} \Delta_{\mathbf{h}_2} (f(2\mathbf{q} - \mathbf{y}) - f(\mathbf{y})) \geq 0,$$

i.e., $\mathbf{y} \mapsto f(2\mathbf{q} - \mathbf{y}) - f(\mathbf{y})$ is a FWNDI of second order, i.e., it is a FWNDI. Now, applying Theorem 3.3.2 and get desired Theorem 3.3.3. □

Remark 3.3.2

(i) If $\dfrac{\int_c^d (2\mathbf{q} - \mathbf{Y}(t)) \, dG(t)}{\int_c^d dG(t)} \in \mathbf{J}$ and $\forall \ y \in c_i \in (d_{i-1}, d_i)$ $(d_0 = c, d_j = d)$ we have either $G(y) \geq G(d)$ or $G(y) \leq G(c)$, then the inequality in (3.3.3) holds for reverse direction.

(ii) By putting $j = 1$, Theorem 3.3.3 gives a special case as Theorem 3.1 of [96].

(iii) We can obtain discrete version of Theorem 3.3.3, using technique as given in Corollary 1 of [100].

Corollary 3.3.4 *Let* $\mathbf{Y}$ *satisfies the assumptions of the Theorem 3.3.3. Then inequalities*

$$0 \leq \left(\int_c^d dG(t) \right)^{k-1} \int_c^d \prod_{i=1}^k Y_i(t) \, dG(t) - \prod_{i=1}^k \int_c^d Y_i(t) \, dG(t)$$

$$\leq \left(\int_c^d dG(t) \right)^{k-1} \int_c^d \prod_{i=1}^k (2q_i - Y_i(t)) \, dG(t) - \prod_{i=1}^k \int_c^d (2q_i - Y_i(t)) \, dG(t)$$

hold, where all $\mathbf{Y}$ *components are non-negative.*

Proof $f(\mathbf{y}) = y_1 \cdots y_k$ is a FWNDI of second and third orders for $\mathbf{y} \in \mathbb{R}_*^k$. Now, applying Theorems 3.3.2 and 3.3.3 and get desired Corollary 3.3.4. □

Remark 3.3.3 By putting $j = 1$, Corollary 3.3.4 gives a special case as Corollary 3.3 (i) of [96].

Theorem 3.3.5 *Let* $G \in BV[c, d]$ *such that (3.3.1) valid and* f *be a continuous FWNDI of third order in the interval* $[\mathbf{p}, \mathbf{q}] \subset \mathbb{R}^k$. *Let* $\mathbf{0} < \mathbf{c} < \mathbf{q} - \mathbf{p}$. *If* $\mathbf{Y} : [c, d] \to$

$[\mathbf{p}, \mathbf{q} - \mathbf{c}]$ *is a monotonic (either non-increasing or nondecreasing) and continuous map in every of j intervals (d_{i-1}, d_i), then the inequality*

$$\frac{\int_c^d f(\mathbf{Y}(t))\, dG(t)}{\int_c^d dG(t)} - f\left(\frac{\int_c^d \mathbf{Y}(t)\, dG(t)}{\int_c^d dG(t)}\right)$$
$$\leq \frac{\int_c^d f(\mathbf{c} + \mathbf{Y}(t))\, dG(t)}{\int_c^d dG(t)} - f\left(\frac{\int_c^d (\mathbf{c} + \mathbf{Y}(t))\, dG(t)}{\int_c^d dG(t)}\right) \quad (3.3.5)$$

holds.

Proof Using (3.3.4) for $\mathbf{h}_3 = \mathbf{c} = $ constant $\in \mathbb{R}^k$, since $\mathbf{y} \mapsto f(\mathbf{c} + \mathbf{y}) - f(\mathbf{y})$ is a FWNDI, now applying Theorem 3.3.2 and get desired Theorem 3.3.5. □

Remark 3.3.4

(i) If $\dfrac{\int_c^d (\mathbf{c} - \mathbf{Y}(t))\, dG(t)}{\int_c^d dG(t)} \in \mathbf{J}$ and $\forall\, y \in c_i \in (d_{i-1}, d_i)$ $(d_0 = c, d_j = d)$ we have either $G(y) \geq G(d)$ or $G(y) \leq G(c)$, then the inequality in (3.3.5) holds for reverse direction.
(ii) By putting $j = 1$, Theorem 3.3.5 gives a special case as Theorem 3.2 of [96].
(iii) We can obtain discrete version of Theorem 3.3.5, using technique as given in Corollary 1 of [100].

Remark 3.3.5 Theorem 3.3.1 and Theorem 3.3.2 both are special cases of the Theorems 3.3.3 and 3.3.5.

Corollary 3.3.6 *Let* $\mathbf{Y}$ *satisfies the assumptions of the Theorem 3.3.5, then inequalities*

$$0 \leq \left(\int_c^d dG(t)\right)^{k-1} \int_c^d \prod_{i=1}^k Y_i(t)\, dG(t) - \prod_{i=1}^k \int_c^d Y_i(t)\, dG(t)$$
$$\leq \left(\int_c^d dG(t)\right)^{k-1} \int_c^d \prod_{i=1}^k (c_i + Y_i(t))\, dG(t) - \prod_{i=1}^k \int_c^d (c_i + Y_i(t))\, dG(t)$$

hold, where all $\mathbf{Y}$ *components are non-negative.*

Proof $f(\mathbf{y}) = y_1 \cdots y_k$ is a FWNDI of second and third orders for $\mathbf{y} \in \mathbb{R}_*^k$. Now, applying Theorems 3.3.2, and 3.3.5 and get desired Corollary 3.3.6. □

Remark 3.3.6 By putting $j = 1$, Corollary 3.3.6 gives a special case as Corollary 3.3 (ii) of [96].

3.3.2 On Jensen-Mercer Type Inequalities

The next theorem is the generalization of Jensen-Mercer inequality using Jensen-Boas inequality, for this purpose using Theorem 3.3.2 for proving the following theorems:

Theorem 3.3.7 *Let $G \in BV[c, d]$ such that (3.3.1) valid and $\mathbf{Y} : [c, d] \to [\mathbf{0}, \mathbf{q}]$, $(\mathbf{q} > \mathbf{0})$ be a monotonic (either non-increasing or nondecreasing) and continuous map in every of j intervals (d_{i-1}, d_i). If f is a continuous FWNDI of third order in the interval $\mathbf{J} = [\mathbf{0}, 2\mathbf{q}] \subset \mathbb{R}^k$ and $J = \int_c^d dG(t) > 0$, then inequality*

$$\frac{1}{J} \int_c^d f(\mathbf{Y}(t)) \, dG(t) - f\left(\frac{1}{J} \int_c^d \mathbf{Y}(t) \, dG(t)\right)$$
$$\leq \frac{1}{J} \int_c^d f(2\mathbf{q} - \mathbf{Y}(t)) \, dG(t) - f\left(\frac{1}{J} \int_c^d (2\mathbf{q} - \mathbf{Y}(t)) \, dG(t)\right) \quad (3.3.6)$$

holds.

Proof Using (3.3.4) for $\mathbf{h}_3 = 2\mathbf{q} - 2\mathbf{y}$, since $\mathbf{y} \mapsto f(2\mathbf{q} - \mathbf{y}) - f(\mathbf{y})$ is a FWNDI of second order, i.e., it is a FWNDI. Now, applying Theorem 3.3.2 and get desired Theorem 3.3.7. □

Remark 3.3.7

(i) If $\frac{1}{J} \int_c^d (2\mathbf{q} - \mathbf{Y}(t)) \, dG(t) \in \mathbf{J}$ and $\forall \, y \in c_i \in (d_{i-1}, d_i)$ $(d_0 = c, d_j = d)$ we have either $G(y) \geq G(d)$ or $G(y) \leq G(c)$, then the inequality in (3.3.6) holds for reverse direction.
(ii) We can obtain discrete version of Theorem 3.3.7, using technique as given in Corollary 1 of [100].

Theorem 3.3.8 *Let $G \in BV[c, d]$ such that (3.3.1) valid and f be a continuous FWNDI of third order in the interval $[\mathbf{p}, \mathbf{q}] \subset \mathbb{R}^k$. Let $\mathbf{0} < \mathbf{c} < \mathbf{q}-\mathbf{p}$. If $\mathbf{Y} : [c, d] \to [\mathbf{p}, \mathbf{q} - \mathbf{c}]$ is a monotonic (either non-increasing or nondecreasing) and continuous map in every of j intervals (d_{i-1}, d_i) and $J = \int_c^d dG(t) > 0$, then the inequality*

$$\frac{1}{J} \int_c^d f(\mathbf{Y}(t)) \, dG(t) - f\left(\frac{1}{J} \int_c^d \mathbf{Y}(t) \, dG(t)\right)$$
$$\leq \frac{1}{J} \int_c^d f(\mathbf{c} - \mathbf{Y}(t)) \, dG(t) - f\left(\frac{1}{J} \int_c^d (\mathbf{c} - \mathbf{Y}(t)) \, dG(t)\right) \quad (3.3.7)$$

holds.

Proof Using (3.3.4) for $\mathbf{h}_3 = \mathbf{c} = $ constant $\in \mathbb{R}^k$, since $\mathbf{y} \mapsto f(\mathbf{c}+\mathbf{y}) - f(\mathbf{y})$ is a FWNDI, now applying Theorem 3.3.2 and get desired Theorem 3.3.8. □

Remark 3.3.8

(i) If $\dfrac{1}{J} \displaystyle\int_c^d (\mathbf{c} - \mathbf{Y}(t))\, dG(t) \in \mathbf{J}$ and $\forall\ y \in c_i \in (d_{i-1}, d_i)$ $(d_0 = c, d_j = d)$ we have either $G(y) \geq G(d)$ or $G(y) \leq G(c)$, then the inequality in (3.3.7) holds for reverse direction.

(ii) We can obtain discrete version of Theorem 3.3.8, using technique as given in Corollary 1 of [100].

Chapter 4
Popoviciu and Čebyšev-Popoviciu Type Identities and Inequalities

> *All analysts spend half their time hunting through the literature for inequalities which they want to use and cannot prove.*
>
> —G. H. Hardy

4.1 Linear Inequalities for Higher Order ∇-Convex and Completely Monotonic Functions

The main aim of this section is to extend the definitions of ∇-convex and completely monotonic functions for two variables. We would construct some examples and applications of completely monotonic functions. In present section, some general identities of Popoviciu type for discrete case for sums $\sum_{i=1}^{M} \sum_{j=1}^{N} p_{ij} f(x_i, y_j)$ and $\sum_{i=1}^{M} \sum_{j=1}^{N} p_{ij} a_{ij}$ have been deduced for function and sequence involving higher order ∇ operator respectively. Then by applying obtained identities, positivity of these expressions will be characterised for higher order ∇-convex functions. Some general identities of Popoviciu type for integral $\int \int P(x, y) f(x, y) dx\, dy$ for differentiable function of higher order with two variables will be deduced by three different methods, then by applying obtained identities, positivity of these expressions will be characterised for higher order ∇-convex and completely monotonic functions. These identities and inequalities would be a generalization of several established results. Some applications in terms of generalized Cauchy means and exponential convexity will also be provided.

Some contents of the current section have been published in years 2020 and 2021 see [88, 125, 126].

We start from some definitions and preliminaries then we would like to discuss discrete identities of two variables function $f(x_i, y_j)$ and sequence a_{ij} involving higher order ∇ operator. Using obtained identities we derive various significant results. We will also discuss the characterisation of Popoviciu type positivity of these discrete sums involving ∇-convex functions.

Over past few decades the notion of completely monotonic functions has gained popularity among researchers in analysis and other related fields due to their interesting properties (see [104]) and higher applicability (see [59]). As it is evident from the following lines taken from a paper with the title "Completely monotone functions: a digest" [128] written by Milan Merkle. He writes "A brief search in MathSciNet reveals total of 286 items that mention this class of functions in the title from 1932 till the end of the year 2011; 98 of them have been published since the beginning of 2006". After some preliminaries, we would like to construct some examples and applications of completely monotonic functions. We would get general integral identity for one variable of Popoviciu type and its positivity for higher order ∇-convex and completely monotonic functions which is extracted from [126]. We will give general integral identity by three different methods of higher order differentiable functions of two variables of Popoviciu type. After getting these identities, we would also discuss the characterisation of Popoviciu type positivity of these general integrals involving ∇-convex and completely monotonic functions. These identities and inequalities would be generalization of several established results. We would give new generalized mean value theorems of Lagrange and Cauchy type and will also discuss exponential convexity by the support of different examples.

Let us recall, few useful definitions and significant results regarding the convex functions extracted from [158] (see also [97]) and also recall a definition from [135]. Throughout the chapter I and J are interval in $\mathbb{R}$ and m, n, M, N are natural numbers.

Definition 4.1.1 Let $E = \{x_1, x_2, \ldots, x_M\} \subset \mathbb{R}$. A function $f : E \to \mathbb{R}$ is known as discrete m-convex function if $[x_i, \ldots, x_{i+m}; f] \geq 0$ holds $\forall$ $(m+1)$ different points $x_i, \ldots, x_{i+m} \in E$ for $i \in \{0, 1, \ldots, m\}$.

Here, we present extension of previously mentioned Definitions 1.0.4 and 3.0.3 for order (m, n). For the sake of this purpose we use interval $I \times J = [a, b] \times [c, d] \subset \mathbb{R}^2$.

Definition 4.1.2 Let $f : I \times J \to \mathbb{R}$, be a function, then (m, n)-*divided difference* or *divided difference of* (m, n)*th order* of f at distinct points $x_i, \ldots, x_{i+m} \in I$, $y_j, \ldots, y_{j+n} \in J$ for some $i, j \in \mathbb{N}$, is stated as

$$\Delta_{(m,n)} f(x_i, y_j) = [x_i, \ldots, x_{i+m}; [y_j, \ldots, y_{j+n}; f]].$$

Definition 4.1.3 The finite difference of a function $f : I \times J \to \mathbb{R}$ of order (m, n), where $h, k \in \mathbb{R}$ and $x \in I, y \in J$, is stated as

$$\Delta_{h,k}^{m,n} f(x, y) = \Delta_h^m(\Delta_k^n f(x, y)) = \Delta_k^n(\Delta_h^m f(x, y))$$

$$= \sum_{i=0}^{m} \sum_{j=0}^{n} (-1)^{m-i+n-j} \binom{m}{i} \binom{n}{j} f(x + ih, y + jk),$$

4.1 Linear Inequalities for Higher Order ∇-Convex and Completely...

where $x+ih \in I$, $y+jk \in J$ for $i \in \{0, 1, \ldots, m\}$ and $j \in \{0, 1, \ldots, n\}$. Moreover, a function $f : I \times J \to \mathbb{R}$ is known as (m, n)-convex, if following condition holds $\Delta_{h,k}^{m,n} f(x, y) \geq 0\, \forall\, x \in I, y \in J$.

Definition 4.1.4 Finite difference and Divided difference of (m, n)th order, of a sequence (a_{ij}) are stated as $\Delta^{m,n} a_{ij} = \Delta_{1,1}^{m,n} f(x_i, y_j)$ and $\Delta_{(m,n)} a_{ij} = \Delta_{(m,n)} f(x_i, y_j)$ respectively, where $i \in \{1, \ldots, m\}$, $j \in \{1, \ldots, n\}$. If $x_i = i$, $y_j = j$, then function f is stated as $f : \{1, \ldots, m\} \times \{1, \ldots, n\} \to \mathbb{R}$ which is $f(i, j) = a_{ij}$. Moreover, a sequence (a_{ij}) is called a (m, n)-convex, if following condition holds $\Delta^{m,n} a_{ij} \geq 0$ for $m, n \geq 0$ and $i, j \in \{1, 2, 3, \ldots\}$.

Definition 4.1.5 A function $f : I \times J \to \mathbb{R}$, known as *convex of (m, n)th order* or (m, n)-*convex*, if $\forall$ different elements $x_i, \ldots, x_{i+m} \in I$ and $y_j, \ldots, y_{j+n} \in J$ we have $\Delta_{(m,n)} f(x_i, y_j) \geq 0$.

Further that the f is convex of (m, n)th order iff $f_{(m,n)} \geq 0$, if partial derivative $\frac{\partial^{m+n} f}{\partial x^m \partial y^n}$ denoted by $f_{(m,n)}$ and exists.

Definition 4.1.6 Let $E = \{x_1, x_2, \ldots, x_M\}$, $F = \{y_1, y_2, \ldots, y_N\} \subset \mathbb{R}$. A function $f : E \times F \to \mathbb{R}$, is known as discrete (m, n)-convex if inequality $[x_i, \ldots, x_{i+m}; [y_j, \ldots, y_{j+n}; f]] \geq 0$, holds $\forall$ $(m + 1)$ different points $x_i, \ldots, x_{i+m} \in E$ and $(n + 1)$ different points $y_j, \ldots, y_{j+n} \in F$.

Further that in this chapter, throughout we would use the following notations, where $I \times J \subset \mathbb{R} \times \mathbb{R}$. For some real sequence (a_m), $m \in \mathbb{N}$ and $n \in \{2, 3, \ldots\}$:

$$\nabla^{(1)} a_m = \nabla a_m = a_m - a_{m+1}, \qquad \nabla^{(n)} a_m = \nabla(\nabla^{(n-1)} a_m).$$

Also for m different real numbers x_i, $i \in \{1, 2, \ldots, m\}$ and $n \geq 0$:

$$(x_k - x_i)^{\{n+1\}} = (x_k - x_i)(x_{k-1} - x_i) \cdots (x_{k-n} - x_i), \qquad (x_k - x_i)^{\{0\}} = 1.$$

Definition 4.1.7 A function $f : I \times J \to \mathbb{R}$ is known as $(m, n) - \nabla$-convex if inequality $\nabla_{(m,n)} f(x_i, y_j) = (-1)^{m+n} \Delta_{(m,n)} f(x_i, y_j) \geq 0$, holds $\forall$ different points $x_i, \ldots, x_{i+m} \in I$, $y_j, \ldots, y_{j+n} \in J$.

Definition 4.1.8 A function $f : I \to \mathbb{R}$ is known as completely monotonic (or totally monotonic) of order m or m-completely monotonic if all its derivatives $f^{(i)}$ exist and satisfy

$$(-1)^i f^{(i)}(x) \geq 0, \quad x \in (0, \infty); \quad i \in \{0, 1, \ldots, m\}.$$

Definition 4.1.9 A function $f : I \times J \to \mathbb{R}$ is known as completely monotonic of order (m, n) or (m, n)-completely monotonic if all its $f_{(i,j)}$ partial derivatives exist and satisfy the condition below:

$$(-1)^{(i+j)} f_{(i,j)}(x, y) \geq 0, \quad x, y, \in (0, \infty); \quad j \in \{0, 1, \ldots, n\}, \; i \in \{0, 1, \ldots, m\}.$$

Remark 4.1.1 It is simple to observe that the notions of completely monotonic function of order m and (m, n) are generalized notions of m-∇-convex function and (m, n)-∇-convex function respectively if there exists differentiability.

Examples of Completely Monotonic Functions

In present subsection, we would use variety of classes of completely monotonic function $F = \{f_v : v \in I \subset \mathbb{R}\}$ and construct examples of completely monotonic function.

Example 4.1.1 Let a family of functions $F_1 = \{\psi_v : \mathbb{R} \to \mathbb{R}_+ | v \in \mathbb{R}_+\}$ which is stated as

$$\psi_v(x) = \frac{e^{-vx}}{v^m}.$$

Since $(-1)^i \frac{d^i}{dx^i} \psi_v(x) > 0$ for $i \in \{0, 1, \ldots, m\}$, therefore the function ψ_v is m-completely monotonic on $\mathbb{R}$, for every $v \in \mathbb{R}_+$.

Example 4.1.2 Let a family of functions $F_2 = \{\phi_v : \mathbb{R}_+ \to \mathbb{R} | v \in \mathbb{R}_+\}$ which is stated as

$$\phi_v(x) = \begin{cases} \frac{v^{-x}}{(\ln v)^m}, & v \neq 1 \\ \frac{(-1)^i x^m}{m!}, & v = 1. \end{cases}$$

Since $(-1)^i \frac{d^i}{dx^i} \phi_v(x) \geq 0$ for $i \in \{0, 1, \ldots, m\}$, therefore the function ϕ_v is m-completely monotonic on $\mathbb{R}_+$ for every $v \in \mathbb{R}_+, x \geq 0$.

Remark 4.1.2 Other examples of completely monotonic functions include:

(i) $f(x) = c$ (a nonnegative real constant), $\forall x \in \mathbb{R}$.
(ii) $f(x) = \frac{1}{(x + \alpha^2)^\beta}$, $\alpha \geq 0$, $\beta \geq 0$, $x > 0$.
(iii) $f(x) = -\ln x$ $\forall x \in \mathbb{R}_*$.

Let us give brief description of two subsections about discrete case as follow, after introduction, preliminaries and examples in the next two subsections, we will get identities for the sums $\sum_{i=1}^{M} \sum_{j=1}^{N} p_{ij} f(x_i, y_j)$ and $\sum_{i=1}^{M} \sum_{j=1}^{N} p_{ij} a_{ij}$ for two dimension involving higher order ∇ operator and investigate the inequality $\sum_{i=1}^{M} \sum_{j=1}^{N} p_{ij} f(x_i, y_j) \geq 0$ for ∇-convex functions of order (m, n) for two variables.

4.1.1 Discrete Identity for Two Dimensional Sequences

Under the given heading, we would consider a discrete sequence of two dimension. Firstly, we will get identities for sequence $\sum_{i=1}^{M} \sum_{j=1}^{N} p_{ij} a_{ij}$ which involve higher

4.1 Linear Inequalities for Higher Order ∇-Convex and Completely...

order ∇ operator. Further that we can split this sequence into two sequences as a special case by using $a_{ij} = a_i b_j$.

In the paper [130] the following result for a real sequence (a_M) was proved:

Theorem 4.1.1 *Let $p_i \in \mathbb{R}$ for $i \in \{1, 2, \ldots, M\}$, then the following identity for any real sequence (a_M) holds:*

$$\sum_{i=1}^{M} p_i a_i = \sum_{k=0}^{m-1} \frac{1}{k!} \nabla^{(k)} a_{M-k} \sum_{i=1}^{M-k} (M-i)^{\{k\}} p_i$$
$$+ \frac{1}{(m-1)!} \sum_{k=1}^{M-m} \left(\sum_{i=1}^{k} (m+k-1-i)^{\{m-1\}} p_i \right) \nabla^{(m)} a_k. \quad (4.1.1)$$

Now we would like to obtain the following theorem for a real sequence (a_{MN}).

Theorem 4.1.2 *Let $p_{ij} \in \mathbb{R}$ and a_{ij} be a sequence, where $i \in \{1, 2, \ldots, M\}$ and $j \in \{1, 2, \ldots, N\}$, then*

$$\sum_{i=1}^{M} \sum_{j=1}^{N} p_{ij} a_{ij}$$
$$= \sum_{k=0}^{n-1} \sum_{t=0}^{m-1} \sum_{s=1}^{M-t} \sum_{r=1}^{N-k} p_{sr} \frac{(M-s)^{\{t\}}}{t!} \frac{(N-r)^{\{k\}}}{k!} \nabla_{(t,k)} a_{(M-t, N-k)}$$
$$+ \sum_{k=0}^{n-1} \sum_{t=1}^{M-m} \sum_{s=1}^{t} \sum_{r=1}^{N-k} p_{sr} \frac{(m+t-1-s)^{\{m-1\}}}{(m-1)!} \frac{(N-r)^{\{k\}}}{k!} \nabla_{(m,k)} a_{(t, N-k)}$$
$$+ \sum_{k=1}^{N-n} \sum_{t=0}^{m-1} \sum_{s=1}^{M-t} \sum_{r=1}^{k} p_{sr} \frac{(M-s)^{\{t\}}}{t!} \frac{(n+k-1-r)^{\{n-1\}}}{(n-1)!} \nabla_{(t,n)} a_{(M-t, k)}$$
$$+ \sum_{k=1}^{N-n} \sum_{t=1}^{M-m} \sum_{s=1}^{t} \sum_{r=1}^{k} p_{sr} \frac{(m+t-1-s)^{\{m-1\}}}{(m-1)!} \frac{(n+k-1-r)^{\{n-1\}}}{(n-1)!} \nabla_{(m,n)} a_{(t,k)}$$

(4.1.2)

holds.

Proof We have

$$\sum_{i=1}^{M} \sum_{j=1}^{N} p_{ij} a_{ij} = \sum_{i=1}^{M} \left(\sum_{j=1}^{N} q_j A_i \right),$$

where $p_{ij} = q_j$ and $A_i : j \mapsto a_{(i,j)}$. Using (4.1.1) in the inner sum we get

$$\sum_{i=1}^{M}\sum_{j=1}^{N} p_{ij}a_{ij} = \sum_{i=1}^{M}\sum_{k=0}^{n-1}\frac{1}{k!}\nabla_{(k)}A_{i(N-k)}\left(\sum_{j=1}^{N-k}q_j(N-j)^{\{k\}}\right)$$

$$+ \sum_{i=1}^{M}\sum_{k=1}^{N-n}\frac{1}{(n-1)!}\nabla_{(n)}A_{i(k)}\left(\sum_{j=1}^{k}q_j(n+k-1-j)^{\{n-1\}}\right)$$

$$= \sum_{k=0}^{n-1}\left(\sum_{i=1}^{M}\frac{1}{k!}\nabla_{(k)}A_{i(N-k)}\left(\sum_{j=1}^{N-k}q_j(N-j)^{\{k\}}\right)\right)$$

$$+ \sum_{k=1}^{N-n}\left(\sum_{i=1}^{M}\frac{1}{(n-1)!}\nabla_{(n)}A_{i(k)}\left(\sum_{j=1}^{k}q_j(n+k-1-j)^{\{n-1\}}\right)\right)$$

$$= \sum_{k=0}^{n-1}\left(\sum_{i=1}^{M}w_i B_i\right) + \sum_{k=1}^{N-n}\left(\sum_{i=1}^{M}v_i C_i\right),$$

where $w_i = \sum_{j=1}^{N-k}q_j(N-j)^{\{k\}} = \sum_{j=1}^{N-k}p_{ij}(N-j)^{\{k\}}$, $v_i = \sum_{j=1}^{k}q_j(n+k-1-j)^{\{n-1\}}$, $B_i = \frac{1}{k!}\nabla_{(k)}A_{i(N-k)}$, and $C_i = \frac{1}{(n-1)!}\nabla_{(n)}A_{i(k)}$.

Using again (4.1.1) in inner sums, then we have

$$\sum_{i=1}^{M}\sum_{j=1}^{N} p_{ij}a_{ij} = \sum_{k=0}^{n-1}\sum_{r=0}^{m-1}\frac{1}{r!}\nabla_{(r)}B_{(M-r)}\left(\sum_{i=1}^{M-r}w_i(M-i)^{\{r\}}\right)$$

$$+ \sum_{k=0}^{n-1}\sum_{r=1}^{M-m}\frac{1}{(m-1)!}\nabla_{(m)}B_{(r)}\left(\sum_{i=1}^{r}w_i(m+r-1-i)^{\{m-1\}}\right)$$

$$+ \sum_{k=1}^{N-n}\sum_{t=0}^{m-1}\frac{1}{t!}\nabla_{(t)}C_{(M-t)}\left(\sum_{i=1}^{M-t}v_i(M-i)^{\{t\}}\right)$$

$$+ \sum_{k=1}^{N-n}\sum_{t=1}^{M-m}\frac{1}{(m-1)!}\nabla_{(m)}C_{(t)}\left(\sum_{i=1}^{t}v_i(m+t-1-i)^{\{m-1\}}\right)$$

$$= \sum_{k=0}^{n-1}\sum_{r=0}^{m-1}\sum_{i=1}^{M-r}\sum_{j=1}^{N-k}p_{ij}\frac{(M-i)^{\{r\}}}{r!}\frac{(N-j)^{\{k\}}}{k!}\nabla_{(r,k)}a_{(M-r,N-k)}$$

$$+ \sum_{k=0}^{n-1}\sum_{r=1}^{M-m}\sum_{i=1}^{r}\sum_{j=1}^{N-k}p_{ij}\frac{(m+r-1-i)^{\{m-1\}}}{(m-1)!}\frac{(N-j)^{\{k\}}}{k!}$$

4.1 Linear Inequalities for Higher Order ∇-Convex and Completely...

$$\times \nabla_{(m,k)} a_{(r,N-k)}$$

$$+ \sum_{k=1}^{N-n} \sum_{t=0}^{m-1} \sum_{i=1}^{M-t} \sum_{j=1}^{k} p_{ij} \frac{(M-i)^{\{t\}}}{t!} \frac{(n+k-1-j)^{\{n-1\}}}{(n-1)!}$$

$$\times \nabla_{(t,n)} a_{(M-t,k)}$$

$$+ \sum_{k=1}^{N-n} \sum_{t=1}^{M-m} \sum_{i=1}^{t} \sum_{j=1}^{k} p_{ij} \frac{(m+t-1-i)^{\{m-1\}}}{(m-1)!} \frac{(n+k-1-j)^{\{n-1\}}}{(n-1)!}$$

$$\times \nabla_{(m,n)} a_{(t,k)}.$$

If we change $i \to s$, $j \to r$ in all sums and put $r \to t$ in first and second sums, then we obtain the required identity (4.1.2). □

Remark 4.1.3 If we simply put $a_{ij} = a_i b_j$ in Theorem 4.1.2, then we obtain similar result for two a_i and b_j sequences as below.

Corollary 4.1.3 Let $p_{ij} \in \mathbb{R}$, $b : j \mapsto b_j$ and $a : i \mapsto a_i$ be two sequences, where $j \in \{1, 2, \ldots, N\}$ and $i \in \{1, 2, \ldots, M\}$, then

$$\sum_{i=1}^{M} \sum_{j=1}^{N} p_{ij} a_i b_j$$

$$= \sum_{k=0}^{n-1} \sum_{t=0}^{m-1} \sum_{s=1}^{M-t} \sum_{r=1}^{N-k} p_{sr} \frac{(M-s)^{\{t\}}}{t!} \nabla_{(t)} a_{(M-t)} \frac{(N-r)^{\{k\}}}{k!} \nabla_{(k)} b_{(N-k)}$$

$$+ \sum_{k=0}^{n-1} \sum_{t=1}^{M-m} \sum_{s=1}^{t} \sum_{r=1}^{N-k} p_{sr} \frac{(m+t-1-s)^{\{m-1\}}}{(m-1)!} \nabla_{(m)} a_{(t)} \frac{(N-r)^{\{k\}}}{k!} \nabla_{(k)} b_{(N-k)}$$

$$+ \sum_{k=1}^{N-n} \sum_{t=0}^{m-1} \sum_{s=1}^{M-t} \sum_{r=1}^{k} p_{sr} \frac{(M-s)^{\{t\}}}{t!} \nabla_{(t)} a_{(M-t)} \frac{(n+k-1-r)^{\{n-1\}}}{(n-1)!} \nabla_{(n)} b_{(k)}$$

$$+ \sum_{k=1}^{N-n} \sum_{t=1}^{M-m} \sum_{s=1}^{t} \sum_{r=1}^{k} p_{sr} \frac{(m+t-1-s)^{\{m-1\}}}{(m-1)!} \nabla_{(m)} a_{(t)} \frac{(n+k-1-r)^{\{n-1\}}}{(n-1)!} \nabla_{(n)} b_{(k)}.$$

4.1.2 Discrete Identity and Inequality for Functions of Two Variables

Under present heading, we would consider a discrete function of two variables that is defined in the interval $I \times J \subset \mathbb{R} \times \mathbb{R}$. Firstly, we will get identities for function $\sum_{i=1}^{M} \sum_{j=1}^{N} p_{ij} f(x_i, y_j)$ in which involves higher order ∇ operator. Moreover, we can split this function into two functions as a special case by

$f(x_i, y_j) = f(x_i)g(y_j)$ and also consider necessary and sufficient conditions of Theorem 4.1.5 for Popoviciu type characterisation of positivity of sums for discrete function of two variables $\sum_{i=1}^{M} \sum_{j=1}^{N} p_{ij} f(x_i, y_j) \geq 0$ holds, for every $(m, n) - \nabla$-convex function.

The following result was proved in [97] for the real function involving ∇ operator and it is a generalization of (4.1.1) which may be stated as:

Theorem 4.1.4 *Let p_i be real numbers for $i \in \{1, 2, \ldots, M\}$, where $M \geq m$. Let f be discrete function and x_i non mutual elements in the interval I for $i \in \{1, 2, \ldots, M\}$, then following identity holds:*

$$\sum_{i=1}^{M} p_i f(x_i) = \sum_{k=0}^{m-1} \nabla_{(k)} f(x_{M-k}) \left(\sum_{j=1}^{M-k} p_j (x_M - x_j)^{\{k\}} \right)$$
$$+ \sum_{k=1}^{M-m} \nabla_{(m)} f(x_k)(x_{m+k} - x_k) \left(\sum_{j=1}^{k} p_j (x_{m+k-1} - x_j)^{\{m-1\}} \right).$$
(4.1.3)

Now we are able to give our main general theorem for discrete function in two dimension.

Theorem 4.1.5 *Let $p_{ij} \in \mathbb{R}$ and $f : I \times J \to \mathbb{R}$ be discrete function, where $i \in \{1, 2, \ldots, M\}$ and $j \in \{1, 2, \ldots, N\}$, then*

$$\sum_{i=1}^{M} \sum_{j=1}^{N} p_{ij} f(x_i, y_j)$$

$$= \sum_{k=0}^{n-1} \sum_{t=0}^{m-1} \sum_{s=1}^{M-t} \sum_{r=1}^{N-k} p_{sr}(y_N - y_r)^{\{k\}} (x_M - x_s)^{\{t\}} \nabla_{(t,k)} f(x_{M-t}, y_{N-k})$$

$$+ \sum_{k=0}^{n-1} \sum_{t=1}^{M-m} \sum_{s=1}^{t} \sum_{r=1}^{N-k} p_{sr}(y_N - y_r)^{\{k\}} (x_{m+t-1} - x_s)^{\{m-1\}}$$
$$\times \nabla_{(m,k)} f(x_t, y_{N-k})(x_{m+t} - x_t)$$

$$+ \sum_{k=1}^{N-n} \sum_{t=0}^{m-1} \sum_{s=1}^{M-t} \sum_{r=1}^{k} p_{sr}(y_{n+k-1} - y_r)^{\{n-1\}} (x_M - x_s)^{\{t\}}$$
$$\times \nabla_{(t,n)} f(x_{M-t}, y_k)(y_{n+k} - y_k)$$

$$+ \sum_{k=1}^{N-n} \sum_{t=1}^{M-m} \sum_{s=1}^{t} \sum_{r=1}^{k} p_{sr}(y_{n+k-1} - y_r)^{\{n-1\}} (x_{m+t-1} - x_s)^{\{m-1\}} \nabla_{(m,n)} f(x_t, y_k) \times$$
$$\times (x_{m+t} - x_t)(y_{n+k} - y_k).$$
(4.1.4)

4.1 Linear Inequalities for Higher Order ∇-Convex and Completely...

holds, where $(x_i, y_j) \in I \times J$ are distinct points.

Proof We have

$$\sum_{i=1}^{M}\sum_{j=1}^{N} p_{ij} f(x_i, y_j) = \sum_{i=1}^{M}\left(\sum_{j=1}^{N} q_j G_i(y_j)\right),$$

where $p_{ij} = q_j$ and $G_i : y \mapsto f(x_i, y)$. Using (4.1.3) in the inner sum we get

$$\sum_{i=1}^{M}\sum_{j=1}^{N} p_{ij} f(x_i, y_j) = \sum_{i=1}^{M}\sum_{k=0}^{n-1} \nabla_{(k)} G_i(y_{N-k}) \left(\sum_{j=1}^{N-k} q_j (y_N - y_j)^{\{k\}}\right)$$

$$+ \sum_{i=1}^{M}\sum_{k=1}^{N-n} \nabla_{(n)} G_i(y_k)(y_{n+k} - y_k)$$

$$\times \left(\sum_{j=1}^{k} q_j (y_{n+k-1} - y_j)^{\{n-1\}}\right)$$

$$= \sum_{k=0}^{n-1}\left(\sum_{i=1}^{M} \nabla_{(k)} G_i(y_{N-k}) \left(\sum_{j=1}^{N-k} q_j (y_N - y_j)^{\{k\}}\right)\right)$$

$$+ \sum_{k=1}^{N-n}\left(\sum_{i=1}^{M} \nabla_{(n)} G_i(y_k)(y_{n+k} - y_k)\right.$$

$$\times \left.\left(\sum_{j=1}^{k} q_j (y_{n+k-1} - y_j)^{\{n-1\}}\right)\right)$$

$$= \sum_{k=0}^{n-1}\left(\sum_{i=1}^{M} w_i F(x_i)\right) + \sum_{k=1}^{N-n}\left(\sum_{i=1}^{M} v_i H(x_i)\right),$$

where $w_i = \sum_{j=1}^{N-k} q_j (y_N - y_j)^{\{k\}} = \sum_{j=1}^{N-k} p_{ij} (y_N - y_j)^{\{k\}}$, $v_i = \sum_{j=1}^{k} q_j (y_{n+k-1} - y_j)^{\{n-1\}}$, $F(x_i) = \nabla_{(k)} G_i(y_{N-k})$, and $H(x_i) = \nabla_{(n)} G_i(y_k)(y_{n+k} - y_k)$.

Using again (4.1.3) in the inner sums, then we have

$$\sum_{i=1}^{M}\sum_{j=1}^{N} p_{ij} f(x_i, y_j) = \sum_{k=0}^{n-1}\sum_{r=0}^{m-1} \nabla_{(r)} F(x_{M-r}) \left(\sum_{i=1}^{M-r} w_i (x_M - x_i)^{\{r\}} \right)$$

$$+ \sum_{k=0}^{n-1}\sum_{r=1}^{M-m} \nabla_{(m)} F(x_r)(x_{m+r} - x_r)$$

$$\times \left(\sum_{i=1}^{r} w_i (x_{m+r-1} - x_i)^{\{m-1\}} \right)$$

$$+ \sum_{k=1}^{N-n}\sum_{t=0}^{m-1} \nabla_{(t)} H(x_{M-t}) \left(\sum_{i=1}^{M-t} v_i (x_M - x_i)^{\{t\}} \right)$$

$$+ \sum_{k=1}^{N-n}\sum_{t=1}^{M-m} \nabla_{(m)} H(x_t)(x_{m+t} - x_t)$$

$$\times \left(\sum_{i=1}^{t} v_i (x_{m+t-1} - x_i)^{\{m-1\}} \right)$$

$$\sum_{i=1}^{M}\sum_{j=1}^{N} p_{ij} f(x_i, y_j) = \sum_{k=0}^{n-1}\sum_{r=0}^{m-1}\sum_{i=1}^{M-r}\sum_{j=1}^{N-k} p_{ij}(y_N - y_j)^{\{k\}}(x_M - x_i)^{\{r\}}$$

$$\times \nabla_{(r,k)} f(x_{M-r}, y_{N-k})$$

$$+ \sum_{k=0}^{n-1}\sum_{r=1}^{M-m}\sum_{i=1}^{r}\sum_{j=1}^{N-k} p_{ij}(y_N - y_j)^{\{k\}}(x_{m+r-1} - x_i)^{\{m-1\}}$$

$$\times \nabla_{(m,k)} f(x_r, y_{N-k})(x_{m+r} - x_r)$$

$$+ \sum_{k=1}^{N-n}\sum_{t=0}^{m-1}\sum_{i=1}^{M-t}\sum_{j=1}^{k} p_{ij}(y_{n+k-1} - y_j)^{\{n-1\}}(x_M - x_i)^{\{t\}}$$

$$\times \nabla_{(t,n)} f(x_{M-t}, y_k)(y_{n+k} - y_k)$$

$$+ \sum_{k=1}^{N-n}\sum_{t=1}^{M-m}\sum_{i=1}^{t}\sum_{j=1}^{k} p_{ij}(y_{n+k-1} - y_j)^{\{n-1\}}$$

$$\times (x_{m+t-1} - x_i)^{\{m-1\}} \nabla_{(m,n)} f(x_t, y_k)(x_{m+t} - x_t)(y_{n+k} - y_k).$$

4.1 Linear Inequalities for Higher Order ∇-Convex and Completely... 223

If we change $i \to s$, $j \to r$ in all sums and put $r \to t$ in first and second sums, then we obtain the required identity (4.1.4). □

Corollary 4.1.6 *Let $p_{ij} \in \mathbb{R}$ and function $f : I^2 \to \mathbb{R}$ be discrete, where $i, j \in \{1, 2, \ldots, M\}$, then following identity holds:*

$$\sum_{i=1}^{M} \sum_{j=1}^{M} p_{ij} f(x_i, y_j)$$

$$= \sum_{k=0}^{n-1} \sum_{t=0}^{m-1} \sum_{s=1}^{M-t} \sum_{r=1}^{M-k} p_{sr} (y_M - y_r)^{\{k\}} (x_M - x_s)^{\{t\}} \nabla_{(t,k)} f(x_{M-t}, y_{M-k})$$

$$+ \sum_{k=0}^{n-1} \sum_{t=1}^{M-m} \sum_{s=1}^{t} \sum_{r=1}^{M-k} p_{sr} (y_M - y_r)^{\{k\}} (x_{m+t-1} - x_s)^{\{m-1\}}$$

$$\times \nabla_{(m,k)} f(x_t, y_{M-k})(x_{m+t} - x_t)$$

$$+ \sum_{k=1}^{M-n} \sum_{t=0}^{m-1} \sum_{s=1}^{M-t} \sum_{r=1}^{k} p_{sr} (y_{n+k-1} - y_r)^{\{n-1\}} (x_M - x_s)^{\{t\}}$$

$$\times \nabla_{(t,n)} f(x_{M-t}, y_k)(y_{n+k} - y_k)$$

$$+ \sum_{k=1}^{M-n} \sum_{t=1}^{M-m} \sum_{s=1}^{t} \sum_{r=1}^{k} p_{sr} (y_{n+k-1} - y_r)^{\{n-1\}} (x_{m+t-1} - x_s)^{\{m-1\}} \nabla_{(m,n)} f(x_t, y_k) \times$$

$$\times (x_{m+t} - x_t)(y_{n+k} - y_k).$$

Remark 4.1.4 If we simply put $f(x_i, y_j) = f(x_i) g(y_j)$ in Theorem 4.1.5, then we obtain similar result for both f and g functions as below.

Corollary 4.1.7 *Let $p_{ij} \in \mathbb{R}$ and functions $f : I \to \mathbb{R}$ and $g : J \to \mathbb{R}$ be both discrete, where $i \in \{1, 2, \ldots, M\}$ and $j \in \{1, 2, \ldots, N\}$, then*

$$\sum_{i=1}^{M} \sum_{j=1}^{N} p_{ij} f(x_i) g(y_j)$$

$$= \sum_{k=0}^{n-1} \sum_{t=0}^{m-1} \sum_{s=1}^{M-t} \sum_{r=1}^{N-k} p_{sr} (x_M - x_s)^{\{t\}} \nabla_{(t)} f(x_{M-t})(y_N - y_r)^{\{k\}} \nabla_{(k)} g(y_{N-k})$$

$$+ \sum_{k=0}^{n-1} \sum_{t=1}^{M-m} \sum_{s=1}^{t} \sum_{r=1}^{N-k} p_{sr} (y_N - y_r)^{\{k\}}$$

$$\times \nabla_{(k)} g(y_{N-k})(x_{m+t-1} - x_s)^{\{m-1\}} \nabla_{(m)} f(x_t)(x_{m+t} - x_t)$$

$$+ \sum_{k=1}^{N-n}\sum_{t=0}^{m-1}\sum_{s=1}^{M-t}\sum_{r=1}^{k} p_{sr}(y_{n+k-1} - y_r)^{\{n-1\}}$$

$$\times \nabla_{(n)} g(y_k)(y_{n+k} - y_k)(x_M - x_s)^{\{t\}} \nabla_{(t)} f(x_{M-t})$$

$$+ \sum_{k=1}^{N-n}\sum_{t=1}^{M-m}\sum_{s=1}^{t}\sum_{r=1}^{k} p_{sr}(y_{n+k-1} - y_r)^{\{n-1\}} \nabla_{(n)} g(y_k)(y_{n+k} - y_k) \times$$

$$\times (x_{m+t-1} - x_s)^{\{m-1\}} \nabla_{(m)} f(x_t)(x_{m+t} - x_t),$$

holds, where $(x_i, y_j) \in I \times J$ *are distinct points.*

Now its time to present necessary and sufficient conditions of Theorem 4.1.5 for Popoviciu type characterisation of positivity of sums for discrete function of two variables involving $(m, n) - \nabla$-convex functions.

Theorem 4.1.8 *Let* $p_{ij} \in \mathbb{R}$ *and* $f : I \times J \to \mathbb{R}$ *be discrete function, where* $i \in \{1, 2, \ldots, M\}$, $j \in \{1, 2, \ldots, N\}$ *and* $I = \{x_{M-r}, x_{M-r+1}, \ldots, x_M\}$, $J = \{y_{N-k}, y_{N-k+1}, \ldots, y_N\}$ *and* $x_{M-r} < \cdots < x_M$, $y_{N-k} < \cdots < y_N$, *then the following inequality holds for all* $(m, n) - \nabla$-*convex function* f

$$\sum_{i=1}^{M}\sum_{j=1}^{N} p_{ij} f(x_i, y_j) \geq 0 \qquad (4.1.5)$$

iff

$$\sum_{s=1}^{M-t}\sum_{r=1}^{N-k} p_{sr}(y_N - y_r)^{\{k\}}(x_M - x_s)^{\{t\}} = 0, \quad \begin{array}{l} k \in \{0, 1, \cdots, n-1\} \\ t \in \{0, 1, \cdots, m-1\} \end{array}$$
(4.1.6)

$$\sum_{s=1}^{t}\sum_{r=1}^{N-k} p_{sr}(y_N - y_r)^{\{k\}}(x_{m+t-1} - x_s)^{\{m-1\}} = 0, \quad \begin{array}{l} k \in \{0, 1, \cdots, n-1\} \\ t \in \{1, 2, \cdots, M-m\} \end{array}$$
(4.1.7)

$$\sum_{s=1}^{M-t}\sum_{r=1}^{k} p_{sr}(y_{n+k-1} - y_r)^{\{n-1\}}(x_M - x_s)^{\{t\}} = 0, \quad \begin{array}{l} k \in \{1, 2, \cdots, N-n\} \\ t \in \{0, 1, \cdots, m-1\} \end{array}$$
(4.1.8)

$$\sum_{s=1}^{t}\sum_{r=1}^{k} p_{sr}(x_{m+t-1} - x_s)^{\{m-1\}}(y_{n+k-1} - y_r)^{\{n-1\}} \geq 0, \quad \begin{array}{l} k \in \{1, 2, \cdots, N-n\} \\ t \in \{1, 2, \cdots, M-m\}. \end{array}$$
(4.1.9)

4.1 Linear Inequalities for Higher Order ∇-Convex and Completely...

Proof If (4.1.6), (4.1.7) and (4.1.8) hold, then 1st, 2nd and 3rd terms are zero in (4.1.4), then by using (4.1.9) we obtain the required inequality (4.1.5).

Conversely, if substitute the following functions in (4.1.5). Then we obtain the required equality (4.1.6)

$$f_1(x_s, y_r) = (y_N - y_r)^{\{k\}}(x_M - x_s)^{\{t\}} \quad \text{and} \quad f_2 = -f_1$$

for $0 \leq t \leq m-1$ and $0 \leq k \leq n-1$ such that $\nabla_{(m,n)} f_j \geq 0$, $j \in \{1, 2\}$

$$\sum_{s=1}^{M-t} \sum_{r=1}^{N-k} p_{sr}(y_N - y_r)^{\{k\}}(x_M - x_s)^{\{t\}} = 0, \quad 0 \leq k \leq n-1; \quad 0 \leq t \leq m-1.$$

In the similar manner, if take the following functions in (4.1.5) for $0 \leq k \leq n-1$ and $1 \leq t \leq M-m$

$$f_3(x_s, y_r) = \begin{cases} (y_N - y_r)^{\{k\}}(x_{m+t-1} - x_s)^{\{m-1\}}, & s < t \\ 0, & s \geq t \end{cases}$$

$$f_4 = -f_3$$

such that $\nabla_{(m,n)} f_j \geq 0$, $j \in \{3, 4\}$, we obtain the equality (4.1.7), i.e.,

$$\sum_{s=1}^{t} \sum_{r=1}^{k} p_{sr}(y_N - y_r)^{\{k\}}(x_{m+t-1} - x_s)^{\{m-1\}} = 0, \quad 0 \leq k \leq n-1; \quad 1 \leq t \leq M-m.$$

Similarly, if take the following functions in (4.1.5) for $0 \leq t \leq m-1$ and $1 \leq k \leq N-n$

$$f_5(x_s, y_r) = \begin{cases} (y_{n+k-1} - y_r)^{\{n-1\}}(x_M - x_s)^{\{t\}}, & r < k \\ 0, & r \geq k \end{cases}$$

$$f_6 = -f_5$$

such that $\nabla_{(m,n)} f_j \geq 0$, $j \in \{5, 6\}$, we obtain the equality (4.1.8) as above, i.e.,

$$\sum_{s=1}^{M-t} \sum_{r=1}^{k} p_{sr}(y_{n+k-1} - y_r)^{\{n-1\}}(x_M - x_s)^{\{t\}} = 0, \quad 0 \leq t \leq m-1; \quad 1 \leq k \leq N-n.$$

We get the last inequality (4.1.9) by considering the following function in (4.1.5) for $1 \leq t \leq M - m$ and $1 \leq k \leq N - n$

$$f_7(x_s, y_r) = \begin{cases} (x_{m+t-1} - x_s)^{\{m-1\}}(y_{n+k-1} - y_r)^{\{n-1\}}, & s < t, \quad r < k; \\ 0, & s \geq t \quad \text{or} \quad r \geq k. \end{cases}$$

□

Remark 4.1.5 Similar remark as given in Remark 4.1.4 and for sequence a_{ij} and split this sequence into two sequences as a special case by using $a_{ij} = a_i b_j$ also holds for this result.

Now, let us briefly explain the format of the following subsections: Firstly we have discussed the discrete case followed by integral case, then in the third subsection we present identity for the integral $\int P(x) f(x) dx$ which involves function of higher order derivatives and also discuss necessary and sufficient conditions of this identity $\forall\, (m+1) - \nabla$-convex function in which $\int P(x) f(x) dx \geq 0$ holds and only necessary condition holds for $(m+1)$-completely monotonic function (see also [126]). In the fourth subsection, we will give general integral identity by three different methods for higher order differentiable function of two variables of Popoviciu type and also deduce positivity of above obtained case for $(M+1, N+1) - \nabla$-convex functions and $(M+1, N+1)$-completely monotonic functions for two variables, then consider an identity of linear functional $\Lambda(f)$ in double integral. In the fifth subsection, we would state some mean value theorems of Lagrange and Cauchy types and consider the nonnegative functional $\Lambda(f)$ and apply this on exponentially convex functions $\psi^{(q)}$ of certain type, and give some properties. At the end of the first section, construct examples of completely monotonic and exponentially convex functions by using various classes of functions.

4.1.3 Integral Identity and Inequality for Functions of One Variable

We give an integral identity of one variable that is analogous to the result of Theorem 4.1.4.

Theorem 4.1.9 *Let $f \in C^{(m+1)}$ and both $p, f : I \to \mathbb{R}$ be integrable functions, then*

$$\int_a^b p(x) f(x) dx = \sum_{i=0}^{m} (-1)^i f^{(i)}(b) \left(\int_a^b p(x) \frac{(b-x)^i}{i!} dx \right)$$
$$+ \int_a^b (-1)^{m+1} f^{(m+1)}(s) \left(\int_a^s p(x) \frac{(s-x)^m}{m!} dx \right) ds.$$

4.1 Linear Inequalities for Higher Order ∇-Convex and Completely...

The following two theorems are proved and extracted from [126] and the following first theorem is generalized form of result (see [151, pp. 121–122]) and (see also [97]).

Theorem 4.1.10 *Let assumption of the Theorem 4.1.9 be valid, then following inequality holds*

$$\int_a^b f(x)p(x)dx \geq 0$$

for all $(m+1) - \nabla$*-convex function f, iff*

$$\int_a^b p(x)\frac{(b-x)^i}{i!}dx = 0, \quad i \in \{1, 2, \ldots, m\};$$

$$\int_a^s p(x)\frac{(s-x)^m}{m!}dx \geq 0, \quad \forall s \in [a,b].$$

Theorem 4.1.11 *Let assumption of the Theorem 4.1.9 be valid, then following inequality holds*

$$\int_a^b p(x)f(x)dx \geq 0$$

for each completely monotonic function f of order $m+1$ if

$$\int_a^b p(x)\frac{(b-x)^i}{i!}dx = 0, \quad i \in \{1, 2, \ldots, m\};$$

$$\int_a^s p(x)\frac{(s-x)^m}{m!}dx \geq 0, \quad \forall s \in [a,b].$$

Remark 4.1.6 Previous result also holds for every exponentially convex function [81]. Moreover, each completely monotonic is log-convex so the stated result also holds for every log-convex function [128].

4.1.4 Integral Identity and Inequality for Functions of Two Variables

Under above heading, we can suppose function in x and y variables which is stated in the interval $I \times J = [a,b] \times [c,d]$. Moreover, $m, n, M, N \in \mathbb{N} \cup \{0\}$ throughout this subsection, and useful notations are:

$$f_{(0,0)} = f \text{ (function itself)}, \quad f_{(1,0)} = \frac{\partial f}{\partial x}, \quad f_{(0,1)} = \frac{\partial f}{\partial y}, \quad f_{(1,1)} = \frac{\partial^2 f}{\partial x \partial y} = \frac{\partial^2 f}{\partial y \partial x},$$

$$f_{(i,j)} = \frac{\partial^{i+j} f}{\partial x^i \partial y^j} = \frac{\partial^{i+j} f}{\partial y^j \partial x^i}.$$

Let $f : I \times J \to \mathbb{R}$ and $p_{i,j}$ be both integrable functions then we also introduce following notations as:

$$p_{(1,1)}(x, y) = \int_x^b \int_y^d p(s, t) dt ds, \quad p_{(m+1,n+1)}(x, y) = \int_x^b \int_y^d p_{(m,n)}(s, t) dt ds$$

and $p_{(m+1,n+1)}(x, y) = \int_x^b \int_y^d p(s, t) \frac{(y-t)^n}{n!} \frac{(x-s)^m}{m!} dt ds.$

To prove our main upcoming theorem, we will have to need Two-dimensional Induction method, for this purpose we required following statement:

Definition 4.1.10 Let $G(M, N)$ denotes a statement involving two variables M and N. Suppose

(i) $G(0, 0)$ is true.
(ii) If $G(m, 0)$ is true for some integer $m \geq 0$, then $G(m + 1, 0)$ is also true.
(iii) If $G(m, n)$ holds for some integers $m, n \geq 0$, then $G(m, n + 1)$ is also true.

Therefore, $G(M, N)$ is true $\forall$ integers $M, N \geq 0$. This process is called Two-dimensional Induction.

Here we need to write following lemma which is a special case of Theorem 4.1.12.

Lemma 4.1.1 Let f has continuous partial derivatives $f_{(0,1)}, f_{(1,0)}, f_{(1,1)}$ and $f, p : I \times J \to \mathbb{R}$ be both integrable functions, then

$$\int_c^d \int_a^b p(x, y) f(x, y) dx dy = \int_d^c \int_b^a p(s, t) f(s, t) ds dt$$

$$= p_{(1,1)}(b, d) f_{(0,0)}(b, d)$$

$$+ \int_b^a p_{(1,1)}(s, d) f_{(1,0)}(s, d) ds$$

$$+ \int_d^c p_{(1,1)}(b, t) f_{(0,1)}(b, t) dt$$

$$+ \int_d^c \int_b^a p_{(1,1)}(s, t) f_{(1,1)}(s, t) ds dt.$$

4.1 Linear Inequalities for Higher Order ∇-Convex and Completely...

Now we recall a result from [97] which would be helpful to prove our upcoming main result:

Lemma 4.1.2 *Let f has continuous partial derivatives $f_{(i,j)}$ and $p, f : I \times J \to \mathbb{R}$ be both integrable functions, where $i \in \{0, 1, \ldots, M+1\}$ and $j \in \{0, 1, \ldots, N+1\}$, then*

$$\int_a^b \int_c^d p(x,y) f(x,y) dy dx$$

$$= \sum_{i=0}^{M} \sum_{j=0}^{N} \int_a^b \int_c^d p(s,t) \frac{(s-a)^i}{i!} \frac{(t-c)^j}{j!} f_{(i,j)}(a,c) dt ds$$

$$+ \sum_{j=0}^{N} \int_a^b \int_x^b \int_c^d p(s,t) \frac{(s-x)^M}{M!} \frac{(t-c)^j}{j!} f_{(M+1,j)}(x,c) dt ds dx$$

$$+ \sum_{i=0}^{M} \int_c^d \int_a^b \int_y^d p(s,t) \frac{(s-a)^i}{i!} \frac{(t-y)^N}{N!} f_{(i,N+1)}(a,y) dt ds dy$$

$$+ \int_a^b \int_c^d \int_x^b \int_y^d p(s,t) \frac{(s-x)^M}{M!} \frac{(t-y)^N}{N!} f_{(M+1,N+1)}(x,y) dt ds dy dx.$$

(4.1.10)

Remark 4.1.7 This above result can also be proved by using Two-dimensional mathematical induction as given in prove of Theorem 4.1.12.

Now we state main integral identity of this subsection using higher order derivatives. We would like to prove following result in three different ways, first by interchanging technique, second by Two-dimensional Induction and third by Taylor expansion.

Theorem 4.1.12 *Let f has continuous partial derivatives $f_{(i,j)}$ and $p, f : I \times J \to \mathbb{R}$ be both integrable functions, where $i \in \{0, 1, \ldots, M+1\}$, $j \in \{0, 1, \ldots, N+1\}$, then*

$$\int_a^b \int_c^d f(x,y) p(x,y) dy\, dx$$

$$= \sum_{i=0}^{M} \sum_{j=0}^{N} \int_a^b \int_c^d (-1)^{i+j} p(x,y) \frac{(b-x)^i}{i!} \frac{(d-y)^j}{j!} f_{(i,j)}(b,d) dy\, dx$$

$$+ \sum_{j=0}^{N} \int_a^b \int_a^s \int_c^d (-1)^{M+j+1} p(x,y) \frac{(s-x)^M}{M!} \frac{(d-y)^j}{j!} f_{(M+1,j)}(s,d) dy\, dx\, ds$$

$$+ \sum_{i=0}^{M} \int_{c}^{d} \int_{a}^{b} \int_{c}^{t} (-1)^{i+N+1} p(x, y) \frac{(b-x)^{i}}{i!} \frac{(t-y)^{N}}{N!} f_{(i,N+1)}(b, t) dy\, dx\, dt$$

$$+ \int_{a}^{b} \int_{c}^{d} \int_{a}^{s} \int_{c}^{t} (-1)^{M+N} p(x, y) \frac{(s-x)^{M}}{M!} \frac{(t-y)^{N}}{N!} f_{(M+1,N+1)}(s, t) dy\, dx\, dt\, ds.$$

(4.1.11)

Proof (Method I) We restate the identity given in Lemma 4.1.2 as follows

$$\int_{A}^{B} \int_{C}^{D} p(x, y) f(x, y) dy\, dx$$

$$= \sum_{i=0}^{M} \sum_{j=0}^{N} \int_{A}^{B} \int_{C}^{D} p(s, t) f_{(i,j)}(A, C) \frac{(s-A)^{i}}{i!} \frac{(t-C)^{j}}{j!} dt\, ds$$

$$+ \sum_{j=0}^{N} \int_{A}^{B} \int_{x}^{B} \int_{C}^{D} p(s, t) f_{(M+1,j)}(x, C) \frac{(s-x)^{M}}{M!} \frac{(t-C)^{j}}{j!} dt\, ds\, dx$$

$$+ \sum_{i=0}^{M} \int_{C}^{D} \int_{A}^{B} \int_{y}^{D} p(s, t) f_{(i,N+1)}(A, y) \frac{(s-A)^{i}}{i!} \frac{(t-y)^{N}}{N!} dt\, ds\, dy$$

$$+ \int_{A}^{B} \int_{C}^{D} \int_{x}^{B} \int_{y}^{D} p(s, t) f_{(M+1,N+1)}(x, y) \frac{(s-x)^{M}}{M!} \frac{(t-y)^{N}}{N!} dt\, ds\, dy\, dx.$$

(4.1.12)

Let us substitute $[A, B] = [b, a]$ and $[C, D] = [d, c]$. Then $\int_{A}^{B} = \int_{b}^{a} = -\int_{a}^{b}$ etc. and we change the variables names $x \leftrightarrow s$, $y \leftrightarrow t$, then

$$\int_{b}^{a} \int_{d}^{c} p(x, y) f(x, y) dy\, dx$$

$$= \sum_{i=0}^{M} \sum_{j=0}^{N} \int_{b}^{a} \int_{d}^{c} f_{(i,j)}(b, d) p(x, y) \frac{(x-b)^{i}}{i!} \frac{(y-d)^{j}}{j!} dy\, dx$$

$$+ \sum_{j=0}^{N} \int_{b}^{a} \int_{s}^{a} \int_{d}^{c} f_{(M+1,j)}(s, d) p(x, y) \frac{(x-s)^{M}}{M!} \frac{(y-d)^{j}}{j!} dy\, dx\, ds$$

$$+ \sum_{i=0}^{M} \int_{d}^{c} \int_{b}^{a} \int_{t}^{c} f_{(i,N+1)}(b, t) p(x, y) \frac{(x-b)^{i}}{i!} \frac{(y-t)^{N}}{N!} dy\, dx\, dt$$

$$+ \int_{b}^{a} \int_{d}^{c} \int_{s}^{a} \int_{t}^{c} f_{(M+1,N+1)}(s, t) p(x, y) \frac{(x-s)^{M}}{M!} \frac{(y-t)^{N}}{N!} dy\, dx\, dt\, ds.$$

(4.1.13)

4.1 Linear Inequalities for Higher Order ∇-Convex and Completely...

Left hand side of (4.1.13) may write

$$\int_b^a \int_d^c p(x,y)f(x,y)dy\,dx = \int_a^b \int_c^d (-1)^2 p(x,y)f(x,y)dy\,dx$$
$$= \int_a^b \int_c^d p(x,y)f(x,y)dy\,dx.$$

First summand on right side may also write

$$\sum_{i=0}^M \sum_{j=0}^N \int_b^a \int_d^c p(x,y) \frac{(x-b)^i}{i!} \frac{(y-d)^j}{j!} f_{(i,j)}(b,d)dy\,dx$$

$$= \sum_{i=0}^M \sum_{j=0}^N \int_a^b \int_c^d (-1)^2 p(x,y)(-1)^i \frac{(b-x)^i}{i!} (-1)^j \frac{(d-y)^j}{j!} f_{(i,j)}(b,d)dy\,dx$$

$$= \sum_{i=0}^M \sum_{j=0}^N \int_a^b \int_c^d (-1)^{i+j} p(x,y) \frac{(b-x)^i}{i!} \frac{(d-y)^j}{j!} f_{(i,j)}(b,d)dy\,dx.$$

Second summand on right side may also write

$$\sum_{j=0}^N \int_b^a \int_s^a \int_d^c p(x,y) \frac{(x-s)^M}{M!} \frac{(y-d)^j}{j!} f_{(M+1,j)}(s,d)dy\,dx\,ds$$

$$= \sum_{j=0}^N \int_a^b \int_a^s \int_c^d (-1)^3 p(x,y)(-1)^M \frac{(s-x)^M}{M!} (-1)^j \frac{(d-y)^j}{j!}$$
$$\times f_{(M+1,j)}(s,d)dy\,dx\,ds$$

$$= \sum_{j=0}^N \int_a^b \int_a^s \int_c^d (-1)^{M+1+j} p(x,y) \frac{(s-x)^M}{M!} \frac{(d-y)^j}{j!} f_{(M+1,j)}(s,d)dy\,dx\,ds.$$

Similarly the third summand is rewritten as

$$\sum_{i=0}^M \int_d^c \int_b^a \int_t^c p(x,y) \frac{(x-b)^i}{i!} \frac{(y-t)^N}{N!} f_{(i,N+1)}(b,t)dy\,dx\,dt$$

$$= \sum_{i=0}^M \int_c^d \int_a^b \int_c^t (-1)^3 p(x,y)(-1)^i \frac{(b-x)^i}{i!} (-1)^N \frac{(t-y)^N}{N!}$$
$$\times f_{(i,N+1)}(b,t)dy\,dx\,dt$$

$$= \sum_{i=0}^M \int_c^d \int_a^b \int_c^t (-1)^{N+1+i} p(x,y) \frac{(b-x)^i}{i!} \frac{(t-y)^N}{N!} f_{(i,N+1)}(b,t)dy\,dx\,dt.$$

Finally, last summand on right side rewritten as

$$\int_b^a \int_d^c \int_s^a \int_t^c p(x,y) \frac{(x-s)^M}{M!} \frac{(y-t)^N}{N!} f_{(M+1,N+1)}(s,t) dy\, dx\, dt\, ds$$

$$= \int_a^b \int_c^d \int_a^s \int_c^t (-1)^4 p(x,y)(-1)^N \frac{(t-y)^N}{N!} (-1)^M \frac{(s-x)^M}{M!}$$

$$\times f_{(M+1,N+1)}(s,t) dy\, dx\, dt\, ds$$

$$= \int_a^b \int_c^d \int_a^s \int_c^t (-1)^{M+N} p(x,y) \frac{(s-x)^M}{M!} \frac{(t-y)^N}{N!}$$

$$\times f_{(M+1,N+1)}(s,t) dy\, dx\, dt\, ds.$$

By substituting all these expression in (4.1.13) we would arrive at our required result. □

Proof (Method II) First we claim that

$$\int_d^c \int_b^a p(x,y) f(x,y) dx\, dy = \sum_{i=0}^M \sum_{j=0}^N f_{(i,j)}(b,d) p_{(i+1,j+1)}(b,d)$$

$$+ \sum_{j=0}^N \int_b^a f_{(M+1,j)}(s,d) p_{(M+1,j+1)}(s,d) ds$$

$$+ \sum_{i=0}^M \int_d^c f_{(i,N+1)}(b,t) p_{(i+1,N+1)}(b,t) dt$$

$$+ \int_d^c \int_b^a f_{(M+1,N+1)}(s,t) p_{(M+1,N+1)}(s,t) ds\, dt.$$
(4.1.14)

Now we prove this equality by using the Definition 4.1.10 of two-dimensional induction and considering base case, i.e., $M = N = 0$

$$\int_c^d \int_a^b p(x,y) f(x,y) dx\, dy = \int_d^c \int_b^a p(s,t) f(s,t) ds\, dt$$

$$= f_{(0,0)}(b,d) p_{(1,1)}(b,d)$$

$$+ \int_b^a f_{(1,0)}(s,d) p_{(1,1)}(s,d) ds$$

$$+ \int_d^c f_{(0,1)}(b,t) p_{(1,1)}(b,t) dt$$

$$+ \int_d^c \int_b^a f_{(1,1)}(s,t) p_{(1,1)}(s,t) ds\, dt,$$

4.1 Linear Inequalities for Higher Order ∇-Convex and Completely...

which is the result of Lemma 4.1.1 so it is proved.

Let us assume that our hypothesis is true for $M = m$ and $N = 0$, i.e.,

$$\int_d^c \int_b^a p(x,y) f(x,y) dx dy$$

$$= \sum_{i=0}^m f_{(i,0)}(b,d) p_{(i+1,1)}(b,d) + \int_b^a f_{(m+1,0)}(s,d) p_{(m+1,1)}(s,d) ds$$

$$+ \sum_{i=0}^m \int_d^c f_{(i,1)}(b,t) p_{(i+1,1)}(b,t) dt + \int_d^c \int_b^a f_{(m+1,1)}(s,t) p_{(m+1,1)}(s,t) ds dt.$$

(4.1.15)

We would show that it is valid for $M = m+1$ and $N = 0$, i.e., following inequality holds

$$\int_d^c \int_b^a p(x,y) f(x,y) dx dy$$

$$= \sum_{i=0}^{m+1} f_{(i,0)}(b,d) p_{(i+1,1)}(b,d) + \int_b^a f_{(m+2,0)}(s,d) p_{(m+2,1)}(s,d) ds$$

$$+ \sum_{i=0}^{m+1} \int_d^c f_{(i,1)}(b,t) p_{(i+1,1)}(b,t) dt + \int_d^c \int_b^a f_{(m+2,1)}(s,t) p_{(m+2,1)}(s,t) ds dt.$$

(4.1.16)

To prove (4.1.16) we consider 2nd term of (4.1.15)

$$\int_b^a f_{(m+1,0)}(s,d) p_{(m+1,1)}(s,d) ds$$

$$= \int_b^a \left(f_{(m+1,0)}(b,d) + \int_b^s f_{(m+2,0)}(\theta,d) d\theta \right) p_{(m+1,1)}(s,d) ds$$

$$= f_{(m+1,0)}(b,d) \int_b^a p_{(m+1,1)}(s,d) ds + \int_b^a \int_b^s f_{(m+2,0)}(\theta,d) p_{(m+1,1)}(s,d) d\theta ds,$$

by using Fubini theorem and interchanging $\theta \leftrightarrow s$

$$\int_b^a f_{(m+1,0)}(s,d) p_{(m+1,1)}(s,d) ds$$

$$= f_{(m+1,0)}(b,d) p_{(m+2,1)}(b,d) + \int_b^a \int_\theta^a f_{(m+2,0)}(\theta,d) p_{(m+1,1)}(s,d) ds d\theta$$

$$= f_{(m+1,0)}(b,d)p_{(m+2,1)}(b,d) + \int_b^a f_{(m+2,0)}(s,d) \int_s^a p_{(m+1,1)}(\theta,d)d\theta ds$$

$$= f_{(m+1,0)}(b,d)p_{(m+2,1)}(b,d) + \int_b^a f_{(m+2,0)}(s,d)p_{(m+2,1)}(s,d)ds. \quad (4.1.17)$$

Now consider 4th term of (4.1.15)

$$\int_d^c \int_b^a f_{(m+1,1)}(s,t)p_{(m+1,1)}(s,t)dsdt$$

$$= \int_d^c \int_b^a \left(f_{(m+1,1)}(b,t) + \int_b^s f_{(m+2,1)}(\theta,t)d\theta \right) p_{(m+1,1)}(s,t)dsdt$$

$$= \int_d^c f_{(m+1,1)}(b,t) \int_b^a p_{(m+1,1)}(s,t)dsdt$$

$$+ \int_d^c \int_b^a \int_b^s f_{(m+2,1)}(\theta,t)p_{(m+1,1)}(s,t)d\theta dsdt$$

$$= \int_d^c f_{(m+1,1)}(b,t)p_{(m+2,1)}(b,t)dt$$

$$+ \int_d^c \int_b^a \int_\theta^a f_{(m+2,1)}(\theta,t)p_{(m+1,1)}(s,t)dsd\theta dt$$

$$= \int_d^c f_{(m+1,1)}(b,t)p_{(m+2,1)}(b,t)dt$$

$$+ \int_d^c \int_b^a f_{(m+2,1)}(s,t) \int_s^a p_{(m+1,1)}(\theta,t)d\theta dsdt$$

$$= \int_d^c f_{(m+1,1)}(b,t)p_{(m+2,1)}(b,t)dt + \int_d^c \int_b^a f_{(m+2,1)}(s,t)p_{(m+2,1)}(s,t)dsdt. \quad (4.1.18)$$

For obtaining desired equality (4.1.16), we will have to substitute values of (4.1.17) and (4.1.18) in (4.1.15) it means that our 1st hypothesis is true.

So, let us set again the hypothesis for $M = m$ and $N = n$, implies that

$$\int_d^c \int_b^a p(x,y)f(x,y)dxdy$$

$$= \sum_{i=0}^m \sum_{j=0}^n f_{(i,j)}(b,d)p_{(i+1,j+1)}(b,d) + \sum_{j=0}^n \int_b^a f_{(m+1,j)}(s,d)p_{(m+1,j+1)}(s,d)ds$$

$$+ \sum_{i=0}^m \int_d^c f_{(i,n+1)}(b,t)p_{(i+1,n+1)}(b,t)dt$$

$$+ \int_d^c \int_b^a f_{(m+1,n+1)}(s,t)p_{(m+1,n+1)}(s,t)dsdt. \quad (4.1.19)$$

4.1 Linear Inequalities for Higher Order ∇-Convex and Completely...

We would show, further it is valid for $M = m$ and $N = n+1$

$$\int_d^c \int_b^a p(x,y) f(x,y) dx dy$$

$$= \sum_{i=0}^m \sum_{j=0}^{n+1} f_{(i,j)}(b,d) p_{(i+1,j+1)}(b,d) + \sum_{j=0}^{n+1} \int_b^a f_{(m+1,j)}(s,d) p_{(m+1,j+1)}(s,d) ds$$

$$+ \sum_{i=0}^m \int_d^c f_{(i,n+2)}(b,t) p_{(i+1,n+2)}(b,t) dt$$

$$+ \int_d^c \int_b^a f_{(m+1,n+2)}(s,t) p_{(m+1,n+2)}(s,t) ds dt. \tag{4.1.20}$$

To prove (4.1.20) we consider 3rd term of (4.1.19)

$$\sum_{i=0}^m \int_d^c f_{(i,n+1)}(b,t) p_{(i+1,n+1)}(b,t) dt$$

$$= \sum_{i=0}^m \int_d^c \left(f_{(i,n+1)}(b,d) + \int_d^t f_{(i,n+2)}(b,\phi) d\phi \right) p_{(i+1,n+1)}(b,t) dt$$

$$= \sum_{i=0}^m \left(f_{(i,n+1)}(b,d) \int_d^c p_{(i+1,n+1)}(b,t) dt \right.$$

$$+ \left. \int_d^c \int_d^t f_{(i,n+2)}(b,\phi) p_{(i+1,n+1)}(b,t) d\phi dt \right)$$

$$= \sum_{i=0}^m \left(f_{(i,n+1)}(b,d) p_{(i+1,n+2)}(b,d) \right.$$

$$+ \left. \int_d^c \int_\phi^c f_{(i,n+2)}(b,\phi) p_{(i+1,n+1)}(b,t) dt d\phi \right)$$

$$= \sum_{i=0}^m \left(f_{(i,n+1)}(b,d) p_{(i+1,n+2)}(b,d) \right.$$

$$+ \left. \int_d^c f_{(i,n+2)}(b,t) \int_t^c p_{(i+1,n+1)}(b,\phi) d\phi dt \right)$$

$$= \sum_{i=0}^m \left(f_{(i,n+1)}(b,d) p_{(i+1,n+2)}(b,d) + \int_d^c f_{(i,n+2)}(b,t) p_{(i+1,n+2)}(b,t) dt \right). \tag{4.1.21}$$

Now consider 4th term of (4.1.19)

$$\int_d^c \int_b^a f_{(m+1,n+1)}(s,t) p_{(m+1,n+1)}(s,t) ds dt$$

$$= \int_b^a \int_d^c f_{(m+1,n+1)}(s,t) p_{(m+1,n+1)}(s,t) dt ds$$

$$= \int_b^a \int_d^c \left(f_{(m+1,n+1)}(s,d) + \int_d^t f_{(m+1,n+2)}(s,\phi) d\phi \right) p_{(m+1,n+1)}(s,t) dt ds$$

$$= \int_b^a f_{(m+1,n+1)}(s,d) \int_d^c p_{(m+1,n+1)}(s,t) dt ds$$

$$+ \int_b^a \int_d^c \int_d^t f_{(m+1,n+2)}(s,\phi) p_{(m+1,n+1)}(s,t) d\phi dt ds$$

$$= \int_b^a f_{(m+1,n+1)}(s,d) p_{(m+1,n+2)}(s,d) ds$$

$$+ \int_b^a \int_d^c \int_\phi^c f_{(m+1,n+2)}(s,\phi) p_{(m+1,n+1)}(s,t) dt d\phi ds$$

$$= \int_b^a f_{(m+1,n+1)}(s,d) p_{(m+1,n+2)}(s,d) ds$$

$$+ \int_b^a \int_d^c f_{(m+1,n+2)}(s,t) \int_t^c p_{(m+1,n+1)}(s,\phi) d\phi dt ds$$

$$= \int_b^a f_{(m+1,n+1)}(s,d) p_{(m+1,n+2)}(s,d) ds$$

$$+ \int_b^a \int_d^c f_{(m+1,n+2)}(s,t) p_{(m+1,n+2)}(s,t) dt ds. \quad (4.1.22)$$

For obtaining desired equality (4.1.20), we will have to substitute values of (4.1.21) and (4.1.22) in (4.1.19) that shows by applying method of induction we have proved our required result.

Here, we need to apply notations that are introduced in starting of subsection for proving further. Considering terms from (4.1.14) separately and find their values such that

$$\sum_{i=0}^M \sum_{j=0}^N f_{(i,j)}(b,d) p_{(i+1,j+1)}(b,d)$$

$$= \sum_{i=0}^M \sum_{j=0}^N \int_b^a \int_d^c f_{(i,j)}(b,d) p(x,y) \frac{(x-b)^i}{i!} \frac{(y-d)^j}{j!} dy dx$$

4.1 Linear Inequalities for Higher Order ∇-Convex and Completely... 237

$$= \sum_{i=0}^{M} \sum_{j=0}^{N} \int_{a}^{b} \int_{c}^{d} (-1)^{i+j} p(x, y) \frac{(b-x)^{i}}{i!} \frac{(d-y)^{j}}{j!} f_{(i,j)}(b, d) dy dx,$$

$$\sum_{j=0}^{N} \int_{b}^{a} f_{(M+1,j)}(s, d) p_{(M+1,j+1)}(s, d) ds$$

$$= \sum_{j=0}^{N} \int_{b}^{a} f_{(M+1,j)}(s, d) \left(\int_{s}^{a} \int_{d}^{c} p(x, y) \frac{(x-s)^{M}}{M!} \frac{(y-d)^{j}}{j!} dy dx \right) ds$$

$$= \sum_{j=0}^{N} \int_{a}^{b} \int_{a}^{s} \int_{c}^{d} (-1)^{M+1+j} p(x, y) \frac{(s-x)^{M}}{M!} \frac{(d-y)^{j}}{j!} f_{(M+1,j)}(s, d) dy dx ds,$$

$$\sum_{i=0}^{M} \int_{d}^{c} f_{(i,N+1)}(b, t) p_{(i+1,N+1)}(b, t) dt$$

$$= \sum_{i=0}^{M} \int_{d}^{c} f_{(i,N+1)}(b, t) \left(\int_{b}^{a} \int_{t}^{c} p(x, y) \frac{(x-b)^{i}}{i!} \frac{(y-t)^{N}}{N!} dy dx \right) dt$$

$$= \sum_{i=0}^{M} \int_{a}^{b} \int_{c}^{d} \int_{c}^{t} (-1)^{i+N+1} p(x, y) \frac{(b-x)^{i}}{i!} \frac{(t-y)^{N}}{N!} f_{(i,N+1)}(b, t) dy dx dt,$$

$$\int_{b}^{a} \int_{d}^{c} f_{(M+1,N+1)}(s, t) p_{(M+1,N+1)}(s, t) ds dt$$

$$= \int_{b}^{a} \int_{d}^{c} f_{(M+1,N+1)}(s, t) \left(\int_{s}^{a} \int_{t}^{c} p(x, y) \frac{(x-s)^{M}}{M!} \frac{(y-t)^{N}}{N!} dy dx \right) ds dt$$

$$= \int_{a}^{b} \int_{c}^{d} \int_{a}^{s} \int_{c}^{t} (-1)^{M+N} p(x, y) \frac{(s-x)^{M}}{M!} \frac{(t-y)^{N}}{N!} f_{(M+1,N+1)}(s, t) dy dx dt ds.$$

Put all these values in (4.1.14) we will get the desired identity (4.1.11) □

Proof (Method III) Let $H(y) = f(x, y)$, i.e., considering $f(x, y)$ as a function of y, where x is fixed. Then H may be written as Taylor expansion

$$f(x, y) = H(y) = \sum_{j=0}^{N} H^{(j)}(d) \frac{(y-d)^{j}}{j!} + \int_{d}^{y} H^{(N+1)}(t) \frac{(y-t)^{N}}{N!} dt$$

$$= \sum_{j=0}^{N} (-1)^{j} \frac{(d-y)^{j}}{j!} f_{(0,j)}(x, d) + \int_{y}^{d} (-1)^{N+1} \frac{(t-y)^{N}}{N!} f_{(0,N+1)}(x, t) dt,$$

where use $H^{(j)}(d) = f_{(0,j)}(x,d)$ and $H^{(N+1)}(t) = f_{(0,N+1)}(x,t)$.

Multiplying above equation by $p(x, y)$ and integrate it by y over the limit c to d, then

$$\int_c^d p(x,y) f(x,y) dy = \sum_{j=0}^N (-1)^j f_{(0,j)}(x,d) \int_c^d p(x,y) \frac{(d-y)^j}{j!} dy$$
$$+ \int_c^d \left(\int_y^d p(x,y)(-1)^{N+1} f_{(0,N+1)}(x,t) \frac{(t-y)^N}{N!} dt \right) dy. \tag{4.1.23}$$

Now we may write functions $x \mapsto f_{(0,j)}(x,d)$, $x \mapsto f_{(0,N+1)}(x,t)$ by applying Taylor expansions:

$$f_{(0,j)}(x,d) = \sum_{i=0}^M (-1)^i \frac{(b-x)^i}{i!} f_{(i,j)}(b,d)$$
$$+ \int_x^b (-1)^{M+1} \frac{(s-x)^M}{M!} f_{(M+1,j)}(s,d) ds,$$

$$f_{(0,N+1)}(x,t) = \sum_{i=0}^M (-1)^i \frac{(b-x)^i}{i!} f_{(i,N+1)}(b,t)$$
$$+ \int_x^b (-1)^{M+1} \frac{(s-x)^M}{M!} f_{(M+1,N+1)}(s,t) ds.$$

Putting these equations in (4.1.23), then

$$\int_c^d p(x,y) f(x,y) dy$$
$$= \sum_{j=0}^N (-1)^j \left(\sum_{i=0}^M (-1)^i \frac{(b-x)^i}{i!} f_{(i,j)}(b,d) \right.$$
$$\left. + \int_x^b (-1)^{M+1} \frac{(s-x)^M}{M!} f_{(M+1,j)}(s,d) ds \right) \int_c^d p(x,y) \frac{(d-y)^j}{j!} dy$$
$$+ \int_c^d \left(\int_y^d (-1)^{N+1} p(x,y) \left(\sum_{i=0}^M (-1)^i \frac{(b-x)^i}{i!} f_{(i,N+1)}(b,t) \right. \right.$$
$$\left. \left. + \int_x^b (-1)^{M+1} \frac{(s-x)^M}{M!} f_{(M+1,N+1)}(s,t) ds \right) \frac{(t-y)^N}{N!} dt \right) dy$$

4.1 Linear Inequalities for Higher Order ∇-Convex and Completely...

$$= \sum_{j=0}^{N} \left(\sum_{i=0}^{M} (-1)^{i+j} \frac{(b-x)^i}{i!} f_{(i,j)}(b,d) \right) \int_{c}^{d} p(x,y) \frac{(d-y)^j}{j!} dy$$

$$+ \sum_{j=0}^{N} \left(\int_{x}^{b} (-1)^{M+1+j} \frac{(s-x)^M}{M!} f_{(M+1,j)}(s,d) ds \right) \int_{c}^{d} p(x,y) \frac{(d-y)^j}{j!} dy$$

$$+ \int_{c}^{d} \int_{y}^{d} p(x,y) \left(\sum_{i=0}^{M} (-1)^{i+N+1} \frac{(b-x)^i}{i!} f_{(i,N+1)}(b,t) \right) \frac{(t-y)^N}{N!} dt\, dy$$

$$+ \int_{c}^{d} \int_{y}^{d} \left(\int_{x}^{b} (-1)^{M+N} p(x,y) \frac{(s-x)^M}{M!} f_{(M+1,N+1)}(s,t) ds \right) \frac{(t-y)^N}{N!} dt\, dy.$$

Now integrate $f(x,y)p(x,y)$ by x over limit a to b and obtain:

$$\int_{a}^{b} \int_{c}^{d} p(x,y) f(x,y) dy\, dx$$

$$= \int_{a}^{b} \left[\sum_{j=0}^{N} \left(\sum_{i=0}^{M} (-1)^{i+j} \frac{(b-x)^i}{i!} f_{(i,j)}(b,d) \right) \int_{c}^{d} p(x,y) \frac{(d-y)^j}{j!} dy \right] dx$$

$$+ \int_{a}^{b} \left[\sum_{j=0}^{N} \left(\int_{x}^{b} (-1)^{M+1+j} \frac{(s-x)^M}{M!} f_{(M+1,j)}(s,d) ds \right) \right.$$

$$\left. \times \int_{c}^{d} p(x,y) \frac{(d-y)^j}{j!} dy \right] dx$$

$$+ \int_{a}^{b} \left[\int_{c}^{d} \int_{y}^{d} p(x,y) \left(\sum_{i=0}^{M} (-1)^{i+N+1} \frac{(b-x)^i}{i!} f_{(i,N+1)}(b,t) \right) \right.$$

$$\left. \times \frac{(t-y)^N}{N!} dt\, dy \right] dx$$

$$+ \int_{a}^{b} \left[\int_{c}^{d} \int_{y}^{d} \left(\int_{x}^{b} (-1)^{M+N} p(x,y) \frac{(s-x)^M}{M!} f_{(M+1,N+1)}(s,t) ds \right) \right.$$

$$\left. \times \frac{(t-y)^N}{N!} dt\, dy \right] dx.$$

Now changing the order of summation in first summand, and using integral's linearity we obtain:

$$\sum_{i=0}^{M}\sum_{j=0}^{N}\int_{a}^{b}\int_{c}^{d}(-1)^{i+j}p(x,y)\frac{(b-x)^{i}}{i!}\frac{(d-y)^{j}}{j!}f_{(i,j)}(b,d)dy\,dx.$$

The second summand is rewritten as:

$$\int_{a}^{b}\left[\sum_{j=0}^{N}\left(\int_{x}^{b}(-1)^{M+1+j}\frac{(s-x)^{M}}{M!}f_{(M+1,j)}(s,d)ds\right)\right.$$
$$\left.\times\int_{c}^{d}p(x,y)\frac{(d-y)^{j}}{j!}dy\right]dx$$
$$=\int_{a}^{b}\left[\sum_{j=0}^{N}\left(\int_{x}^{b}\int_{c}^{d}(-1)^{M+1+j}p(x,y)\frac{(d-y)^{j}}{j!}\right.\right.$$
$$\left.\left.\times\frac{(s-x)^{M}}{M!}f_{(M+1,j)}(s,d)dy\,ds\right)\right]dx$$
$$=\sum_{j=0}^{N}\int_{a}^{b}\int_{x}^{b}\int_{c}^{d}(-1)^{M+1+j}p(x,y)\frac{(s-x)^{M}}{M!}\frac{(d-y)^{j}}{j!}f_{(M+1,j)}(s,d)dy\,ds\,dx$$
$$=\sum_{j=0}^{N}\int_{a}^{b}\int_{a}^{s}\int_{c}^{d}(-1)^{M+1+j}p(x,y)\frac{(s-x)^{M}}{M!}\frac{(d-y)^{j}}{j!}f_{(M+1,j)}(s,d)dy\,dx\,ds.$$

Here, Fubini theorem for variables s and x is applied in the last step. The variable x has been changed from $a \to b$ and s from $x \to b$. The order of integration is then changed followed by changing s from $a \to b$ and x from $a \to s$. Similarly the third summand can be rewritten as:

$$\int_{a}^{b}\left[\int_{c}^{d}\int_{y}^{d}p(x,y)\left(\sum_{i=0}^{M}(-1)^{i+N+1}\frac{(b-x)^{i}}{i!}f_{(i,N+1)}(b,t)\right)\frac{(t-y)^{N}}{N!}dt\,dy\right]dx$$
$$=\sum_{i=0}^{M}\int_{a}^{b}\int_{c}^{d}\int_{y}^{d}(-1)^{i+N+1}p(x,y)\frac{(b-x)^{i}}{i!}\frac{(t-y)^{N}}{N!}f_{(i,N+1)}(b,t)dt\,dy\,dx$$
$$=\sum_{i=0}^{M}\int_{a}^{b}\int_{c}^{d}\int_{c}^{t}(-1)^{i+N+1}p(x,y)\frac{(b-x)^{i}}{i!}\frac{(t-y)^{N}}{N!}f_{(i,N+1)}(b,t)dy\,dt\,dx$$
$$=\sum_{i=0}^{M}\int_{c}^{d}\int_{a}^{b}\int_{c}^{t}(-1)^{i+N+1}p(x,y)\frac{(b-x)^{i}}{i!}\frac{(t-y)^{N}}{N!}f_{(i,N+1)}(b,t)dy\,dx\,dt.$$

4.1 Linear Inequalities for Higher Order ∇-Convex and Completely... 241

In the above step Fubini theorem has been applied twice. First, t and y have been changed and then t and x. Therefore, last summand is rewritten as:

$$\int_a^b \left[\int_c^d \int_y^d \left(\int_x^b (-1)^{M+N} p(x,y) \frac{(s-x)^M}{M!} f_{(M+1,N+1)}(s,t) ds \right) \frac{(t-y)^N}{N!} dt\, dy \right] dx$$

$$= \int_a^b \int_c^d \int_y^d \int_x^b (-1)^{M+N} p(x,y) \frac{(s-x)^M}{M!} \frac{(t-y)^N}{N!} f_{(M+1,N+1)}(s,t) ds\, dt\, dy\, dx$$

$$= \int_a^b \int_c^d \int_a^s \int_c^t (-1)^{M+N} p(x,y) \frac{(s-x)^M}{M!} \frac{(t-y)^N}{N!} f_{(M+1,N+1)}(s,t) dy\, dx\, dt\, ds.$$

The above steps involve application of Fubini theorem various times. The variables are changed in the following order, first t and y followed by y and s, after that s and t, then s and x and finally t and x. The required identity by using all these results.

$$\int_a^b \int_c^d p(x,y) f(x,y) dy\, dx$$

$$= \sum_{i=0}^M \sum_{j=0}^N \int_a^b \int_c^d (-1)^{i+j} p(x,y) \frac{(b-x)^i}{i!} \frac{(d-y)^j}{j!} f_{(i,j)}(b,d) dy\, dx$$

$$+ \sum_{j=0}^N \int_a^b \int_a^s \int_c^d (-1)^{M+1+j} p(x,y) \frac{(s-x)^M}{M!} \frac{(d-y)^j}{j!} f_{(M+1,j)}(s,d) dy\, dx\, ds$$

$$+ \sum_{i=0}^M \int_c^d \int_a^b \int_c^t (-1)^{i+N+1} p(x,y) \frac{(b-x)^i}{i!} \frac{(t-y)^N}{N!} f_{(i,N+1)}(b,t) dy\, dx\, dt$$

$$+ \int_a^b \int_c^d \int_a^s \int_c^t (-1)^{M+N} p(x,y) \frac{(s-x)^M}{M!} \frac{(t-y)^N}{N!} f_{(M+1,N+1)}(s,t) dy\, dx\, dt\, ds.$$

□

Remark 4.1.8 We may also obtain corollary of Theorem 4.1.12 for I^2 by changing the variables names on right side $x \leftrightarrow s$, $y \leftrightarrow t$.

Corollary 4.1.13 *Let both p, $f : I^2 \to \mathbb{R}$ be functions, $f \in C^{(M+1,N+1)}(I^2)$ and p is an integrable, where $j \in \{0, 1, \ldots, N+1\}$ and $i \in \{0, 1, \ldots, M+1\}$, then*

$$\int_a^b \int_a^b p(x,y) f(x,y) dy\, dx$$

$$= \sum_{i=0}^M \sum_{j=0}^N \int_a^b \int_a^b (-1)^{i+j} p(s,t) \frac{(b-s)^i}{i!} \frac{(b-t)^j}{j!} f_{(i,j)}(b,b) dt\, ds$$

$$+ \sum_{j=0}^{N} \int_{a}^{b} \int_{a}^{x} \int_{a}^{b} (-1)^{M+1+j} p(s,t) \frac{(x-s)^M}{M!} \frac{(b-t)^j}{j!} f_{(M+1,j)}(x,b) dt\, ds\, dx$$

$$+ \sum_{i=0}^{M} \int_{a}^{b} \int_{a}^{b} \int_{a}^{y} (-1)^{i+N+1} p(s,t) \frac{(b-s)^i}{i!} \frac{(y-t)^N}{N!} f_{(i,N+1)}(b,y) dt\, ds\, dy$$

$$+ \int_{a}^{b} \int_{a}^{b} \int_{a}^{x} \int_{a}^{y} (-1)^{M+N} p(s,t) \frac{(x-s)^M}{M!} \frac{(y-t)^N}{N!} f_{(M+1,N+1)}(x,y) dt\, ds\, dy\, dx$$

holds.

Remark 4.1.9 If we replace $f(x,y)$ by $f(x)g(y)$ into Theorem 4.1.12, then we obtain the below statement.

Corollary 4.1.14 *Let $f \in C^{(M+1)}(I)$, $g \in C^{(N+1)}(J)$, be two different functions and function $p : I \times J \to \mathbb{R}$ be integrable, then*

$$\int_{a}^{b} \int_{c}^{d} f(x,y) p(x,y) dy\, dx$$

$$= \sum_{i=0}^{M} \sum_{j=0}^{N} \int_{a}^{b} \int_{c}^{d} (-1)^{i+j} p(x,y) \frac{(d-y)^j}{j!} g^{(j)}(d) \frac{(b-x)^i}{i!} f^{(i)}(b) dy\, dx$$

$$+ \sum_{j=0}^{N} \int_{a}^{b} \int_{a}^{s} \int_{c}^{d} (-1)^{M+1+j} p(x,y) \frac{(d-y)^j}{j!} g^{(j)}(d) \frac{(s-x)^M}{M!} f^{(M+1)}(x) dy\, dx\, ds$$

$$+ \sum_{i=0}^{M} \int_{c}^{d} \int_{a}^{b} \int_{c}^{t} (-1)^{N+1+i} p(x,y) \frac{(t-y)^N}{N!} g^{(N+1)}(y) \frac{(b-x)^i}{i!} f^{(i)}(b) dy\, dx\, dt$$

$$+ \int_{a}^{b} \int_{c}^{d} \int_{a}^{s} \int_{c}^{t} (-1)^{N+M} p(x,y) \frac{(t-y)^N}{N!} g^{(N+1)}(y) \frac{(s-x)^M}{M!} f^{(M+1)}(x) dy\, dx\, dt\, ds.$$

We obtain necessary and sufficient conditions by using results of previous theorem that $\Lambda(f) \geq 0$ holds $\forall \ (M+1, N+1) - \nabla$-convex function and only necessary condition $\forall \ (M+1, N+1)$-completely monotonic function for two variables function.

Theorem 4.1.15 *Let all the assumptions of Theorem 4.1.12 be valid, then following inequality holds*

$$\Lambda(f) = \int_{a}^{b} \int_{c}^{d} p(x,y) f(x,y) dy\, dx \geq 0 \tag{4.1.24}$$

4.1 Linear Inequalities for Higher Order ∇-Convex and Completely... 243

$\forall\,(M+1, N+1)-\nabla$-convex function f on $I \times J$, iff

$$\int_a^b \int_c^d p(x, y)\frac{(b-x)^i}{i!}\frac{(d-y)^j}{j!} dy\, dx = 0, \ i \in \{0, 1, \ldots, M\}; \ j \in \{0, 1, \ldots, N\};$$
(4.1.25)

$$\int_a^s \int_c^d p(x, y)\frac{(s-x)^M}{M!}\frac{(d-y)^j}{j!} dy\, dx = 0, \quad j \in \{0, 1, \ldots, N\}; \ \forall s \in [a, b];$$
(4.1.26)

$$\int_a^b \int_c^t p(x, y)\frac{(b-x)^i}{i!}\frac{(t-y)^N}{N!} dy\, dx = 0, \quad i \in \{0, 1, \ldots, M\}; \ \forall t \in [c, d];$$
(4.1.27)

$$\int_a^s \int_c^t p(x, y)\frac{(s-x)^M}{M!}\frac{(t-y)^N}{N!} dy\, dx \geq 0, \quad \forall s \in [a, b]; \ \forall t \in [c, d]. \quad (4.1.28)$$

Proof If (4.1.25), (4.1.26) and (4.1.27) hold, then first, second and third sums are zero in (4.1.11), then using (4.1.28) we get desired inequality (4.1.24).

Conversely, if substitute the following functions in (4.1.24). Then

$$f^1(x, y) = \frac{(b-x)^m}{m!}\frac{(d-y)^n}{n!} \quad \text{and} \quad f^2 = -f^1$$

for $0 \leq n \leq N;\ 0 \leq m \leq M$ such that $(-1)^{M+N} f^j_{(M+1, N+1)} \geq 0,\ j \in \{1, 2\}$, then obtain the desired equation (4.1.25), i.e.,

$$\int_a^b \int_c^d p(x, y)\frac{(b-x)^m}{m!}\frac{(d-y)^n}{n!} dy\, dx = 0, \quad 0 \leq m \leq M;\ 0 \leq n \leq N.$$

In the similar manner, if take the following functions in (4.1.24) $\forall\ s \in [a, b]$ and $0 \leq n \leq N$

$$f^3(x, y) = \begin{cases} \frac{(s-x)^M}{M!}\frac{(d-y)^n}{n!}, & x < s \\ 0, & x \geq s \end{cases} \quad \text{and} \quad f^4 = -f^3$$

such that $(-1)^{N+M} f^j_{(M+1, N+1)} \geq 0,\ j \in \{3, 4\}$, we obtain desired equation (4.1.26), i.e.,

$$\int_a^s \int_c^d p(x, y)\frac{(s-x)^M}{M!}\frac{(d-y)^n}{n!} dy\, dx = 0, \quad 0 \leq n \leq N;\ \forall\ s \in [a, b].$$

Similarly, if take the following functions in (4.1.24) $\forall\ t \in [c, d]$ and $0 \leq m \leq M$

$$f^5(x, y) = \begin{cases} \frac{(b-x)^m}{m!} \frac{(t-y)^N}{N!}, & y < t \\ 0, & y \geq t \end{cases} \quad \text{and} \quad f^6 = -f^5$$

such that $(-1)^{N+M} f^j_{(M+1,N+1)} \geq 0$, $j \in \{5, 6\}$, we can obtain above equation (4.1.27), i.e.,

$$\int_a^b \int_c^t p(x, y) \frac{(b-x)^m}{m!} \frac{(t-y)^N}{N!} dy\, dx = 0, \quad 0 \leq m \leq M;\ \forall\ t \in [c, d].$$

By considering the below function in (4.1.24), obtain the desired last inequality (4.1.28) for $s \in [a, b]$, $t \in [c, d]$

$$f^7(x, y) = \begin{cases} \frac{(s-x)^M}{M!} \frac{(t-y)^N}{N!}, & x < s,\quad y < t \\ 0, & x \geq s\ \text{ or }\ y \geq t. \end{cases}$$

$\square$

Theorem 4.1.16 *Let all the assumptions of Theorem 4.1.12 be valid, then following inequality holds*

$$\Lambda(f) = \int_a^b \int_c^d p(x, y) f(x, y) dy\, dx \geq 0 \qquad (4.1.29)$$

for all completely monotonic function f of $(M+1, N+1)$th order on $I \times J$ if

$$\int_a^b \int_c^d p(x, y) \frac{(b-x)^i}{i!} \frac{(d-y)^j}{j!} dy\, dx = 0, i \in \{0, 1, \ldots, M\};\ j \in \{0, 1, \ldots, N\}; \qquad (4.1.30)$$

$$\int_a^s \int_c^d p(x, y) \frac{(s-x)^M}{M!} \frac{(d-y)^j}{j!} dy\, dx = 0, \quad j \in \{0, 1, \ldots, N\};\ \forall\ s \in [a, b]; \qquad (4.1.31)$$

$$\int_a^b \int_c^t p(x, y) \frac{(b-x)^i}{i!} \frac{(t-y)^N}{N!} dy\, dx = 0,\quad i \in \{0, 1, \ldots, M\};\ \forall\ t \in [c, d]; \qquad (4.1.32)$$

$$\int_a^s \int_c^t p(x, y) \frac{(s-x)^M}{M!} \frac{(t-y)^N}{N!} dy\, dx \geq 0,\ \forall\ s \in [a, b];\ \forall\ t \in [c, d]. \qquad (4.1.33)$$

Proof If (4.1.30), (4.1.31) and (4.1.32) hold, then first, second and third sums are zero in (4.1.11), then by using (4.1.33) obtain desired inequality (4.1.29). $\square$

4.1 Linear Inequalities for Higher Order ∇-Convex and Completely...

Remark 4.1.10 If we simply put $f(x, y) = f(x)g(y)$ in previous two results, then we obtain similar results for functions f and g.

4.1.5 Mean Value Theorems and Exponential Convexity

Mean Value Theorems

The mean value theorem is valuable tool to obtain interesting and important results of classical real analysis. In the field of differential calculus, Lagrange and Cauchy mean value theorems are the most demanding theorems. For more materials on this topic (see [167]). Here, we would give some generalized mean value theorems of Lagrange and Cauchy type.

Theorem 4.1.17 *Let function $p : I \times J \to \mathbb{R}$ be integrable and $f \in C^{(M+1,N+1)}(I \times J)$, be a $(M+1, N+1) - \nabla$-convex function in the interval $I \times J$. Let Λ be a linear functional as stated in (4.1.24) and conditions (4.1.25), (4.1.26), (4.1.27) and (4.1.28) be valid for function p in Theorem 4.1.15, then $\exists (\eta, \zeta) \in I \times J$, such that*

$$\Lambda(f) = \Lambda(G_0) f_{(M+1,N+1)}(\eta, \zeta), \qquad (4.1.34)$$

where $G_0(x, y) = (-1)^{M+N} \dfrac{x^{M+1}}{(M+1)!} \dfrac{y^{N+1}}{(N+1)!}$.

Proof Let $U = \max\limits_{(x,y) \in I \times J} (-1)^{M+N} f_{(M+1,N+1)}(x, y)$,
$J = \min\limits_{(x,y) \in I \times J} (-1)^{M+N} f_{(M+1,N+1)}(x, y)$.
So the function

$$G(x, y) = U(-1)^{M+N} \frac{y^{N+1}}{(N+1)!} \frac{x^{M+1}}{(M+1)!} - f(x, y) = U G_0(x, y) - f(x, y),$$

gives us

$$(-1)^{M+N} G_{(M+1,N+1)}(x, y) = U - (-1)^{M+N} f_{(M+1,N+1)}(x, y) \geq 0.$$

i.e., G is ∇-convex function of $(M+1, N+1)$th order in the interval $I \times J$. Hence $0 \leq \Lambda(G)$ using Theorem 4.1.15 and would summarize

$$\Lambda(f) \leq U \Lambda(G_0).$$

Similarly

$$L\Lambda(G_0) \leq \Lambda(f).$$

Now, we can write the above two inequalities as:

$$L\Lambda(G_0) \leq \Lambda(f) \leq U\Lambda(G_0),$$

which gives the required result (4.1.34). □

Theorem 4.1.18 *Let* $g, f \in C^{(M+1,N+1)}(I \times J)$, *be two* ∇-*convex functions of order* $(M+1, N+1)$ *in the interval and* $p : I \times J \to R$ *be integrable function. Let* Λ *be a linear functional as stated in* (4.1.24) *and conditions* (4.1.25), (4.1.26), (4.1.27) *and* (4.1.28) *be valid for function p in Theorem* 4.1.15, *then* $\exists\, (\eta, \zeta) \in I \times J$, *such that*

$$\frac{f_{(M+1,N+1)}(\eta, \zeta)}{g_{(M+1,N+1)}(\eta, \zeta)} = \frac{\Lambda(f)}{\Lambda(g)},$$

provided that the denominators are non zero.

Proof Let $u \in C^{(M+1,N+1)}$ be a ∇-convex function of order $(M + 1, N + 1)$ in the interval $I \times J$, stated as:

$$u = \Lambda(g)f - \Lambda(f)g.$$

Applying Theorem 4.1.17 $\exists\, (\eta, \zeta)$, such that

$$0 = \Lambda(u) = u_{(M+1,N+1)}(\eta, \zeta)\Lambda(G_0)$$

or

$$[\Lambda(g)f_{(M+1,N+1)}(\eta, \zeta) - \Lambda(f)g_{(M+1,N+1)}(\eta, \zeta)]\Lambda(G_0) = 0,$$

that gives desired consequence. □

Corollary 4.1.19 *Let* Λ *be a linear functional as stated in* (4.1.24) *and the conditions* (4.1.25), (4.1.26), (4.1.27) *and* (4.1.28) *be valid for function p with* $N = M$ *in Theorem* 4.1.15, *then* $\exists\, (\eta, \zeta) \in I \times J$, *such that*

$$(\eta\zeta)^{t-s} = \frac{[(s+1)s(s-1)\cdots(s-M+2)(s-M+1)]^2 \Lambda((xy)^{t+1})}{[(t+1)t(t-1)\cdots(t-M+2)(t-M+1)]^2 \Lambda((xy)^{s+1})},$$

where $t, s \notin \{-1, 0, 1, \ldots, M-1\}$, *but lie on* $-\infty < t \neq s < +\infty$.

Proof Let

$$f(x, y) = (xy)^{t+1}, \qquad g(x, y) = (xy)^{s+1}.$$

4.1 Linear Inequalities for Higher Order ∇-Convex and Completely... 247

Then

$$f_{(M+1,M+1)} = [(t+1)t(t-1)\cdots(t-M+2)(t-M+1)]^2(xy)^{t-M}$$

and

$$g_{(M+1,M+1)} = [(s+1)s(s-1)\cdots(s-M+2)(s-M+1)]^2(xy)^{s-M}$$

$$\frac{f_{(M+1,M+1)}(\eta,\zeta)}{g_{(M+1,M+1)}(\eta,\zeta)} = \frac{[(t+1)t(t-1)\cdots(t-M+2)(t-M+1)]^2(\eta\zeta)^{t-M}}{[(s+1)s(s-1)\cdots(s-M+2)(s-M+1)]^2(\eta\zeta)^{s-M}}$$

$$= \frac{[(t+1)t(t-1)\cdots(t-M+2)(t-M+1)]^2}{[(s+1)s(s-1)\cdots(s-M+2)(s-M+1)]^2}(\eta\zeta)^{t-s}.$$

We have

$$\Lambda(f) = \Lambda((xy)^{t+1}).$$

So if we put all these equality in Theorem 4.1.18, we get

$$\frac{\Lambda(f)}{\Lambda(g)} = \frac{f_{(M+1,M+1)}(\eta,\zeta)}{g_{(M+1,M+1)}(\eta,\zeta)}$$

$$\frac{\Lambda((xy)^{t+1})}{\Lambda((xy)^{s+1})} = \frac{[(t+1)t(t-1)\cdots(t-M+2)(t-M+1)]^2}{[(s+1)s(s-1)\cdots(s-M+2)(s-M+1)]^2}(\eta\zeta)^{t-s}$$

$$(\eta\zeta)^{t-s} = \frac{[(s+1)s(s-1)\cdots(s-M+2)(s-M+1)]^2\Lambda((xy)^{t+1})}{[(t+1)t(t-1)\cdots(t-M+2)(t-M+1)]^2\Lambda((xy)^{s+1})}.$$

□

Remark 4.1.11 Here we observe that for the case $M = N$ the $(M+1, N+1) - \nabla$-convex function becomes $(M+1, M+1)$-convex function and hence we retrieve the results from [97].

Exponential Convexity

For definition and detailed discussion on exponential convexity we refer the reader to Chap. 1 of this book. Here, in this subsection $J = (a, b) \subset \mathbb{R}$.

Example 4.1.3 A function $x \mapsto ce^{kx}$ is an example of exponentially convex function for $k \in \mathbb{R}$ and constant $c \geq 0$.

Theorem 4.1.20 ([13]) *Let $\omega : J \to \mathbb{R}$, then given statements are similar:*

(i) ω is exponentially convex in an open interval J.

(ii) ω continuous and satisfying the following condition

$$\sum_{i,j=1}^{m} \rho_i \rho_j \omega\left(\frac{x_i + x_j}{2}\right) \geq 0,$$

$\forall \rho_i, \rho_j \in \mathbb{R}$ *and every* $x_i, x_j \in J$; $i, j \in \{1, 2, \ldots, m\}$.

□

Let $D = \{\psi^{(q)} : (0, \infty) \times (0, \infty) \to \mathbb{R} | q \in \mathbb{R}\}$ is a family of functions, stated as:

$$\psi^{(q)}(x, y) = \begin{cases} \dfrac{(x+k_1)^q (y-k_2)^q}{[q(q-1)\cdots(q-M)]^2}, & q \notin \{0, 1, \ldots, M\} \\ \dfrac{[\log(x+k_1)(y-k_2)]^2 (x+k_1)^q (y-k_2)^q}{2[q!(M-q)!]^2}, & q \in \{0, 1, \ldots, M\}. \end{cases}$$

Clearly $\psi_{(M+1,M+1)}^{(q)}(x, y) = [(x+k_1)(y-k_2)]^{q-M-1} = e^{(q-M-1)\log[(x+k_1)(y-k_2)]}$ for $(x+k_1, y-k_2) \in (0, \infty) \times (0, \infty)$ so $\psi^{(q)}$ is completely monotonic function of $(M+1, M+1)$th order since $(-1)^{2i} \psi_{(i+1,i+1)}^{(q)}(x, y) \geq 0$ and $q \mapsto \psi_{(M+1,M+1)}^{(q)}$ is an exponentially convex function f on real numbers. We can say that every positive exponentially convex is log-convex function, by using the Corollary 1.2.19.

Now, at this stage we can give next theorem which is stated as:

Theorem 4.1.21 *Let Λ be a linear functional stated in (4.1.24) and conditions (4.1.25), (4.1.26), (4.1.27) and (4.1.28) be valid for function p in Theorem 4.1.15 and $\psi^{(q)}$ be a completely monotonic function, then following points hold:*

(i) $q \mapsto \Lambda(\psi^{(q)})$ is continuous in the interval $\mathbb{R}$.
(ii) $q \mapsto \Lambda(\psi^{(q)})$ is an exponentially convex function on $\mathbb{R}$.
(iii) If function $q \mapsto \Lambda(\psi^{(q)})$ is positive in the interval $\mathbb{R}$, then $q \mapsto \Lambda(\psi^{(q)})$ is log-convex on $\mathbb{R}$. Moreover, following inequality

$$[\Lambda(\psi^{(s)})]^{t-p} \leq [\Lambda(\psi^{(p)})^{t-s} [\Lambda(\psi^{(t)})]^{s-p} \quad (4.1.35)$$

holds, where $p < s < t$; $p, s, t \in J$.
(iv) $\forall\, m \in \mathbb{N}$ and $q_1, q_2, \ldots, q_m \in \mathbb{R}$, then following matrix is positive semi-definite.

$$\left[\Lambda(\psi^{(\frac{q_i+q_j}{2})})\right]_{i,j=1}^{m}.$$

Particularly,

$$\det\left[\Lambda(\psi^{(\frac{q_i+q_j}{2})})\right]_{i,j=1}^{m} \geq 0.$$

(v) If $q \mapsto \Lambda(\psi^{(q)})$ is differentiable in the interval $\mathbb{R}$, and $\forall s, t, u, v \in \mathbb{R}$, such that $u \geq s$ and $v \geq t$, then

$$\mathfrak{M}_{s,t}(x, y) \leq \mathfrak{M}_{u,v}(x, y), \tag{4.1.36}$$

where

$$\mathfrak{M}_{s,t}(x, y) = \begin{cases} \left(\dfrac{\Lambda(\psi^{(s)})}{\Lambda(\psi^{(t)})}\right)^{\frac{1}{s-t}}, & s \neq t \\ \exp\left(\dfrac{\frac{d}{ds}\Lambda(\psi^{(s)})}{\Lambda(\psi^{(s)})}\right), & s = t, \end{cases}$$

for $\psi^{(s)}, \psi^{(t)} \in D$.

Proof

(i) For fixed $M \in \mathbb{N} \cup \{0\}$, using L' Hopital rule twice then apply limit, we obtain

$$\lim_{q \to 0} \Lambda(\psi^{(q)}) = \lim_{q \to 0} \frac{\int_a^b \int_a^b (y - k_2)^q (x + k_1)^q p(x, y) dy dx}{[q(q-1) \cdots (q-M)]^2}$$

$$= \frac{\int_a^b \int_a^b p(x, y)[\log(x + k_1)(y - k_2)]^2 dy dx}{2[M!]^2}$$

$$= \Lambda(\psi^{(0)}).$$

Similarly, we can show

$$\lim_{q \to k} \Lambda(\psi^{(q)}) = \Lambda(\psi^{(k)}), \quad k \in \{1, 2, \ldots, M\}.$$

(ii) For the proof of 2nd part of this theorem, we define function as:

$$\eta(x, y) = \sum_{i,j=1}^{k} \alpha_i \alpha_j \psi^{(\frac{q_i+q_j}{2})}(x, y),$$

$q_i \in \mathbb{R}, \alpha_i \in \mathbb{R}$ and $i \in \{1, 2, \ldots, k\}$.
Since the function $q \mapsto \psi^{(q)}_{(M+1,M+1)}$ is exponentially convex function, then can write

$$\eta_{(M+1,M+1)} = \sum_{i,j=1}^{k} \alpha_i \alpha_j \psi^{(\frac{q_i+q_j}{2})}_{(M+1,M+1)} \geq 0,$$

that implies function η is ∇-convex of order $(M + 1, M + 1)$ in the interval $\mathbb{R}_+ \times \mathbb{R}_+$ and we know that $\Lambda(\eta) \geq 0$. Therefore

$$\sum_{i,j=1}^{k} \alpha_i \alpha_j \Lambda(\psi^{(\frac{q_i+q_j}{2})}) \geq 0.$$

On the behalf of above working we can summarize that function $q \to \Lambda(\psi^{(q)})$ is exponentially convex on real numbers.

(iii) It follows from (ii) and Corollary 1.2.19. As function $t \mapsto \Lambda(\psi^{(t)})$ is log-convex, i.e., $\ln \Lambda(\psi^{(t)})$ is convex. Now applying the Theorem 1.0.1 of convex function from Chap. 1, we have

$$(x_2 - x_1) f(x_3) + (x_1 - x_3) f(x_2) + (x_3 - x_2) f(x_1) \geq 0$$

for each $x_i \in I$, where $i = 1, 2, 3$ such that $x_1 < x_2 < x_3$, which gives (4.1.35), i.e.,

$$\ln[\Lambda(\psi^{(p)})]^{t-s} + \ln[\Lambda(\psi^{(t)})]^{s-p} \geq \ln[\Lambda(\psi^{(s)})]^{t-p}.$$

(iv) It is consequence of Corollary 1.2.18.

(v) We recall an another definition of convex function φ from [158]

$$\frac{\varphi(u) - \varphi(v)}{u - v} \geq \frac{\varphi(s) - \varphi(t)}{s - t}, \quad (4.1.37)$$

$\forall s, t, u, v \in J$, such that $u \geq s$, $v \geq t$, $u \neq v$, $s \neq t$.

We know that from part (iii), $\Lambda(\psi^{(q)})$ is log-convex, then by setting $\varphi(x) = \log \Lambda(\psi^{(x)})$ in (4.1.37) we have

$$\frac{\log \Lambda(\psi^{(u)}) - \log \Lambda(\psi^{(v)})}{u - v} \geq \frac{\log \Lambda(\psi^{(s)}) - \log \Lambda(\psi^{(t)})}{s - t} \quad (4.1.38)$$

for $u \geq s$, $v \geq t$, $u \neq v$, $s \neq t$, that is similar to (4.1.36). The cases for $v = u$ or/and $t = s$ simply getting from (4.1.38) using respective limits. □

Examples of Completely Monotonic and Exponentially Convex Functions

In the last heading of our first section, we construct various examples of classes of completely monotonic function and exponentially convex function by using different classes of functions $F = \{f_q : q \in I \subset \mathbb{R}\}$. For this purpose, we consider the following examples.

Example 4.1.4 Let $F_1 = \{\zeta_q : \mathbb{R}_+ \times \mathbb{R}_+ \to \mathbb{R}_* | q \in \mathbb{R}\}$ be family of functions which is stated as

$$\zeta_q(x,y) = \begin{cases} \frac{e^{q(x+y)}}{q^{2M+2}}, & q \neq 0; \\ \frac{(x+y)^{2M+2}}{(2M+2)!}, & q = 0. \end{cases}$$

Clearly $\zeta_{q(M+1,M+1)}(x,y) = e^{q(x+y)} > 0$, the function $\zeta_q(x,y)$ is $(M+1, M+1)$-completely monotonic on $\mathbb{R}_+ \times \mathbb{R}_+$ since $(-1)^{2i}\zeta_{q(i+1,i+1)}(x,y) \geq 0$ for $i \in \{0, 1, \ldots, M\}$, $\forall\, q \in \mathbb{R}$ and $q \to \zeta_{q(M+1,M+1)}(x,y)$ is exponentially convex by definition.

Example 4.1.5 Let $F_2 = \{\phi_q : \mathbb{R}_+ \times \mathbb{R}_+ \to \mathbb{R}_+ | q \in \mathbb{R}_+\}$ be family of functions which is stated as

$$\phi_q(x,y) = \begin{cases} \frac{(xy)^q}{[q(q-1)\cdots(q-M)]^2}, & q \notin \{0, 1, \ldots, M\}; \\ \frac{(xy)^q \ln(xy)^2}{2[q!(M-q)!]^2}, & q \in \{0, 1, \ldots, M\}. \end{cases}$$

Clearly $\phi_{q(M+1,M+1)}(x,y) = e^{(q-M-1)\ln(xy)} > 0$, the function $\phi_q(x,y)$ is $(M+1, M+1)$-completely monotonic on $\mathbb{R}_+ \times \mathbb{R}_+$ since $(-1)^{2i}\phi_{q(i+1,i+1)}(x,y) \geq 0$ for $i \in \{0, 1, \ldots, M\}$, $\forall\, q \in \mathbb{R}$ and $q \to \phi_{q(M+1,M+1)}(x,y)$ is exponentially convex by definition.

4.2 Generalized Čebyšev and Ky Fan Identities and Inequalities for ∇-Convex Functions

In the current section we would establish two generalizations, first generalization of discrete Čebyšev identity for function of higher order ∇ operator of two variables and give its special case as sequence of higher order ∇ operator and would also deduce results of discrete inequality of Čebyšev involving higher order ∇-convex function. We have plan to give generalization for integral Čebyšev and integral Ky Fan identities for function of higher order derivative and will discuss its inequalities using ∇-convex function.

Some contents of the present section are extracted from [127].

Over past few decades, there were some reviewers by Mitrinović and Vasić [133] and Mitrinović and Pečarić [132] respectively, which traced completely the chronological and historical development of Čebyšev identities and its connected inequalities. These research works are remarkable due to many instances incorrect quotations of consequences, sometimes by change of several mathematical scholars–have been uncritically transferred paper to paper and book to book. It is well known that the famous Čebyšev functional is applied in many fields such as numerical quadrature, probability, transform theory, special functions and statistical problems (see [33]).

This current section is divided into four parts. In the first part, we introduce inequality of Čebyšev and give its some related results and will define some notations as well. In the second part, we will discuss the generalization of discrete identity and inequality of Čhebyšev type. In third and fourth parts, we would deduce the generalization of integral identities and inequalities of Čebyšev and Ky Fan type respectively.

We start this section from a significant result of Čebyšev [31, 32] may be stated as (see also [158, p. 197]):

Theorem 4.2.1 *Let functions $f, h : [a, b] \to \mathbb{R}$ be both integrable and $p : [a, b] \to \mathbb{R}_+$, where p is also integrable. If f and h are monotone in the same direction, then the inequality*

$$\int_a^b p(x)dx \int_a^b f(x)h(x)p(x)dx \geq \int_a^b h(x)p(x)dx \int_a^b f(x)p(x)dx \quad (4.2.1)$$

holds, there exists integrability. If $f(x)$ and $h(x)$ are monotone in the opposite directions, then (4.2.1) is also valid for reverse inequality. Equality holds in (4.2.1) in the both cases, iff either h or f is constant function almost everywhere.

A discrete form of above theorem can be present as following (see [158]).

Theorem 4.2.2 *Let p be a nonnegative m-tuple and $\mathbf{a}$ and $\mathbf{b}$ be both real m-tuples monotone in the same direction. Then*

$$\sum_{i=1}^M p_i a_i b_i \sum_{i=1}^M p_i \geq \sum_{i=1}^M p_i b_i \sum_{i=1}^M p_i a_i \quad (4.2.2)$$

holds. If $\mathbf{a}$ and $\mathbf{b}$ are monotone in the opposite directions, then (4.2.2) is also valid for reverse inequality. Equality holds in (4.2.2) in the both cases, iff either $a_1 = a_2 = \cdots = a_m$ or $b_1 = b_2 = \cdots = b_m$.

For more discussion about the Čebyšev inequality, we suggest [120, 135, 158].

In [142] Ostrowski obtained the result which is related to inequality of Čebyšev as follows:

Theorem 4.2.3 *Let $p : I \to \mathbb{R}_+$ be an integrable function and $f, h \in C^{(1)}(I)$ be both monotone functions. Then $\exists \, v, \zeta \in I$, such that*

$$T(f, h, p) = f'(v)h'(\zeta)T(x - a, x - a, p), \quad (4.2.3)$$

where

$$T(f, h, p) = \int_a^b p(x)f(x)h(x)dx \int_a^b p(x)dx - \int_a^b p(x)h(x)dx \int_a^b p(x)f(x)dx. \quad (4.2.4)$$

4.2 Generalized Čebyšev and Ky Fan Identities and Inequalities for ∇-Convex...

For other generalizations of Theorem 4.2.3, (see [149]). By using the functional, Pečarić has given the main generalization of Theorem 4.2.3 in [150] which is as follows:

$$\mathcal{C}(f, p) = \int_a^b \int_a^b p(x, y) f(x, x) dy\, dx - \int_a^b \int_a^b p(x, y) f(x, y) dy\, dx, \tag{4.2.5}$$

where the functions f and p are integrable.

Theorem 4.2.4 *Let p be integrable function and stated as $p : I^2 \to \mathbb{R}$, such that*

$$Y(x, x) = \overline{Y}(x, x) \quad \text{for every } x \text{ belongs to } I$$

and let either

$$Y(x, y) \geq 0, \quad a \leq y \leq x \leq b; \quad \overline{Y}(x, y) \geq 0, \quad a \leq x \leq y \leq b$$

or their reverse inequalities be valid, where

$$Y(x, y) = \int_x^b \int_a^y p(s, t)\, dt\, ds$$

and

$$\overline{Y}(x, y) = \int_a^x \int_y^b p(s, t)\, dt\, ds.$$

If function $f : I^2 \to \mathbb{R}$ has continuous partial derivatives, i.e., $f_{(0,1)} = \frac{\partial}{\partial y} f(x, y)$, $f_{(1,0)} = \frac{\partial}{\partial x} f(x, y)$ and $f_{(1,1)} = \frac{\partial^2}{\partial x \partial y} f(x, y)$, then there is existence of v, $\zeta \in [a, b]$, such that

$$\mathcal{C}(f, p) = \mathcal{C}((x - a)(y - a), p) f_{(1,1)}(v, \zeta). \tag{4.2.6}$$

Now its time to describe discrete Čebyšev identity and inequality which are stated as following [150]. Let

$$\mathcal{C}^\Delta(a, p) = \sum_{i=1}^M \sum_{j=1}^M p_{ij} a_{ii} - \sum_{i=1}^M \sum_{j=1}^M p_{ij} a_{ij}, \tag{4.2.7}$$

where $p_{ij}, a_{ij} \in \mathbb{R}; i, j \in \{1, \ldots, M\}$.

Theorem 4.2.5 *The following given inequality*

$$\mathcal{C}^\Delta(a, p) \geq 0 \tag{4.2.8}$$

holds, $\forall$ real numbers a_{ij}, for $i, j \in \{1, 2, \ldots, M\}$ such that $\Delta^{(1,1)}a_{ij} \geq 0$ for $i, j \in \{1, 2, \ldots, M-1\}$, iff

$$Y_{j+1,j} = \overline{Y}_{j,j+1} \quad j \in \{1, 2, \ldots, M-1\}$$

and

$$Y_{ij} \geq 0, \quad i \in \{j+1, \ldots, m\} \quad for \quad j \in \{1, 2, \ldots, M-1\}$$
$$\overline{Y}_{ij} \geq 0, \quad i \in \{1, 2, \ldots, j-1\} \quad for \quad j \in \{2, 3, \ldots, M\}$$

hold. The reverse inequality of above (4.2.8) is also valid for $i, j \in \{1, 2, \ldots, M-1\}$, if $\Delta^{(1,1)}a_{ij} \leq 0$, where

$$Y_{ij} = \sum_{r=i}^{M}\sum_{s=1}^{j} p_{rs} \quad and \quad \overline{Y}_{ij} = \sum_{r=1}^{i}\sum_{s=j}^{M} p_{rs}.$$

Ky Fan [58] proposed the following result in 1952, as a problem (see also [133]):

Theorem 4.2.6 *Let $(x, y) \mapsto v(x, y)$ be a function of non-negative Lebesgue integrable over square $\{(x, y) : a \leq x \leq b \ ; \ a \leq y \leq b\}$ and let D be positive constant such that $\int_a^b v(x, y)dx \leq D$ and $\int_a^b v(x, y)dy \leq D$ for almost all y and $x \in [a, b]$ respectively. If f and h finite valued functions and both are non-increasing and non-negative in the interval $[a, b]$, then inequality*

$$D\int_a^b f(x)h(x)dx \geq \int_a^b \int_a^b v(x, y)f(x)h(y)dx\,dy \qquad (4.2.9)$$

holds.

Remark 4.2.1 If put $v(x, y) = constant$, then (4.2.9) becomes special case of inequality (4.2.1).

In [150] J. Pečarić considered the following expression for f, p and q integrable functions for generalization of result of K. Fan

$$R(f, p, q) = \int_a^b q(x)f(x, x)dx - \int_a^b \int_a^b p(x, y)f(x, y)dx\,dy \qquad (4.2.10)$$

and gave the result as follows.

Theorem 4.2.7 *Let $q : I \to \mathbb{R}$ and $p : I^2 \to \mathbb{R}$ be both integrable functions, such that*

$$P_1(x, y) \leq S_1(max\{x, y\}); \quad P_1(x, a) = S_1(x), \quad P_1(a, y) = S_1(y), \quad \forall x, y \in [a, b],$$

where $S_1(x) = \int_x^b q(t)dt, \quad P_1(x, y) = \int_x^b \int_y^b p(s, t)dt\,ds.$

4.2 Generalized Čebyšev and Ky Fan Identities and Inequalities for ∇-Convex...

If $f : I^2 \to \mathbb{R}$ has $f_{(0,1)}$, $f_{(1,0)}$, and $f_{(1,1)}$ continuous partial derivatives on I^2. Then there is existence of $(v, \zeta) \in I^2$, such that

$$R(f, p, q) = f_{(1,1)}(v, \zeta) R((x-a)(y-a), p, q) \quad \text{for} \quad v, \zeta \in [a, b].$$

We give following theorem for our chapter from above theorem.

Theorem 4.2.8 Let $q : I \to \mathbb{R}$ and $p : I^2 \to \mathbb{R}$ be both integrable functions, such that

$$P(x, y) \leq S(\max\{x, y\}); \quad P(x, b) = S(x), \quad P(b, y) = S(y), \quad \forall x, y \in [a, b],$$

where $S(x) = \int_a^x q(t)dt$, $P(x, y) = \int_a^x \int_a^y p(s, t) dt\, ds$.
If $f \in C''(I^2)$, then there exists $(v, \zeta) \in I^2$ such that

$$\overline{R}(f, p, q) = f_{(1,1)}(v, \zeta) R((b-x)(b-y), p, q). \tag{4.2.11}$$

Under the assumptions of Theorem 4.2.8, we would like to use some notations for easy to present the statements for following upcoming theorems:

$$P^{(i,j)}(x, y) = \int_a^x \int_a^y p(s, t) \frac{(x-s)^i}{i!} \frac{(y-t)^j}{j!} dt\, ds, \tag{4.2.12}$$

$$\overline{P}^{(i,j)}(x, y) = \int_a^x \int_a^y p(s, t) \frac{(x-s)^i}{i!} \frac{(y-s)^j}{j!} dt\, ds, \tag{4.2.13}$$

$$S^{(i,j)}(x) = \int_a^x q(s) \frac{(x-s)^i}{i!} \frac{(b-s)^j}{j!} ds, \tag{4.2.14}$$

$$\Upsilon(x, y) = \int_a^{\max\{x,y\}} \int_a^b p(s, t) \frac{(x-s)^M}{M!} \frac{(y-s)^N}{N!} dt\, ds$$

$$- \int_a^x \int_a^y p(s, t) \frac{(x-s)^M}{M!} \frac{(y-t)^N}{N!} dt\, ds, \tag{4.2.15}$$

$$\overline{\Upsilon}(x, y) = \int_a^{\max\{x,y\}} q(s) \frac{(x-s)^M}{M!} \frac{(y-s)^N}{N!} ds$$

$$- \int_a^x \int_a^y p(s, t) \frac{(x-s)^M}{M!} \frac{(y-t)^N}{N!} dt\, ds. \tag{4.2.16}$$

Now at this stage we require useful definition which is taken from [180] and for this purpose we also require few notations that are follow: let $B = [a, b] \times [c, d]$ represents rectangle in two dimension, $S(B)$ represents the all rectangles system $[x_1, x_2] \times [y_1, y_2]$ contained in B and provided that a function $\omega : B \to \mathbb{R}$, we put $E_\omega([x_1, x_2] \times [y_1, y_2]) = \omega(x_2, y_2) - \omega(x_2, y_1) - \omega(x_1, y_2) + \omega(x_1, y_1)$ for $[x_1, x_2] \times [y_1, y_2] \in S(B)$. The function of rectangles $E_\omega : S(B) \to \mathbb{R}$ is just stated as a function of rectangles associated with ω.

Definition 4.2.1 A function $\omega : B \to \mathbb{R}$ is known as absolutely continuous in the Carathéodory's sense on B, if the following statements hold:

(a) The rectangles function E_ω associated with ω is absolutely continuous, i.e. $\forall\; \epsilon > 0, \exists\; \delta > 0$ such that, if $P_1, P_2, \ldots, P_j \in S(B)$ are rectangles (mutually non-overlapped) with property $\sum_{j=1}^{i} |P_j| \leq \delta$, where $|\cdot|$ denotes the rectangle area, then $\sum_{j=1}^{i} E_\omega(P_j) \leq \epsilon$.

(b) The functions $\omega(\cdot, c) : [c, d] \to \mathbb{R}$ and $\omega(a, \cdot) : [a, b] \to \mathbb{R}$ are absolutely continuous.

If function $\omega : B \to \mathbb{R}$ is absolutely continuous in the Carathéodory's sense on B, then it admits the representation of integral $\forall\; (x, y) \in B$

$$\omega(x, y) = \omega(a, c) + \int_a^x \omega_{(1,0)}(s, c)\, ds + \int_c^y \omega_{(0,1)}(a, t)\, dt + \int_a^x \int_c^y \omega_{(1,1)}(s, t)\, dt\, ds, \tag{4.2.17}$$

provided that the existence of partial derivatives almost everywhere in the above equation (see more details in [133]).

Let $q : I \to \mathbb{R}$ and $f, p : I^2 \to \mathbb{R}$ be functions and in which p, q are integrable and there should be existence of $f_{(M,N)}$ with absolutely continuous in the Carathéodory's sense, then for this chapter $\overline{R}(f, p, q)$ and $\overline{C}(f, p)$ are defined as:

$$\overline{C}(f, p) = C(f, p) - \sum_{i=0}^{M} \sum_{j=0}^{N} \left[\overline{P}^{(i,j)}(b, b) - P^{(i,j)}(b, b) \right] f_{(i,j)}(b, b)$$

$$- \sum_{j=0}^{N} \int_a^b \left[\overline{P}^{(M,j)}(x, b) - P^{(M,j)}(x, b) \right] f_{(M+1,j)}(x, b)\, dx$$

$$- \sum_{i=0}^{M} \int_a^b \left[\overline{P}^{(i,N)}(b, y) - P^{(i,N)}(b, y) \right] f_{(i,N+1)}(b, y)\, dy, \tag{4.2.18}$$

where $C(f, p)$ is stated in (4.2.5).

$$\overline{R}(f, p, q) = R(f, p, q) - \sum_{j=0}^{N} \sum_{i=0}^{M} \left[S^{(i,j)}(b) - P^{(i,j)}(b, b) \right] f_{(i,j)}(b, b)$$

$$- \sum_{j=0}^{N} \int_a^b \left[S^{(M,j)}(x) - P^{(M,j)}(x, b) \right] f_{(M+1,j)}(x, b)\, dx$$

$$- \sum_{i=0}^{M} \int_a^b \left[S^{(i,N)}(y) - P^{(i,N)}(b, y) \right] f_{(i,N+1)}(b, y)\, dy, \tag{4.2.19}$$

where $R(f, p, q)$ is stated in (4.2.10).

4.2.1 Generalized Discrete Čebyšev Identity and Inequality

In present subsection we will get discrete identity and inequality of Čebyšev in sequential manner.

In this subsection we present some important identities followed by some inequalities. For that purpose we would require Theorem 4.1.4 and Corollary 4.1.6 from Sect. 4.1.2.

Now we would obtain our main theorem about discrete Čebyšev's identity in the following.

Theorem 4.2.9 *Let $f : I^2 \to \mathbb{R}$ be function and $(x_i, y_j) \in I^2 = [a, b] \times [a, b]$ be mutually different points, where $i, j \in \{1, 2, \ldots, M\}$. Let p_{ij} be real numbers. Then*

$$C^\nabla(f, p) = \sum_{i=1}^{M} \sum_{j=1}^{M} p_{ij} f(x_i, y_i) - \sum_{i=1}^{M} \sum_{j=1}^{M} p_{ij} f(x_i, y_j)$$

$$= \sum_{k=0}^{n-1} \sum_{t=0}^{m-1} \nabla_{(t,k)} f(x_{M-t}, y_{M-k}) \times$$

$$\times \left[\sum_{s=1}^{M-\max\{t,k\}} \sum_{r=1}^{M-k} p_{sr}(y_M - y_s)^{\{k\}}(x_M - x_s)^{\{t\}} \right.$$

$$\left. - \sum_{s=1}^{M-t} \sum_{r=1}^{M-k} p_{sr}(y_M - y_r)^{\{k\}}(x_M - x_s)^{\{t\}} \right]$$

$$+ \sum_{k=0}^{n-1} \sum_{t=1}^{M-m} \nabla_{(m,k)} f(x_t, y_{M-k})(x_{m+t} - x_t) \times$$

$$\times \left[\sum_{s=1}^{\max\{t,k\}} \sum_{r=1}^{M-k} p_{sr}(y_M - y_s)^{\{k\}}(x_{m+t-1} - x_s)^{\{m-1\}} \right.$$

$$\left. - \sum_{s=1}^{t} \sum_{r=1}^{M-k} p_{sr}(y_M - y_r)^{\{k\}}(x_{m+t-1} - x_s)^{\{m-1\}} \right]$$

$$+ \sum_{k=1}^{M-n} \sum_{t=0}^{m-1} \nabla_{(t,n)} f(x_{M-t}, y_k)(y_{n+k} - y_k) \times$$

$$\times \left[\sum_{s=1}^{M-\max\{t,k\}} \sum_{r=1}^{k} p_{sr}(y_{n+k-1} - y_s)^{\{n-1\}}(x_M - x_s)^{\{t\}} \right.$$

$$-\sum_{s=1}^{M-t}\sum_{r=1}^{k} p_{sr}(y_{n+k-1}-y_r)^{\{n-1\}}(x_M-x_s)^{\{t\}}\Bigg]$$

$$+\sum_{k=1}^{M-n}\sum_{t=1}^{M-m} \nabla_{(m,n)} f(x_t, y_k)(x_{m+t}-x_t)(y_{n+k}-y_k) \times$$

$$\times\Bigg[\sum_{s=1}^{\max\{t,k\}}\sum_{r=1}^{k} p_{sr}(y_{n+k-1}-y_s)^{\{n-1\}}(x_{m+t-1}-x_s)^{\{m-1\}}$$

$$-\sum_{s=1}^{t}\sum_{r=1}^{k} p_{sr}(y_{n+k-1}-y_r)^{\{n-1\}}(x_{m+t-1}-x_s)^{\{m-1\}}\Bigg] \qquad (4.2.20)$$

holds, where $\nabla_{(m,n)} f(x, y) = \Delta_{(m,n)} f(x, y)(-1)^{m+n}$.

Proof By considering the following expression we begin the proof of this theorem

$$\sum_{i=1}^{M}\sum_{j=1}^{M} \tilde{p}_{ij} f(x_i, y_i),$$

where $\tilde{p}_{ij}$ is defined as

$$\tilde{p}_{ij} = \begin{cases} \sum_{r=1}^{M} p_{ir}, & i = j, \\ 0, & i \neq j. \end{cases}$$

$$\sum_{i=1}^{M}\sum_{j=1}^{M} \tilde{p}_{ij} f(x_i, y_i) = \sum_{i=1}^{M}\sum_{j=1}^{M} p_{ij} f(x_i, y_i).$$

We have

$$\sum_{i=1}^{M}\sum_{j=1}^{M} p_{ij} f(x_i, y_i) = \sum_{i=1}^{M}\left(\sum_{j=1}^{M} q_j G_i(y_i)\right),$$

where $p_{ij} = q_j$ and $G_i : y \mapsto f(x_i, y)$. Using (4.1.3) in the inner sum we get

$$\sum_{i=1}^{M}\sum_{j=1}^{M} p_{ij} f(x_i, y_i) = \sum_{i=1}^{M}\sum_{k=0}^{n-1} \nabla_{(k)} G_i(y_{M-k}) \left(\sum_{j=1}^{M-k} q_j (y_M - y_i)^{\{k\}}\right)$$

$$+ \sum_{i=1}^{M}\sum_{k=1}^{M-n} \nabla_{(n)} G_i(y_k)(y_{n+k}-y_k) \left(\sum_{j=1}^{k} q_j (y_{n+k-1} - y_i)^{\{n-1\}}\right)$$

4.2 Generalized Čebyšev and Ky Fan Identities and Inequalities for ∇-Convex... 259

$$= \sum_{k=0}^{n-1}\left(\sum_{i=1}^{M}\nabla_{(k)}G_i(y_{M-k})\left(\sum_{j=1}^{M-k}q_j(y_M-y_i)^{\{k\}}\right)\right)$$

$$+ \sum_{k=1}^{M-n}\left(\sum_{i=1}^{M}\nabla_{(n)}G_i(y_k)(y_{n+k}-y_k)\left(\sum_{j=1}^{k}q_j(y_{n+k-1}-y_i)^{\{n-1\}}\right)\right)$$

$$= \sum_{k=0}^{n-1}\left(\sum_{i=1}^{M}w_iF(x_i)\right) + \sum_{k=1}^{M-n}\left(\sum_{i=1}^{M}v_iH(x_i)\right),$$

where $w_i = \sum_{j=1}^{M-k}q_j(y_M-y_i)^{\{k\}} = \sum_{j=1}^{M-k}p_{ij}(y_M-y_i)^{\{k\}}$, $v_i = \sum_{j=1}^{k}q_j(y_{n+k-1}-y_i)^{\{n-1\}}$, $F(x_i) = \nabla_{(k)}G_i(y_{M-k})$, and $H(x_i) = \nabla_{(n)}G_i(y_k)(y_{n+k}-y_k)$.

Using again (4.1.3) in the inner sums, then we have

$$\sum_{i=1}^{M}\sum_{j=1}^{M}p_{ij}f(x_i,y_i)$$

$$= \sum_{k=0}^{n-1}\sum_{r=0}^{m-1}\nabla_{(r)}F(x_{M-r})\left(\sum_{i=1}^{M-r}w_i(x_M-x_i)^{\{r\}}\right)$$

$$+ \sum_{k=0}^{n-1}\sum_{r=1}^{M-m}\nabla_{(m)}F(x_r)(x_{m+r}-x_r)\left(\sum_{i=1}^{r}w_i(x_{m+r-1}-x_i)^{\{m-1\}}\right)$$

$$+ \sum_{k=1}^{M-n}\sum_{t=0}^{m-1}\nabla_{(t)}H(x_{M-t})\left(\sum_{i=1}^{M-t}v_i(x_M-x_i)^{\{t\}}\right)$$

$$+ \sum_{k=1}^{M-n}\sum_{t=1}^{M-m}\nabla_{(m)}H(x_t)(x_{m+t}-x_t)\left(\sum_{i=1}^{t}v_i(x_{m+t-1}-x_i)^{\{m-1\}}\right)$$

$$= \sum_{k=0}^{n-1}\sum_{r=0}^{m-1}\sum_{i=1}^{M-\max\{r,k\}}\sum_{j=1}^{M-k}p_{ij}(y_M-y_i)^{\{k\}}(x_M-x_i)^{\{r\}}\nabla_{(r,k)}f(x_{M-r},y_{M-k})$$

$$+ \sum_{k=0}^{n-1}\sum_{r=1}^{M-m}\sum_{i=1}^{\max\{r,k\}}\sum_{j=1}^{M-k}p_{ij}(y_M-y_i)^{\{k\}}(x_{m+r-1}-x_i)^{\{m-1\}}\nabla_{(m,k)}f(x_r,y_{M-k})\times$$

$$\times (x_{m+r}-x_r)$$

$$+ \sum_{k=1}^{M-n}\sum_{t=0}^{m-1}\sum_{i=1}^{M-\max\{t,k\}}\sum_{j=1}^{k}p_{ij}(y_{n+k-1}-y_i)^{\{n-1\}}(x_M-x_i)^{\{t\}}\nabla_{(t,n)}f(x_{M-t},y_k)\times$$

$$\times (y_{n+k} - y_k)$$

$$+ \sum_{k=1}^{M-n} \sum_{t=1}^{M-m} \sum_{i=1}^{\max\{t,k\}} \sum_{j=1}^{k} p_{ij}(y_{n+k-1} - y_i)^{\{n-1\}}(x_{m+t-1} - x_i)^{\{m-1\}} \times$$

$$\times \nabla_{(m,n)} f(x_t, y_k)(x_{m+t} - x_t)(y_{n+k} - y_k).$$

If we change $i \to s$, $j \to r$ in all sums and put $r \to t$ in first and second sums, then we obtain the required result by putting values of $\sum_{i=1}^{M} \sum_{j=1}^{M} p_{ij} f(x_i, y_i)$ and Corollary 4.1.6 in $\mathcal{C}^{\nabla}(f, p) = \sum_{i=1}^{M} \sum_{j=1}^{M} p_{ij} f(x_i, y_i) - \sum_{i=1}^{M} \sum_{j=1}^{M} p_{ij} f(x_i, y_j)$. □

Remark 4.2.2 If we put $x_i = i$, $y_j = j$ and $f(x_i, y_j) = f(i, j) = a_{ij}$ in Theorem 4.2.9, then get following corollary.

Corollary 4.2.10 Let p_{ij} and $a_{ij} \in \mathbb{R}$ for $i, j \in \{1, 2, \ldots, M\}$. Then following identity holds

$$\mathcal{C}^{\nabla}(a, p) = \sum_{i=1}^{M} \sum_{j=1}^{M} p_{ij} a_{ii} - \sum_{i=1}^{M} \sum_{j=1}^{M} p_{ij} a_{ij}$$

$$= \sum_{k=0}^{n-1} \sum_{t=0}^{m-1} \nabla_{(t,k)} a_{(M-t, M-k)} \left[\sum_{s=1}^{M-\max\{t,k\}} \sum_{r=1}^{M-k} p_{sr} \frac{(M-s)^{\{t\}}}{t!} \frac{(M-s)^{\{k\}}}{k!} \right.$$

$$\left. - \sum_{s=1}^{M-t} \sum_{r=1}^{M-k} p_{sr} \frac{(M-s)^{\{t\}}}{t!} \frac{(M-r)^{\{k\}}}{k!} \right]$$

$$+ \sum_{k=0}^{n-1} \sum_{t=1}^{M-m} \nabla_{(m,k)} a_{(t, M-k)} \left[\sum_{s=1}^{\max\{t,k\}} \sum_{r=1}^{M-k} p_{sr} \frac{(M-s)^{\{k\}}}{k!} \frac{(m+t-1-s)^{\{m-1\}}}{(m-1)!} \right.$$

$$\left. - \sum_{s=1}^{t} \sum_{r=1}^{M-k} p_{sr} \frac{(M-r)^{\{k\}}}{k!} \frac{(m+t-1-s)^{\{m-1\}}}{(m-1)!} \right]$$

$$+ \sum_{k=1}^{M-n} \sum_{t=0}^{m-1} \nabla_{(t,n)} a_{(M-t, k)} \left[\sum_{s=1}^{M-\max\{t,k\}} \sum_{r=1}^{k} p_{sr} \frac{(M-s)^{\{t\}}}{t!} \frac{(n+k-1-s)^{\{n-1\}}}{(n-1)!} \right.$$

$$\left. - \sum_{s=1}^{M-t} \sum_{r=1}^{k} p_{sr} \frac{(M-s)^{\{t\}}}{t!} \frac{(n+k-1-r)^{\{n-1\}}}{(n-1)!} \right]$$

4.2 Generalized Čebyšev and Ky Fan Identities and Inequalities for ∇-Convex...

$$+ \sum_{k=1}^{M-n} \sum_{t=1}^{M-m} \nabla_{(m,n)} a_{(t,k)} \left[\sum_{s=1}^{\max\{t,k\}} \sum_{r=1}^{k} p_{sr} \frac{(n+k-1-s)^{\{n-1\}}}{(n-1)!} \frac{(m+t-1-s)^{\{m-1\}}}{(m-1)!} \right.$$

$$\left. - \sum_{s=1}^{t} \sum_{r=1}^{k} p_{sr} \frac{(n+k-1-r)^{\{n-1\}}}{(n-1)!} \frac{(m+t-1-s)^{\{m-1\}}}{(m-1)!} \right]$$

Before starting the next theorem, we would like to state few notations, under assumptions of Theorem 4.2.9:

$$\overline{C}^\nabla(f,p) = C^\nabla(f,p) - \sum_{k=0}^{n-1}\sum_{t=0}^{m-1} \nabla_{(t,k)} f(x_{M-t}, y_{M-k}) \times$$

$$\times \left[\sum_{s=1}^{M-\max\{t,k\}} \sum_{r=1}^{M-k} p_{sr}(y_M - y_s)^{\{k\}}(x_M - x_s)^{\{t\}} \right.$$

$$\left. - \sum_{s=1}^{M-t}\sum_{r=1}^{M-k} p_{sr}(y_M - y_r)^{\{k\}}(x_M - x_s)^{\{t\}} \right]$$

$$- \sum_{k=0}^{n-1}\sum_{t=1}^{M-m} \nabla_{(m,k)} f(x_t, y_{M-k})(x_{m+t} - x_t) \times$$

$$\times \left[\sum_{s=1}^{\max\{t,k\}}\sum_{r=1}^{M-k} p_{sr}(y_M - y_s)^{\{k\}}(x_{m+t-1} - x_s)^{\{m-1\}} \right.$$

$$\left. - \sum_{s=1}^{t}\sum_{r=1}^{M-k} p_{sr}(y_M - y_r)^{\{k\}}(x_{m+t-1} - x_s)^{\{m-1\}} \right]$$

$$- \sum_{k=1}^{M-n}\sum_{t=0}^{m-1} \nabla_{(t,n)} f(x_{M-t}, y_k)(y_{n+k} - y_k) \times$$

$$\times \left[\sum_{s=1}^{M-\max\{t,k\}}\sum_{r=1}^{k} p_{sr}(y_{n+k-1} - y_s)^{\{n-1\}}(x_M - x_s)^{\{t\}} \right.$$

$$\left. - \sum_{s=1}^{M-t}\sum_{r=1}^{k} p_{sr}(y_{n+k-1} - y_r)^{\{n-1\}}(x_M - x_s)^{\{t\}} \right], \qquad (4.2.21)$$

$$\Upsilon^\nabla(t,k) = \left[\sum_{s=1}^{\max\{t,k\}} \sum_{r=1}^{k} p_{sr}(y_{n+k-1} - y_s)^{\{n-1\}}(x_{m+t-1} - x_s)^{\{m-1\}} \right.$$

$$\left. - \sum_{s=1}^{t} \sum_{r=1}^{k} p_{sr}(y_{n+k-1} - y_r)^{\{n-1\}}(x_{m+t-1} - x_s)^{\{m-1\}} \right]. \quad (4.2.22)$$

Theorem 4.2.11 *Let (x_i) and (y_j) for $i, j \in \{1, 2, \ldots, M\}$ be real sequences and monotonic in the same sense and f is ∇-convex function of order (m, n) and $p_{ij} \in \mathbb{R}$ for $i, j \in \{1, 2, \ldots, M\}$. Then*

$$\overline{C}^\nabla(f, p) \geq 0 \quad \text{if} \quad \Upsilon^\nabla(t, k) \geq 0; \quad t \in \{m+1, \ldots, M\}, \quad k \in \{n+1, \ldots, M\}.$$

Where $\overline{C}^\nabla(f, p)$ and $\Upsilon^\nabla(t, k)$ are stated respectively in (4.2.21) and (4.2.22).

Proof This result can easily obtain using (4.2.20). □

Remark 4.2.3 If we put $x_i = i$, $y_j = j$ and $f(x_i, y_j) = f(i, j) = a_{ij}$ in last theorem for $m = n = 1$ then can obtain similar result for ∇-convex function of Theorem 3 of paper [150] and therefore in this result for $a_{ij} = f(a_i, b_j)$ we can also get similar result for ∇-convex function of Corollary 2 of paper [150].

Theorem 4.2.12 *Let $(x_i, y_j) \in I^2 = [a, b] \times [a, b]$ where $i, j \in \{1, 2, \ldots, M\}$, be mutually different elements, $p_{ij} \in \mathbb{R}$ for $i, j \in \{1, 2, \ldots, M\}$ and suppose that $f, h : I^2 \to \mathbb{R}$ be $(m, n) - \nabla$-convex functions, such that inequalities*

$$\Upsilon^\nabla(t, k) \geq 0; \quad t \in \{m+1, \ldots, M\}, \quad k \in \{n+1, \ldots, M\} \quad (4.2.23)$$

and

$$L\nabla_{(m,n)} h(x_i, y_j) \leq \nabla_{(m,n)} f(x_i, y_j) \leq U\nabla_{(m,n)} h(x_i, y_j) \quad (4.2.24)$$

hold, then below are valid

$$L\overline{C}^\nabla(h, p) \leq \overline{C}^\nabla(f, p) \leq U\overline{C}^\nabla(h, p), \quad (4.2.25)$$

where $\Upsilon^\nabla(t, k)$ is stated in (4.2.22) and U, L are some real constants.

Proof Let functions $F_1(x_i, y_j) = f(x_i, y_j) - Lh(x_i, y_j)$ and $F_2(x_i, y_j) = Uh(x_i, y_j) - f(x_i, y_j)$, then $\nabla_{(m,n)} F_1(x_i, y_j) \geq 0$ and $\nabla_{(m,n)} F_2(x_i, y_j) \geq 0$, now using Theorem 4.2.11 we get our required result. □

Remark 4.2.4 If reverse inequalities hold in (4.2.23) and (4.2.24), then inequalities in (4.2.25) remain hold. Further that the reverse inequalities in (4.2.25) are also valid, if reverse of inequality holds in (4.2.23).

4.2 Generalized Čebyšev and Ky Fan Identities and Inequalities for ∇-Convex...

Remark 4.2.5 If put $x_i = i$, $y_j = j$ and $f(x_i, y_j) = f(i,j) = a_{ij}$ and $h(i,j) = b_{ij}$ in previous theorem then we get similar result for ∇-convex function of Theorem 4 of paper [150].

4.2.2 Generalized Integral Čebyšev Identity and Inequality

Now we would obtain our main theorem about integral Čebyšev's identity in the following, for the sake of this purpose we apply the Corollary 4.1.13 from Chap. 4 of this book.

Theorem 4.2.13 *Let $p, f : I^2 \to \mathbb{R}$ be both functions, where p is an integrable and there should be existence of partial derivatives $f_{(M+1,N)}$ and $f_{(M,N+1)}$ those are absolutely continuous, then*

$$\mathcal{C}(f, p) = \int_a^b \int_a^b p(x,y) f(x,x) dy\, dx - \int_a^b \int_a^b p(x,y) f(x,y) dy\, dx$$

$$= \sum_{i=0}^{M} \sum_{j=0}^{N} (-1)^{i+j} \left[\overline{P}^{(i,j)}(b,b) - P^{(i,j)}(b,b) \right] f_{(i,j)}(b,b)$$

$$+ \sum_{j=0}^{N} \int_a^b (-1)^{M+1+j} \left[\overline{P}^{(M,j)}(x,b) - P^{(M,j)}(x,b) \right] f_{(M+1,j)}(x,b)\, dx$$

$$+ \sum_{i=0}^{M} \int_a^b (-1)^{i+N+1} \left[\overline{P}^{(i,N)}(b,y) - P^{(i,N)}(b,y) \right] f_{(i,N+1)}(b,y)\, dy$$

$$+ \int_a^b \int_a^b (-1)^{M+N} \Upsilon(x,y) f_{(M+1,N+1)}(x,y)\, dy\, dx, \qquad (4.2.26)$$

where $P^{(i,j)}$, $\overline{P}^{(i,j)}$, and $\Upsilon(x,y)$ are stated in (4.2.12), (4.2.13), and (4.2.15) respectively.

Proof For fixed x we define a function $f(x,y) = F_x(y)$. Now we write Taylor expansion of $F_x(y)$ as follows:

$$f(x,y) = F_x(y) = \sum_{j=0}^{N} F^{(j)}(b) \frac{(y-b)^j}{j!} + \int_b^y F^{(N+1)}(t) \frac{(y-t)^N}{N!} dt$$

$$= \sum_{j=0}^{N} (-1)^j \frac{(b-y)^j}{j!} f_{(0,j)}(x,b) + \int_y^b (-1)^{N+1} \frac{(t-y)^N}{N!} f_{(0,N+1)}(x,t)\, dt,$$

where $F^{(j)}(b) = f_{(0,j)}(x,b)$ and $F^{(N+1)}(t) = f_{(0,N+1)}(x,t)$.

Now, for $y = x$ we have

$$f(x, x) = \sum_{j=0}^{N}(-1)^j \frac{(b-x)^j}{j!} f_{(0,j)}(x, b) + \int_x^b (-1)^{N+1} \frac{(t-x)^N}{N!} f_{(0,N+1)}(x, t)\, dt,$$

Multiplying above equation by $p(x, y)$ and integrate it by y over the limit a to b, then

$$\int_a^b p(x, y) f(x, x)\, dy = \sum_{j=0}^{N}(-1)^j f_{(0,j)}(x, b) \int_a^b p(x, y) \frac{(b-x)^j}{j!}\, dy$$

$$+ \int_a^b \left(\int_x^b (-1)^{N+1} p(x, y) \frac{(t-x)^N}{N!} f_{(0,N+1)}(x, t)\, dt \right) dy.$$
(4.2.27)

Now we use further representation of functions $x \mapsto f_{(0,j)}(x, b)$ and $x \mapsto f_{(0,N+1)}(x, t)$ by Taylor expansions:

$$f_{(0,j)}(x, b) = \sum_{i=0}^{M}(-1)^i \frac{(b-x)^i}{i!} f_{(i,j)}(b, b)$$

$$+ \int_x^b (-1)^{M+1} \frac{(s-x)^M}{M!} f_{(M+1,j)}(s, b)\, ds,$$

$$f_{(0,N+1)}(x, t) = \sum_{i=0}^{M}(-1)^i \frac{(b-x)^i}{i!} f_{(i,N+1)}(b, t)$$

$$+ \int_x^b (-1)^{M+1} \frac{(s-x)^M}{M!} f_{(M+1,N+1)}(s, t)\, ds.$$

Putting these above formulae in (4.2.27), then

$$\int_a^b p(x, y) f(x, x)\, dy$$

$$= \sum_{j=0}^{N}(-1)^j \left(\sum_{i=0}^{M}(-1)^i \frac{(b-x)^i}{i!} f_{(i,j)}(b, b) \right.$$

$$\left. + \int_x^b (-1)^{M+1} \frac{(s-x)^M}{M!} f_{(M+1,j)}(s, b)\, ds \right) \int_a^b p(x, y) \frac{(b-x)^j}{j!}\, dy$$

$$+ \int_a^b \left(\int_x^b (-1)^{N+1} p(x, y) \left(\sum_{i=0}^{M}(-1)^i \frac{(b-x)^i}{i!} f_{(i,N+1)}(b, t) \right. \right.$$

4.2 Generalized Čebyšev and Ky Fan Identities and Inequalities for ∇-Convex...

$$+ \int_x^b (-1)^{M+1} \frac{(s-x)^M}{M!} f_{(M+1,N+1)}(s,t) ds \bigg) \frac{(t-x)^N}{N!} dt \bigg) dy$$

$$= \sum_{j=0}^{N} \left(\sum_{i=0}^{M} (-1)^{i+j} \frac{(b-x)^i}{i!} f_{(i,j)}(b,b) \right) \int_a^b p(x,y) \frac{(b-x)^j}{j!} dy$$

$$+ \sum_{j=0}^{N} \left(\int_x^b (-1)^{M+1+j} \frac{(s-x)^M}{M!} f_{(M+1,j)}(s,b) ds \right) \int_a^b p(x,y) \frac{(b-x)^j}{j!} dy$$

$$+ \int_a^b \int_x^b p(x,y) \left(\sum_{i=0}^{M} (-1)^{i+N+1} \frac{(b-x)^i}{i!} f_{(i,N+1)}(b,t) \right) \frac{(t-x)^N}{N!} dt\, dy$$

$$+ \int_a^b \int_x^b \left(\int_x^b (-1)^{M+N} p(x,y) \frac{(s-x)^M}{M!} f_{(M+1,N+1)}(s,t) ds \right) \frac{(t-x)^N}{N!} dt\, dy.$$

Now integrate $p(x,y)f(x,x)$ by x over the limit a to b and obtain:

$$\int_a^b \int_a^b p(x,y)f(x,x) dy\, dx$$

$$= \int_a^b \left[\sum_{j=0}^{N} \left(\sum_{i=0}^{M} (-1)^{i+j} \frac{(b-x)^i}{i!} f_{(i,j)}(b,b) \right) \int_a^b p(x,y) \frac{(b-x)^j}{j!} dy \right] dx$$

$$+ \int_a^b \left[\sum_{j=0}^{N} \left(\int_x^b (-1)^{M+1+j} \frac{(s-x)^M}{M!} f_{(M+1,j)}(s,b) ds \right) \int_a^b p(x,y) \frac{(b-x)^j}{j!} dy \right] dx$$

$$+ \int_a^b \left[\int_a^b \int_x^b p(x,y) \left(\sum_{i=0}^{M} (-1)^{i+N+1} \frac{(b-x)^i}{i!} f_{(i,N+1)}(b,t) \right) \frac{(t-x)^N}{N!} dt\, dy \right] dx$$

$$+ \int_a^b \left[\int_a^b \int_x^b \left(\int_x^b (-1)^{M+N} p(x,y) \frac{(s-x)^M}{M!} f_{(M+1,N+1)}(s,t) ds \right) \frac{(t-x)^N}{N!} dt\, dy \right] dx.$$

Now changing the order of summation in first summand, and use integral linearity property and obtain:

$$\sum_{i=0}^{M} \sum_{j=0}^{N} \int_a^b \int_a^b (-1)^{i+j} p(x,y) \frac{(b-x)^i}{i!} \frac{(b-x)^j}{j!} f_{(i,j)}(b,b) dy\, dx.$$

The second summand is rewritten as:

$$\int_a^b \left[\sum_{j=0}^N \left(\int_x^b (-1)^{M+1+j} \frac{(s-x)^M}{M!} f_{(M+1,j)}(s,b) ds \right) \int_a^b p(x,y) \frac{(b-y)^j}{j!} dy \right] dx$$

$$= \int_a^b \left[\sum_{j=0}^N \left(\int_x^b \int_a^b (-1)^{M+1+j} p(x,y) \frac{(b-y)^j}{j!} \frac{(s-x)^M}{M!} f_{(M+1,j)}(s,b) dy\, ds \right) \right] dx$$

$$= \sum_{j=0}^N \int_a^b \int_x^b \int_a^b (-1)^{M+1+j} p(x,y) \frac{(s-x)^M}{M!} \frac{(b-y)^j}{j!} f_{(M+1,j)}(s,b) \, dy\, ds\, dx$$

$$= \sum_{j=0}^N \int_a^b \int_a^s \int_a^b (-1)^{M+1+j} p(x,y) \frac{(s-x)^M}{M!} \frac{(b-y)^j}{j!} f_{(M+1,j)}(s,b) \, dy\, dx\, ds.$$

Applying Fubini theorem for variables s and x in the last step. Let us first, the change of variable x from $a \to b$ while changing of variable s from $x \to b$. After the change of order of integration, s is changed $a \to b$ while x is changed $a \to s$. In the similar manner the third summand may be rewritten as:

$$\int_a^b \left[\int_a^b \int_x^b p(x,y) \left(\sum_{i=0}^M (-1)^{i+N+1} \frac{(b-x)^i}{i!} f_{(i,N+1)}(b,t) \right) \frac{(t-x)^N}{N!} dt\, dy \right] dx$$

$$= \sum_{i=0}^M \int_a^b \int_a^b \int_x^b (-1)^{i+N+1} p(x,y) \frac{(b-x)^i}{i!} \frac{(t-x)^N}{N!} f_{(i,N+1)}(b,t) \, dt\, dy\, dx$$

$$= \sum_{i=0}^M \int_a^b \int_a^b \int_a^t (-1)^{i+N+1} p(x,y) \frac{(b-x)^i}{i!} \frac{(t-x)^N}{N!} f_{(i,N+1)}(b,t) \, dy\, dt\, dx$$

$$= \sum_{i=0}^M \int_a^b \int_a^b \int_a^t (-1)^{i+N+1} p(x,y) \frac{(b-x)^i}{i!} \frac{(t-x)^N}{N!} f_{(i,N+1)}(b,t) \, dy\, dx\, dt.$$

In above using Fubini theorem twice. First, changing t and y, then changing t and x. Therefore, the last summand may be rewritten as:

$$\int_a^b \left[\int_a^b \int_x^b \left(\int_x^b (-1)^{M+N} p(x,y) \frac{(s-x)^M}{M!} f_{(M+1,N+1)}(s,t) ds \right) \right.$$

$$\left. \times \frac{(t-x)^N}{N!} dt\, dy \right] dx$$

$$= \int_a^b \int_a^b \int_x^b \int_x^b (-1)^{M+N} p(x,y) \frac{(s-x)^M}{M!} \frac{(t-x)^N}{N!} f_{(M+1,N+1)}(s,t) \, ds\, dt\, dy\, dx$$

4.2 Generalized Čebyšev and Ky Fan Identities and Inequalities for ∇-Convex...

$$= \int_a^b \int_a^b \int_a^{\max\{s,t\}} \int_a^b (-1)^{M+N} p(x,y) \frac{(s-x)^M}{M!} \frac{(t-x)^N}{N!}$$
$$\times f_{(M+1,N+1)}(s,t)\, dy\, dx\, dt\, ds.$$

Now, adding up all these summand results to obtain:

$$\int_a^b \int_a^b p(x,y) f(x,x)\, dy\, dx$$

$$= \sum_{i=0}^M \sum_{j=0}^N \int_a^b \int_a^b (-1)^{i+j} p(x,y) \frac{(b-x)^i}{i!} \frac{(b-x)^j}{j!} f_{(i,j)}(b,b)\, dy\, dx$$

$$+ \sum_{j=0}^N \int_a^b \int_a^s \int_a^b (-1)^{M+1+j} p(x,y) \frac{(s-x)^M}{M!} \frac{(b-x)^j}{j!} f_{(M+1,j)}(s,b)\, dy\, dx\, ds$$

$$+ \sum_{i=0}^M \int_a^b \int_a^b \int_a^t (-1)^{i+N+1} p(x,y) \frac{(b-x)^i}{i!} \frac{(t-x)^N}{N!} f_{(i,N+1)}(b,t)\, dy\, dx\, dt$$

$$+ \int_a^b \int_a^b \int_a^{\max\{s,t\}} \int_a^b (-1)^{M+N} p(x,y) \frac{(s-x)^M}{M!} \frac{(t-x)^N}{N!}$$
$$\times f_{(M+1,N+1)}(s,t)\, dy\, dx\, dt\, ds.$$

After changing $x \leftrightarrow s$, $y \leftrightarrow t$ on right side, then get

$$\int_a^b \int_a^b p(x,y) f(x,x)\, dy\, dx$$

$$= \sum_{i=0}^M \sum_{j=0}^N \int_a^b \int_a^b (-1)^{i+j} p(s,t) \frac{(b-s)^{i+j}}{i!j!} f_{(i,j)}(b,b)\, dt\, ds$$

$$+ \sum_{j=0}^N \int_a^b \int_a^x \int_a^b (-1)^{M+1+j} p(s,t) \frac{(x-s)^M}{M!} \frac{(b-s)^j}{j!} f_{(M+1,j)}(x,b)\, dt\, ds\, dx$$

$$+ \sum_{i=0}^M \int_a^b \int_a^b \int_a^y (-1)^{i+N+1} p(s,t) \frac{(b-s)^i}{i!} \frac{(y-s)^N}{N!} f_{(i,N+1)}(b,y)\, dt\, ds\, dy$$

$$+ \int_a^b \int_a^b \int_a^{\max\{x,y\}} \int_a^b (-1)^{M+N} p(s,t) \frac{(x-s)^M}{M!} \frac{(y-s)^N}{N!}$$
$$\times f_{(M+1,N+1)}(x,y)\, dt\, ds\, dy\, dx.$$

Now by defined notations, finally we obtain

$$\int_a^b \int_a^b p(x,y) f(x,x) dy\, dx$$

$$= \sum_{i=0}^M \sum_{j=0}^N (-1)^{i+j} \overline{P}^{(i,j)}(b,b) f_{(i,j)}(b,b)$$

$$+ \sum_{j=0}^N \int_a^b (-1)^{M+1+j} \overline{P}^{(M,j)}(x,b) f_{(M+1,j)}(x,b)\, dx$$

$$+ \sum_{i=0}^M \int_a^b (-1)^{i+N+1} \overline{P}^{(i,N)}(b,y) f_{(i,N+1)}(b,y)\, dy$$

$$+ \int_a^b \int_a^b (-1)^{N+M} f_{(M+1,N+1)}(x,y) \int_a^{\max\{x,y\}} \int_a^b p(s,t) \frac{(x-s)^M}{M!}$$

$$\times \frac{(y-s)^N}{N!} dt\, ds\, dy\, dx,$$

where $\overline{P}^{(i,j)}$ is stated in (4.2.13). Use above expression for $\int_a^b \int_a^b p(x,y) f(x,x) dy\, dx$ and Corollary 4.1.13 in the following

$$C(f,p) = \int_a^b \int_a^b p(x,y) f(x,x) dy\, dx - \int_a^b \int_a^b p(x,y) f(x,y) dy\, dx,$$

we get our required identity. □

Remark 4.2.6 If put $f(x,y) = f(x)h(y)$ and $p(x,y) = p(x)p(y)$ in Theorem 4.2.13, then we can give corollary as:

Corollary 4.2.14 *Let $p, h, f : I \to \mathbb{R}$ be three functions such that p is an integrable and there should be existence of derivatives $f^{(M)}$ and $h^{(N)}$ with absolutely continuous, then*

$$T(f,h,p) = T(P_M(f), P_N(h), p) + T(\Upsilon_M(f), P_N(h), p) + T(P_M(f), \Upsilon_N(h), p)$$

$$+ \int_a^b p(x)\, dx \int_a^b \int_a^b \int_a^{\max\{x,y\}} (-1)^{M+N} \frac{(x-s)^M f^{(M+1)}(x)}{M!} \frac{(y-s)^N h^{(N+1)}(y)}{N!}$$

$$\times p(s)\, ds\, dy\, dx - \int_a^b (-1)^{M+N} p(x) \Upsilon_M(f)(x)\, dx \int_a^b p(x) \Upsilon_N(h)(x)\, dx,$$

(4.2.28)

where $P_k(g)(x) = \sum_{i=0}^k \frac{(b-x)^i g^{(i)}(b)}{i!}$, $\Upsilon_k(g)(x) = \int_a^x \frac{(x-s)^M g^{(M+1)}(s)}{M!}\, ds$, $k \in \mathbb{N}$, and g is a function and $T(f,h,p)$ is stated in (4.2.4).

Proof Applying Taylor formula for f and h, then we can easily obtain (4.2.28). □

Corollary 4.2.15 *Let $p, f : I^2 \to \mathbb{R}$ be both functions, p is an integrable and there should be existence of partial derivatives $f_{(M+1,N)}$ and $f_{(M,N+1)}$ with absolutely continuous, then for $\frac{1}{s} + \frac{1}{t} = 1$; $s, t > 1$; we have*

$$|\overline{C}(f, p)| \leq \left(\int_a^b \int_a^b |\Upsilon(x,y)|^t \, dy \, dx \right)^{\frac{1}{t}}$$

$$\times \left(\int_a^b \int_a^b |(-1)^{M+N} f_{(M+1,N+1)}(x,y)|^s \, dy \, dx \right)^{\frac{1}{s}}, \qquad (4.2.29)$$

where $\overline{C}(f, p)$ and $\Upsilon(x, y)$ are stated in (4.2.18) and (4.2.15) respectively.

Proof Applying Hölder inequality for integrals on Theorem 4.2.13, we may obtain (4.2.29). □

Theorem 4.2.16 *Let $p, f : I^2 \to \mathbb{R}$ be functions, p is integrable and f is $(M+1, N+1) - \nabla$-convex, then*

$$\overline{C}(f, p) \geq 0 \quad \text{if} \quad \Upsilon(x, y) \geq 0 \quad \forall x, y \in [a, b],$$

where $\Upsilon(x, y)$ and $\overline{C}(f, p)$ are stated in (4.2.15) and (4.2.18) respectively.

Proof If function f is ∇-convex function of $(M+1, N+1)$th order and on domain, it can be approximated uniformly by polynomials containing non-negative $(M+1, N+1)$th order partial derivatives. From polynomials of Bernstein

$$B^{m,n}(x, y) = \sum_{i=0}^{m} \sum_{j=0}^{n} \binom{n}{j}\binom{m}{i} f(a_i, b_j)(x-a)^i (y-a)^j (b-y)^{n-j}(b-x)^{m-i},$$

(where $k = (b-a)/n$ and $h = (b-a)/m$) converge to function f uniformly in the domain I^2 limits as $n \to \infty$, $m \to \infty$ provided that function is continuous. Furthermore, if function f $(M+1, N+1)$th order ∇-convex function, where polynomial containing non-negative $(M+1, N+1)$th order partial derivatives, that is, $(-1)^{M+N} B^{m,n}_{(M+1,N+1)} \geq 0$, applying following formula it may be proved by using method of induction:

$$(-1)^{M+N} B^{m,n}_{(M+1,N+1)}(x, y)$$

$$= (N+1)!(M+1)! \binom{n}{N+1}\binom{m}{M+1} \sum_{i=0}^{m-M-1} \sum_{j=0}^{n-N-1}$$

$$\times \binom{n-N-1}{j}\binom{m-M-1}{i} \times$$

$$\times \left(\Delta_{h,k}^{(M+1,N+1)} f(a+ih, a+jk)\right)(y-a)^j(b-y)^{n-N-1-j}$$

$$\times (x-a)^i(b-x)^{m-M-1-i}$$

$$= ((N+1)!)^2((M+1)!)^2 h^{M+1} k^{N+1} \binom{n}{N+1}\binom{m}{M+1}^{m-M-1}\sum_{i=0}$$

$$\times \sum_{j=0}^{n-N-1} \binom{m-M-1}{i}$$

$$\times \binom{n-N-1}{j}\left(\nabla_{(M+1,N+1)} f(a_i, b_j)\right)(y-a)^j(b-y)^{n-N-1-j}$$

$$\times (x-a)^i(b-x)^{m-M-1-i},$$

where $a_i = a+ih$, $b_j = a+jk$ and as (a_i) and (b_j) increasing sequences.

Since f is $(M+1, N+1)th$ order ∇-convex, so $0 \leq \nabla_{(M+1,N+1)} f$. Since $\Upsilon(x, y)$ is continuous and $(-1)^{M+N} B_{(M+1,N+1)}^{m,n} \geq 0$ in the domain I^2 so by (4.2.18), we get

$$\overline{C}(B^{m,n}, p)$$
$$= \int_a^b \int_a^b (-1)^{N+M} B_{(M+1,N+1)}^{m,n}(x, y)$$
$$\times \left[\int_a^{max\{x,y\}} \int_a^b p(s, t)\frac{(y-s)^N}{N!}\frac{(x-s)^M}{M!} dt\, ds\right.$$
$$\left. - \int_a^x \int_a^y p(s, t)\frac{(x-s)^M}{M!}\frac{(y-t)^N}{N!} dt\, ds\right] dy\, dx \geq 0$$

or we can write

$$\overline{C}(B^{m,n}, p) = \int_a^b \int_a^b \Upsilon(x, y)(-1)^{M+N} B_{(M+1,N+1)}^{m,n}(x, y) dy\, dx \geq 0. \tag{4.2.30}$$

Now convergence of $B_{(M+1,N+1)}^{m,n}$ uniformly to $f_{(M+1,N+1)}$ by supposing $n, m \to \infty$ through an appropriate sequence, provides required result. □

Theorem 4.2.17 *Let $p, f : I^2 \to \mathbb{R}$ be two functions, where $f \in C^{(M+1,N+1)}$ is $(M+1, N+1)th$ order ∇-convex function in the interval I^2 and p is integrable. If*

$$\Upsilon(x, y) \geq 0$$

4.2 Generalized Čebyšev and Ky Fan Identities and Inequalities for ∇-Convex...

holds, $\forall x, y$ belong to $[a, b]$, $\exists v, \zeta$ belong to $[a, b]$, such that

$$\overline{C}(f, p) = (-1)^{M+N} C(G_0, p)\, f_{(M+1,N+1)}(v, \zeta), \qquad (4.2.31)$$

where

$$G_0(x, y) = (-1)^{M+N} \frac{(b-x)^{M+1}(b-y)^{N+1}}{(M+1)!(N+1)!} \qquad (4.2.32)$$

and $\overline{C}(f, p)$, $\Upsilon(x, y)$ are stated in (4.2.18) and (4.2.15) respectively.

Proof Since

$$\overline{C}(f, p) = \int_a^b \int_a^b (-1)^{N+M} \Upsilon(x, y) f_{(M+1,N+1)}(x, y) dy\, dx,$$

by applying mean value theorem for the purpose of double integrals, then obtain

$$\overline{C}(f, p) = (-1)^{N+M} f_{(M+1,N+1)}(v, \zeta) \int_a^b \int_a^b \Upsilon(x, y) dy\, dx.$$

In above equation, if put $f(x, y) = G_0(x, y)$ then we can write as:

$$\overline{C}(G_0, p) = C(G_0, p) = \int_a^b \int_a^b \Upsilon(x, y)\, dy\, dx$$

and hence we get what we wanted. □

Remark 4.2.7 By putting $f(x, y) = f(x)h(y)$ and $p(x, y) = p(x)p(y)$ in Theorem 4.2.17 with $M = N = 0$, then can obtain similar result for ∇-convex function of (4.2.3).

Theorem 4.2.18 *Let $h, f : I^2 \to \mathbb{R}$ be both functions and $p : I^2 \to \mathbb{R}$ is integrable, such that f is $(M+1, N+1) - \nabla$-convex function and $h \in C^{(M+1,N+1)}(I^2)$ with $h_{(M+1,N+1)} \neq 0$ on I^2. If*

$$\Upsilon(x, y) \geq 0 \qquad \forall x, y \text{ belong to } [a, b]$$

holds, $\exists v, \zeta$ belong to $[a, b]$, such that

$$\overline{C}(f, p) = \frac{f_{(M+1,N+1)}(v, \zeta)}{h_{(M+1,N+1)}(v, \zeta)} \overline{C}(h, p),$$

where $\Upsilon(x, y)$ and $\overline{C}(f, p)$ are stated in (4.2.15) and (4.2.18) respectively.

Proof (Method I) Applying mean value theorem of integral and (4.2.31), then

$$\overline{C}(f,p) = \int_a^b \int_a^b (-1)^{M+N} \frac{f_{(M+1,N+1)}(x,y)}{h_{(M+1,N+1)}(x,y)} h_{(M+1,N+1)}(x,y) \Upsilon(x,y) dy\,dx$$

$$= \frac{f_{(M+1,N+1)}(\nu,\zeta)}{h_{(M+1,N+1)}(\nu,\zeta)} \int_a^b \int_a^b (-1)^{N+M} \Upsilon(x,y) h_{(M+1,N+1)}(x,y) dy\,dx$$

$$= \frac{f_{(M+1,N+1)}(\nu,\zeta)}{h_{(M+1,N+1)}(\nu,\zeta)} \overline{C}(h,p).$$

$\square$

Proof (Method II) Let $u \in C^{(M+1,N+1)}$ be $(M+1, N+1)$th order ∇-convex function in the interval $I \times J$, is stated as:

$$u = \overline{C}(h,p)f - \overline{C}(f,p)h,$$

by applying Theorem 4.2.17, $\exists\, \nu, \zeta \in I$, such that

$$0 = \overline{C}(u,p) = (-1)^{N+M} \overline{C}(G_0,p) u_{(M+1,N+1)}(\nu,\zeta)$$

or

$$[\overline{C}(h,p) f_{(M+1,N+1)}(\nu,\zeta) - \overline{C}(f,p) h_{(M+1,N+1)}(\nu,\zeta)] \overline{C}(G_0,p) = 0.$$

This gives required result. $\square$

Remark 4.2.8 If we put $M = N = 0$ in Theorem 4.2.18, then we get similar result for ∇-convex function of Theorem 2 of [150].

Theorem 4.2.19 *Let $p, f : I^2 \to \mathbb{R}$ be two functions, f is $(M+1, N+1)$th order ∇-convex and p is integrable. $\exists\, \nu, \zeta \in [a,b]$, such that*

$$\overline{C}(f,p) = (-1)^{M+N} \Upsilon(\nu,\zeta) \big(f_{(M,N)}(b,b) - f_{(M,N)}(b,a)$$
$$- f_{(M,N)}(a,b) + f_{(M,N)}(a,a)\big),$$

where $\Upsilon(x,y)$ and $\overline{C}(f,p)$ are stated in (4.2.15) and (4.2.18).

Proof Since $(-1)^{M+N} B^{m,n}_{(M+1,N+1)} \geq 0$ in the interval I^2 and $\Upsilon(x,y)$ is continuous, here $B^{m,n}$ is polynomial of Bernstein, using same statement that was applied in the proof of the Theorem 4.2.17, we start from (4.2.30), we obtain

$$\overline{C}(B^{m,n},p) = \int_a^b \int_a^b (-1)^{M+N} \Upsilon(x,y) B^{m,n}_{(M+1,N+1)}(x,y) dy\,dx$$

$$= (-1)^{M+N} \Upsilon(\nu_{m,n}, \zeta_{m,n}) \int_a^b \int_a^b B^{m,n}_{(M+1,N+1)}(x,y) dy\,dx$$

$$= (-1)^{N+M} \Upsilon (v_{m,n}, \zeta_{m,n}) \left(B^{m,n}_{(M,N)}(b,b) - B^{m,n}_{(M,N)}(a,b) \right.$$
$$\left. - B^{m,n}_{(M,N)}(b,a) + B^{m,n}_{(M,N)}(a,a) \right).$$

The points $x_{m,n} = (v_{m,n}, \zeta_{m,n})$ have a limit point (v, ζ) on I^2 as $m, n \to \infty$, so the uniform convergence of $B^{m,n}_{(M,N)}$ to $f_{(M,N)}$ by letting $m, n \to \infty$ through an appropriate sequence, gives our desired result. □

Remark 4.2.9 For case $M = N = 0$ in Theorem 4.2.19, we may get similar result for ∇-convex function of Theorem 6 of [150].

4.2.3 Generalized Integral Ky Fan Identity and Inequality

According to MathSciNet, Ky Fan (1914–2010) published 126 research papers and books. The contributions of Ky Fan in mathematics, have provided many of influence in development of convex analysis, nonlinear analysis, linear algebra, operator theory, mathematical economics, approximation theory and mathematical programming (see [108]). In literature, there are different kinds of inequalities due to Ky Fan worked in several fields; cf. [26].

Now in this subsection we have to obtain some important identities and inequalities as below:

Theorem 4.2.20 *Let $q : I \to \mathbb{R}$ and $f, p : I^2 \to \mathbb{R}$ be functions, such that q and p are integrable and there should be existence of partial derivatives $f_{(M+1,N)}$ and $f_{(M,N+1)}$ with absolutely continuous, then*

$$R(f, p, q) = \sum_{j=0}^{N} \sum_{i=0}^{M} (-1)^{i+j} \left[S^{(i,j)}(b) - P^{(i,j)}(b,b) \right] f_{(i,j)}(b,b)$$
$$+ \sum_{j=0}^{N} \int_a^b (-1)^{M+1+j} \left[S^{(M,j)}(x) - P^{(M,j)}(x,b) \right] f_{(M+1,j)}(x,b)\, dx$$
$$+ \sum_{i=0}^{M} \int_a^b (-1)^{i+N+1} \left[S^{(i,N)}(y) - P^{(i,N)}(b,y) \right] f_{(i,N+1)}(b,y)\, dy$$
$$+ \int_a^b \int_a^b (-1)^{M+N} \overline{\Upsilon}(x,y) f_{(M+1,N+1)}(x,y)\, dy\, dx,$$

where $S^{(i,j)}$, $P^{(i,j)}$, and $\overline{\Upsilon}(x,y)$ are stated in (4.2.14), (4.2.12), and (4.2.16) respectively.

Proof By applying the substitution

$$\int_a^b p(x, y) dy = q(x),$$

we may prove this theorem in similar manner as done in Theorem's 4.2.13 proof. □

Remark 4.2.10 By putting $f(x, y) = f(x)h(y)$ and $p(x, y) = \frac{q(x)q(y)}{\int_a^b q(t)dt}$ in Theorem 4.2.20, here q is integrable and $\int_a^b q(t)\,dt \neq 0$, then we can give corollary as below.

Corollary 4.2.21 *Let $f, h, q : I \to \mathbb{R}$ be functions, such that q is integrable, where $\int_a^b q(t)\,dt \neq 0$ and there should be existence of derivatives $f^{(M)}$ and $g^{(N)}$ with absolutely continuous. Then*

$$T(f, h, q) = T(P_M(f), P_N(h), q) + T(\Upsilon_M(f), P_N(h), q) + T(P_M(f), \Upsilon_N(h), q)$$
$$+ \int_a^b \int_a^b \int_a^{max\{x,y\}} (-1)^{M+N} \frac{(x-s)^M f^{(M+1)}(x)}{M!} \frac{(y-s)^N h^{(N+1)}(y)}{N!}$$
$$\times q(s)\,ds\,dy\,dx - \int_a^b (-1)^{M+N} \Upsilon_M(f)(x)q(x)\,dx \int_a^b \Upsilon_N(h)(x)q(x)\,dx,$$

where $P_k(g)(x) = \sum_{i=0}^k \frac{(b-x)^i g^{(i)}(b)}{i!}$, $\Upsilon_k(g)(x) = \int_a^x \frac{(x-s)^M g^{(M+1)}(s)}{M!}\,ds$, $k \in \mathbb{N}$, and here g is a function and $T(f, h, p)$ is defined in (4.2.4).

Corollary 4.2.22 *Let $q : I \to \mathbb{R}$ and $f, p : I^2 \to \mathbb{R}$ be functions and also p and q are integrable and there should be existence of partial derivatives $f_{(M+1,N)}$ and $f_{(M,N+1)}$ with absolutely continuous. Then for $\frac{1}{s} + \frac{1}{t} = 1; s, t > 1;$ we have*

$$|\overline{R}(f, p, q)| \leq \left(\int_a^b \int_a^b |(-1)^{M+N} f_{(M+1,N+1)}(x, y)|^s\,dy\,dx \right)^{\frac{1}{s}}$$
$$\times \left(\int_a^b \int_a^b |\overline{\Upsilon}(x, y)|^t\,dy\,dx \right)^{\frac{1}{t}},$$

where $\overline{R}(f, p, q)$ and $\overline{\Upsilon}(x, y)$ are stated in (4.2.19) and (4.2.16) respectively.

Theorem 4.2.23 *Let $q : I \to \mathbb{R}$ and $f, p : I^2 \to \mathbb{R}$ be three functions and also p and q are integrable and function f is $(M+1, N+1)$th order ∇-convex. Then*

$$\overline{R}(f, p, q) \geq 0 \quad \text{if} \quad \overline{\Upsilon}(x, y) \geq 0, \quad \forall x, y \in [a, b],$$

where $\overline{R}(f, p, q)$ and $\overline{\Upsilon}(x, y)$ are stated in (4.2.19) and (4.2.16) respectively.

Remark 4.2.11 We can prove Theorem 4.2.23 in similar manner as done in Theorem's 4.2.16 proof.

Theorem 4.2.24 *Let $q : I \to \mathbb{R}$ and $f, p : I^2 \to \mathbb{R}$ be functions and also p and q are integrable and function f is $(M+1, N+1)$th order ∇-convex and assuming $\forall x, y$ belong to $[a, b]$*

$$\overline{\Upsilon}(x, y) \geq 0.$$

$\exists \, \nu, \zeta \in [a, b]$, *such that*

$$\overline{R}(f, p, q) = (-1)^{N+M} f_{(M+1,N+1)}(\nu, \zeta) R(G_0, p, q),$$

where $\overline{R}(f, p, q)$ and G_0 are stated in (4.2.19) and (4.2.32) respectively.

Remark 4.2.12 We can give proof of Theorem 4.2.24 in similar way as done in Theorem's 4.2.17 proof. Further that we obtain Theorem 4.2.8 from Theorem 4.2.24 by putting $M = N = 0$.

Theorem 4.2.25 *Let $q : I \to \mathbb{R}$ and $f, h, p : I^2 \to \mathbb{R}$ be four functions and also q and p are integrable and function f is $(M+1, N+1) - \nabla$-convex and $h \in C^{(M+1,N+1)}(I^2)$ with $h_{(M+1,N+1)} \neq 0$ in the interval I^2 and assuming $\forall x, y$ belong to $[a, b]$*

$$\overline{\Upsilon}(x, y) \geq 0.$$

Then $\exists \, \nu, \zeta \in [a, b]$, such that

$$\overline{R}(f, p, q) = \frac{f_{(M+1,N+1)}(\nu, \zeta)}{h_{(M+1,N+1)}(\nu, \zeta)} \overline{R}(h, p, q),$$

where $\overline{R}$ and $\overline{\Upsilon}$ are stated in (4.2.19) and (4.2.16) respectively.

Remark 4.2.13 We can give proof of Theorem 4.2.25 in similar ways as done in Theorem's 4.2.18 proof by two different methods. Further, if put $M = N = 0$ in Theorem 4.2.25, then we get similar result for ∇-convex function of Theorem 16 of [150].

4.3 Weighted Montgomery Identities for Higher Order Differentiable Function of Two Variables and Related Inequalities

We would provide double weighted integrals identities of Montgomery for differentiable function of higher order for two variables and by help of those identities we would get generalization of Ostrowski and Grüss type inequalities for weighted integrals for differentiable functions of higher order for two variables.

The some contents of the current section have published in 2019 see [86].

In this section we would deduce the generalizations of weighted integral identities of Montgomery for differentiable function of higher order for two variables and their consequences. Above said identities recapture other known and important identities and inequalities like Ostrowski type, Čebyšev type and Grüss type inequalities.

The topic of Montgomery's identities have many applications and cover other important known identities and inequalities in which involve Ostrowski and Grüss type inequalities (also see in present section). There are many applications of Ostrowski inequalities in the field of numerical integrations and probability theory (see [23, 48, 56, 110, 123, 138, 176]). We can also get special means using such inequalities (see [9, 10]). The special case of Ostrowski type inequalities is Čebyšev inequality which is very popular (see [149, 150]). Also there are many applications of Grüss type inequalities in the numerical integrations and other different fields (see [28, 34, 45, 175]).

Now-a-days, due to rapid advancement of these types of inequalities are very popular. In current section, several generalizations of identities of Montgomery and inequalities of Ostrowski type, Grüss type also proposed by us for differentiable function of higher order. All these identities and inequalities generalize several consequences that we can see in [16, 52, 54, 66, 141, 156] etc.

In this section, we use interval $I \times J = [a, b] \times [c, d] \subset \mathbb{R}^2$. We recall Montgomery identity (1.1.1) from Chap. 1. We present here the generalization of Montgomery identity which is collected from [146].

Theorem 4.3.1 *Let f be function and provided f' is continuous in the interval I. Then*

$$f(x) = \int_a^b v(r)f(r)dr + \int_a^b p_v(x,r)f'(r)dr,$$

holds for weighted Peano kernel p_v, defined as

$$p_v(x,r) = \begin{cases} V(r), & a \leq r \leq x; \\ V(r) - 1, & x < r \leq b; \end{cases}$$

where $v : I \to \mathbb{R}_*$ is some probability density function, i.e., it is a function that satisfies $\int_a^b v(r)dr = 1$ and

$$V(r) = \begin{cases} 0, & r < a; \\ \int_a^r v(v)dv, & r \in [a,b]; \\ 1, & r > b. \end{cases}$$

4.3 Weighted Montgomery Identities for Higher Order Differentiable Function...

The following generalized identities are obtained from [16] and [54] for functions with 2 independent variables.

Theorem 4.3.2 *Let f be function and provided $f_{(1,1)}$ is continuous in the interval $I \times J$. Then*

$$(d-c)(b-a)f(x,y) = -\int_a^b \int_c^d f(r,s)\,ds\,dr + (d-c)\int_a^b f(r,y)\,dr$$

$$+(b-a)\int_c^d f(x,s)\,ds + \int_a^b \int_c^d p(x,r)q(y,s)f_{(1,1)}(r,s)\,ds\,dr$$

and $(d-c)(b-a)f(x,y) = \int_a^b \int_c^d f(r,s)\,ds\,dr + \int_a^b \int_c^d p(x,r)f_{(1,0)}(r,s)\,ds\,dr$

$$+ \int_a^b \int_c^d q(y,s)f_{(0,1)}(r,s)\,ds\,dr + \int_a^b \int_c^d p(x,r)\,q(y,s)f_{(1,1)}(r,s)\,ds\,dr$$

hold, here p and q are Peano kernels as defined above.

In [156] authors gave the identities of weighted Montgomery for two variables functions.

Theorem 4.3.3 *Let function $p : I \times J \to \mathbb{R}$ be integrable and P is stated as*

$$P(x,y) = \int_x^b \int_y^d p(v,\zeta)\,d\zeta\,dv. \tag{4.3.1}$$

If f is a function and provided that $f_{(1,1)}$ is continuous in the interval $I \times J$. Then

$$P(a,c)f(x,y) = \int_a^b \int_c^d p(r,s)f(r,s)\,ds\,dr + \int_a^b \hat{P}(x,r)f_{(1,0)}(r,y)\,dr$$

$$+ \int_c^d \tilde{P}(y,s)f_{(0,1)}(x,s)\,ds - \int_a^b \int_c^d \bar{P}(x,r,y,s)f_{(1,1)}(r,s)\,ds\,dr \tag{4.3.2}$$

holds,

$$\text{where} \quad \hat{P}(x,r) = \begin{cases} \int_a^r \int_c^d p(v,\zeta)\,d\zeta\,dv, & a \leq r \leq x; \\ -P(r,c), & x < r \leq b; \end{cases}$$

$$\tilde{P}(y,s) = \begin{cases} \int_a^b \int_c^t p(v,\zeta)\,d\zeta\,dv, & c \leq s \leq y; \\ -P(a,s), & y < s \leq d; \end{cases}$$

and $\bar{P}(x,r,y,s) = \begin{cases} \int_a^r \int_c^s p(v,\zeta)d\zeta dv, & a \le r \le x, \ c \le s \le y; \\ -\int_r^b \int_c^s p(v,\zeta)d\zeta dv, & x < r \le b, \ c \le s \le y; \\ -\int_a^r \int_s^d p(v,\zeta)d\zeta dv, & a \le r \le x, \ y < s \le d; \\ P(r,s), & x < r \le b, \ y < s \le d. \end{cases}$

Theorem 4.3.4 *Let the assumptions of Theorem 4.3.3 be valid, then identity*

$$P(a,c)f(x,y) = -\int_a^b \int_c^d p(r,s)f(r,s)\,ds\,dr + \int_a^b \int_c^d p(r,s)f(r,y)\,ds\,dr$$

$$+ \int_a^b \int_c^d p(r,s)f(x,s)\,ds\,dr + \int_a^b \int_c^d \bar{P}(x,r,y,s)f_{(1,1)}(r,s)\,ds\,dr \qquad (4.3.3)$$

holds, where $\bar{P}$ is stated in Theorem 4.3.3.

Theorem 4.3.5 *Let the assumptions of Theorem 4.3.3 be valid, then*

$$[P(a,c)]^2 f(x,y) = P(a,c)\int_a^b \int_c^d p(r,s)f(r,s)\,ds\,dr$$

$$+ \int_a^b \left(\int_a^b \int_c^d p(v,s)\hat{P}(x,r)f_{(1,0)}(r,s)\,ds\,dr \right) dv$$

$$+ \int_c^d \left(\int_a^b \int_c^d p(r,\zeta)\tilde{P}(y,s)f_{(0,1)}(r,s)\,ds\,dr \right) d\zeta$$

$$+ \int_a^b \int_c^d \check{P}(x,r,y,s)f_{(1,1)}(r,s)\,ds\,dr$$

holds, where $\hat{P}$, $\tilde{P}$ and $\bar{P}$ are defined in Theorem 4.3.3 and

$$\check{P}(x,r,y,s) = 2\hat{P}(x,r)\tilde{P}(y,s) - P(a,c)\bar{P}(x,r,y,s).$$

The present section is divided into four parts. The 1st part consists on introduction and preliminaries. In the 2nd part, we would give generalized identities of Montgomery for differentiable function of higher order with two independent variables. In 3rd and 4th parts respectively we get the generalization of Ostrowski and Grüss type inequalities for differentiable functions of higher order for two independent variables by applying identities which were proved in 2nd part. These identities and inequalities give the generalization of many important results (see [16, 52, 54, 66, 141, 156] etc.).

4.3.1 Montgomery Identities for Double Weighted Integrals of Higher Order Differentiable Functions

In starting of this part, we would like to state some notations for simplification of the lengthy expressions as:

$$P^{(i,j)}_{\binom{a}{c}\Rightarrow\binom{b}{d}}(x,y) = \int_a^b \int_c^d p(v,\zeta) \frac{(x-v)^i}{i!} \frac{(y-\eta)^j}{j!} d\zeta\, dv, \qquad (4.3.4)$$

$$P^{(0,j)}_{\binom{a}{c}\Rightarrow\binom{b}{d}}(y) = \int_a^b \int_c^d p(v,\zeta) \frac{(y-\zeta)^j}{j!} d\zeta\, dv, \qquad (4.3.5)$$

$$P^{(i,0)}_{\binom{a}{c}\Rightarrow\binom{b}{d}}(x) = \int_a^b \int_c^d p(v,\zeta) \frac{(x-v)^i}{i!} d\zeta\, dv, \qquad (4.3.6)$$

$$R(f;x,y) = -\sum_{i=1}^M \sum_{j=1}^N (-1)^{i+j} f_{(i,j)}(x,y) P^{(i,j)}_{\binom{a}{c}\Rightarrow\binom{b}{d}}(x,y)$$

$$-\sum_{j=1}^N (-1)^j f_{(0,j)}(x,y) P^{(0,j)}_{\binom{a}{c}\Rightarrow\binom{b}{d}}(y)$$

$$-\sum_{i=1}^M (-1)^i f_{(i,0)}(x,y) P^{(i,0)}_{\binom{a}{c}\Rightarrow\binom{b}{d}}(x). \qquad (4.3.7)$$

For our next main theorem of this recent part, we use a Theorem 4.1.12, by using new notations as follows:

Lemma 4.3.1 *Let f has continuous partial derivatives $f_{(i,j)}$ and p, $f : I \times J \to \mathbb{R}$ be both integrable functions, where $i \in \{0, 1, \ldots, M+1\}$, $j \in \{0, 1, \ldots, N+1\}$, then*

$$\int_a^b \int_c^d p(x,y) f(x,y) dy\, dx = \sum_{i=0}^M \sum_{j=0}^N P^{(i,j)}_{\binom{a}{c}\Rightarrow\binom{b}{d}}(b,d)(-1)^{i+j} f_{(i,j)}(b,d)$$

$$+ \sum_{j=0}^N \int_a^b P^{(M,j)}_{\binom{a}{c}\Rightarrow\binom{s}{d}}(s,d)(-1)^{M+1+j} f_{(M+1,j)}(s,d)\, ds$$

$$+ \sum_{i=0}^M \int_c^d P^{(i,N)}_{\binom{a}{c}\Rightarrow\binom{b}{t}}(b,t)(-1)^{i+N+1} f_{(i,N+1)}(b,t)\, dt$$

$$+ \int_a^b \int_c^d P^{(M,N)}_{\binom{a}{c}\Rightarrow\binom{s}{t}}(s,t)(-1)^{M+N} f_{(M+1,N+1)}(s,t)\, dt\, ds.$$
$$(4.3.8)$$

Now, at this stage ready to give our important new theorems as:

Theorem 4.3.6 *Let function f has continuous partial derivatives $f_{(i,j)}$ and $f, p : I \times J \to \mathbb{R}$ be both integrable functions, where $i \in \{0, 1, \ldots, M+1\}$, $j \in \{0, 1, \ldots, N+1\}$, then*

$$f(x,y)P(b,d) = R(f; x, y) + \int_a^b \int_c^d p(s,t) f(s,t) \, dt \, ds$$

$$+ \sum_{j=0}^{N} \int_a^b (-1)^{M+1+j} \hat{P}^{(M,j)}(x,s,y) f_{(M+1,j)}(s,y) \, ds$$

$$+ \sum_{i=0}^{M} \int_c^d (-1)^{i+N+1} \tilde{P}^{(i,N)}(x,y,t) f_{(i,N+1)}(x,t) \, dt$$

$$- \int_a^b \int_c^d (-1)^{M+N} \bar{P}^{(M,N)}(x,s,y,t) f_{(M+1,N+1)}(s,t) \, dt \, ds,$$

(4.3.9)

where

$$\hat{P}^{(M,j)}(x,s,y) = \begin{cases} -P^{(M,j)}_{\binom{a}{c} \rightrightarrows \binom{s}{d}}(s,y), & a \leq s \leq x; \\ P^{(M,j)}_{\binom{s}{c} \rightrightarrows \binom{b}{d}}(s,y), & x < s \leq b; \end{cases}$$

$$\tilde{P}^{(i,N)}(x,y,t) = \begin{cases} -P^{(i,N)}_{\binom{a}{c} \rightrightarrows \binom{b}{t}}(x,t), & c \leq t \leq y; \\ P^{(i,N)}_{\binom{a}{t} \rightrightarrows \binom{b}{d}}(x,t), & y < t \leq d; \end{cases}$$

and

$$\bar{P}^{(M,N)}(x,s,y,t) = \begin{cases} P^{(M,N)}_{\binom{a}{c} \rightrightarrows \binom{s}{t}}(s,t), & a \leq s \leq x, \ c \leq t \leq y; \\ P^{(M,N)}_{\binom{b}{c} \rightrightarrows \binom{s}{t}}(s,t), & x < s \leq b, \ c \leq t \leq y; \\ P^{(M,N)}_{\binom{a}{d} \rightrightarrows \binom{s}{t}}(s,t), & a \leq s \leq x, \ y < t \leq d; \\ P^{(M,N)}_{\binom{b}{d} \rightrightarrows \binom{s}{t}}(s,t), & x < s \leq b, \ y < t \leq d; \end{cases}$$

where $P^{(M,j)}_{() \rightrightarrows ()}(s,y)$, $P^{(M,N)}_{() \rightrightarrows ()}(s,t)$, $P^{(i,N)}_{() \rightrightarrows ()}(x,t)$ and $R(f; x, y)$ are defined in (4.3.5), (4.3.4), (4.3.6) and (4.3.7) respectively.

4.3 Weighted Montgomery Identities for Higher Order Differentiable Function...

Proof Apply Lemma 4.3.1 for $[a, x] \times [c, y]$ and obtain

$$\int_a^x \int_c^y p(s,t) f(s,t) dt \, ds$$

$$= \sum_{i=0}^{M} \sum_{j=0}^{N} P^{(i,j)}_{\binom{a}{c} \Rightarrow \binom{x}{y}}(x,y) (-1)^{i+j} f_{(i,j)}(x,y)$$

$$+ \sum_{j=0}^{N} \int_a^x P^{(M,j)}_{\binom{a}{c} \Rightarrow \binom{s}{y}}(s,y) (-1)^{M+1+j} f_{(M+1,j)}(s,y) \, ds$$

$$+ \sum_{i=0}^{M} \int_c^y P^{(i,N)}_{\binom{a}{c} \Rightarrow \binom{x}{t}}(x,t) (-1)^{i+N+1} f_{(i,N+1)}(x,t) \, dt$$

$$+ \int_a^x \int_c^y P^{(M,N)}_{\binom{a}{c} \Rightarrow \binom{s}{t}}(s,t) (-1)^{M+N} f_{(M+1,N+1)}(s,t) \, dt \, ds$$

$$= \sum_{i=0}^{M} \sum_{j=0}^{N} (-1)^{i+j} f_{(i,j)}(x,y) \left[P^{(i,j)}_{\binom{b}{d} \Rightarrow \binom{a}{c}}(x,y) - P^{(i,j)}_{\binom{b}{d} \Rightarrow \binom{a}{y}}(x,y) - P^{(i,j)}_{\binom{b}{d} \Rightarrow \binom{x}{c}}(x,y) \right.$$

$$\left. + P^{(i,j)}_{\binom{b}{d} \Rightarrow \binom{x}{y}}(x,y) \right] + \sum_{j=0}^{N} \int_a^x (-1)^{M+1+j} f_{(M+1,j)}(s,y)$$

$$\times \left[P^{(M,j)}_{\binom{b}{d} \Rightarrow \binom{a}{c}}(s,y) - P^{(M,j)}_{\binom{b}{d} \Rightarrow \binom{a}{y}}(s,y) \right.$$

$$\left. - P^{(M,j)}_{\binom{b}{d} \Rightarrow \binom{s}{c}}(s,y) + P^{(M,j)}_{\binom{b}{d} \Rightarrow \binom{s}{y}}(s,y) \right] ds + \sum_{i=0}^{M} \int_c^y (-1)^{i+N+1} f_{(i,N+1)}(x,t)$$

$$\times \left[P^{(i,N)}_{\binom{b}{d} \Rightarrow \binom{a}{c}}(x,t) - P^{(i,N)}_{\binom{b}{d} \Rightarrow \binom{a}{t}}(x,t) - P^{(i,N)}_{\binom{b}{d} \Rightarrow \binom{x}{c}}(x,t) + P^{(i,N)}_{\binom{b}{d} \Rightarrow \binom{x}{t}}(x,t) \right] dt$$

$$+ \int_a^x \int_c^y (-1)^{N+M} f_{(M+1,N+1)}(s,t) \left[P^{(M,N)}_{\binom{b}{d} \Rightarrow \binom{a}{c}}(s,t) - P^{(M,N)}_{\binom{b}{d} \Rightarrow \binom{a}{t}}(s,t) \right.$$

$$\left. - P^{(M,N)}_{\binom{b}{d} \Rightarrow \binom{s}{c}}(s,t) + P^{(M,N)}_{\binom{b}{d} \Rightarrow \binom{s}{t}}(s,t) \right] dt \, ds.$$

In similar manner for $[x, b] \times [c, y]$ obtain

$$\int_x^b \int_c^y p(s,t)f(s,t)dt\,ds = -\int_b^x \int_c^y p(s,t)f(s,t)dt\,ds$$

$$= -\sum_{i=0}^M \sum_{j=0}^N (-1)^{i+j} f_{(i,j)}(x,y) \left[-P^{(i,j)}_{\binom{b}{d} \Rightarrow \binom{x}{c}}(x,y) + P^{(i,j)}_{\binom{b}{d} \Rightarrow \binom{x}{y}}(x,y) \right]$$

$$+ \sum_{j=0}^N \int_x^b (-1)^{M+1+j} f_{(M+1,j)}(s,y) \left[-P^{(M,j)}_{\binom{b}{d} \Rightarrow \binom{s}{c}}(s,y) + P^{(M,j)}_{\binom{b}{d} \Rightarrow \binom{s}{y}}(s,y) \right] ds$$

$$- \sum_{i=0}^M \int_c^y (-1)^{i+N+1} f_{(i,N+1)}(x,t) \left[-P^{(i,N)}_{\binom{b}{d} \Rightarrow \binom{x}{c}}(x,t) + P^{(i,N)}_{\binom{b}{d} \Rightarrow \binom{x}{t}}(x,t) \right] dt$$

$$+ \int_x^b \int_c^y (-1)^{N+M} f_{(M+1,N+1)}(s,t) \left[-P^{(M,N)}_{\binom{b}{d} \Rightarrow \binom{s}{c}}(s,t) + P^{(M,N)}_{\binom{b}{d} \Rightarrow \binom{s}{t}}(s,t) \right] dt\,ds.$$

For $[a, x] \times [y, d]$

$$\int_a^x \int_y^d p(s,t)f(s,t)dt\,ds = -\int_a^x \int_d^y p(s,t)f(s,t)dt\,ds$$

$$= -\sum_{i=0}^M \sum_{j=0}^N (-1)^{i+j} f_{(i,j)}(x,y) \left[-P^{(i,j)}_{\binom{b}{d} \Rightarrow \binom{a}{y}}(x,y) + P^{(i,j)}_{\binom{b}{d} \Rightarrow \binom{x}{y}}(x,y) \right]$$

$$- \sum_{j=0}^N \int_a^x (-1)^{M+1+j} f_{(M+1,j)}(s,y) \left[-P^{(M,j)}_{\binom{b}{d} \Rightarrow \binom{a}{y}}(s,y) + P^{(M,j)}_{\binom{b}{d} \Rightarrow \binom{s}{y}}(s,y) \right] ds$$

$$+ \sum_{i=0}^M \int_y^d (-1)^{i+N+1} f_{(i,N+1)}(x,t) \left[-P^{(i,N)}_{\binom{b}{d} \Rightarrow \binom{a}{t}}(x,t) + P^{(i,N)}_{\binom{b}{d} \Rightarrow \binom{x}{t}}(x,t) \right] dt$$

$$+ \int_a^x \int_y^d (-1)^{N+M} f_{(M+1,N+1)}(s,t) \left[-P^{(M,N)}_{\binom{b}{d} \Rightarrow \binom{a}{t}}(s,t) + P^{(M,N)}_{\binom{b}{d} \Rightarrow \binom{s}{t}}(s,t) \right] dt\,ds.$$

Finally $[x, b] \times [y, d]$

$$\int_x^b \int_y^d p(s,t)f(s,t)dt\,ds = \int_b^x \int_d^y p(s,t)f(s,t)dt\,ds$$

$$= \sum_{i=0}^M \sum_{j=0}^N (-1)^{i+j} f_{(i,j)}(x,y) P^{(i,j)}_{\binom{b}{d} \Rightarrow \binom{x}{y}}(x,y)$$

4.3 Weighted Montgomery Identities for Higher Order Differentiable Function...

$$-\sum_{j=0}^{N}\int_{x}^{b}(-1)^{M+1+j}f_{(M+1,j)}(s,y)P_{\binom{b}{d}\Rightarrow\binom{s}{y}}^{(M,j)}(s,y)\,ds$$

$$-\sum_{i=0}^{M}\int_{y}^{d}(-1)^{i+N+1}f_{(i,N+1)}(x,t)P_{\binom{b}{d}\Rightarrow\binom{x}{t}}^{(i,N)}(x,t)\,dt$$

$$+\int_{x}^{b}\int_{y}^{d}(-1)^{M+N}f_{(M+1,N+1)}(s,t)P_{\binom{b}{d}\Rightarrow\binom{s}{t}}^{(M,N)}(s,t)\,dt\,ds.$$

Adding above 4 expressions and obtain desired result. □

Theorem 4.3.7 *Let f has continuous partial derivatives $f_{(M+1,N+1)}$ and $f, p : I \times J \to \mathbb{R}$ be both integrable functions. Then following identity*

$$f(x,y)P(b,d) = R(f;x,y) - \int_{a}^{b}\int_{c}^{d}p(s,t)f(s,t)\,dt\,ds$$

$$+ \int_{a}^{b}\int_{c}^{d}p(s,t)f(s,y)\,dt\,ds + \int_{a}^{b}\int_{c}^{d}p(s,t)f(x,t)\,dt\,ds$$

$$+ \sum_{j=1}^{N}\int_{a}^{b}\int_{c}^{d}(-1)^{j}p(s,\zeta)\frac{(\zeta-y)^{j}}{j!}f_{(0,j)}(s,y)\,d\zeta\,ds$$

$$+ \sum_{i=1}^{M}\int_{a}^{b}\int_{c}^{d}(-1)^{i}p(\nu,t)\frac{(\nu-x)^{i}}{i!}f_{(i,0)}(x,t)\,dt\,d\nu$$

$$+ \int_{a}^{b}\int_{c}^{d}(-1)^{M+N}\bar{P}^{(M,N)}(x,s,y,t)f_{(M+1,N+1)}(s,t)\,dt\,ds$$

(4.3.10)

holds, where $\bar{P}^{(M,N)}(x,s,y,t)$ is given in Theorem 4.3.6, and $P(b,d)$ and $R(f;x,y)$ are stated as (4.3.1) and (4.3.6) respectively.

Proof Firstly, finding expression for the following

$$\int_{a}^{b}(-1)^{M+1+j}\hat{P}^{(M,j)}(x,s,y)f_{(M+1,j)}(s,y)\,ds,$$

using integration by parts as:

$$\int_{a}^{b}(-1)^{M+1+j}\hat{P}^{(M,j)}(x,s,y)f_{(M+1,j)}(s,y)\,ds$$

$$= -\int_{a}^{x}(-1)^{M+1+j}P_{\binom{a}{c}\Rightarrow\binom{s}{d}}^{(M,j)}(s,y)f_{(M+1,j)}(s,y)\,ds$$

$$+ \int_x^b (-1)^{M+1+j} P^{(M,j)}_{\binom{s}{c} \Rightarrow \binom{b}{d}}(s, y) f_{(M+1,j)}(s, y) \, ds$$

$$= -\left(\int_a^x (-1)^{M+1+j} P^{(M,j)}_{\binom{a}{c} \Rightarrow \binom{s}{d}}(s, y) f_{(M+1,j)}(s, y) \, ds \right.$$

$$\left. + \int_x^b (-1)^{M+1+j} P^{(M,j)}_{\binom{b}{c} \Rightarrow \binom{s}{d}}(s, y) f_{(M+1,j)}(s, y) \, ds \right)$$

$$= P^{(M,j)}_{\binom{a}{c} \Rightarrow \binom{x}{d}}(x, y)(-1)^{M+j} f_{(M,j)}(x, y)$$

$$+ \int_a^x P^{(M-1,j)}_{\binom{a}{c} \Rightarrow \binom{s}{d}}(s, y)(-1)^{M+j} f_{(M,j)}(s, y) \, ds$$

$$+ P^{(M,j)}_{\binom{x}{c} \Rightarrow \binom{b}{d}}(x, y)(-1)^{M+j} f_{(M,j)}(x, y)$$

$$+ \int_x^b P^{(M-1,j)}_{\binom{b}{c} \Rightarrow \binom{s}{d}}(s, y)(-1)^{M+j} f_{(M,j)}(s, y) \, ds$$

$$= P^{(M,j)}_{\binom{a}{c} \Rightarrow \binom{b}{d}}(x, y)(-1)^{M+j} f_{(M,j)}(x, y)$$

$$+ \int_a^x P^{(M-1,j)}_{\binom{a}{c} \Rightarrow \binom{s}{d}}(s, y)(-1)^{M+j} f_{(M,j)}(s, y) \, ds$$

$$+ \int_x^b P^{(M-1,j)}_{\binom{b}{c} \Rightarrow \binom{s}{d}}(s, y)(-1)^{M+j} f_{(M,j)}(s, y) \, ds$$

$$= P^{(M,j)}_{\binom{a}{c} \Rightarrow \binom{b}{d}}(x, y)(-1)^{M+j} f_{(M,j)}(x, y)$$

$$+ \int_a^b (-1)^{M+j} \hat{P}^{(M-1,j)}(x, s, y) f_{(M,j)}(s, y) \, ds,$$

continuing in this way we finally get

$$\int_a^b (-1)^{M+1+j} \hat{P}^{(M,j)}(x, s, y) f_{(M+1,j)}(s, y) \, ds$$

$$= \int_a^b \int_c^d p(v, \zeta) \frac{(\zeta - y)^j}{j!} \left[\sum_{i=0}^M \frac{(v - x)^i}{i!} (-1)^{i+j} f_{(i,j)}(x, y) \right] d\zeta \, dv$$

$$- \int_a^b \int_c^d p(s, \zeta) \frac{(\zeta - y)^j}{j!} (-1)^j f_{(0,j)}(s, y) \, d\zeta \, ds. \qquad (4.3.11)$$

4.3 Weighted Montgomery Identities for Higher Order Differentiable Function...

Similarly

$$\int_c^d (-1)^{i+N+1} \tilde{P}^{(i,N)}(x, y, t) f_{(i,N+1)}(x, t) dt$$

$$= \int_a^b \int_c^d p(v, \zeta) \frac{(v-x)^i}{i!} \left[\sum_{k=0}^N \frac{(\zeta-y)^k}{k!} (-1)^{i+k} f_{(i,k)}(x, y) \right] d\zeta \, dv$$

$$- \int_a^b \int_c^d p(v, t) \frac{(v-x)^i}{i!} (-1)^i f_{(i,0)}(x, t) \, dt \, dv. \tag{4.3.12}$$

By putting all above values in (4.3.9) and after some cancellation and rearrangements, we get desired identity. $\square$

Theorem 4.3.8 *Let $f : I \times J \to \mathbb{R}$ be function, has $f \in C^{(2N+1,2M+1)}$ and $\forall$ $(x, y) \in I \times J$, we have*

$$f(x, y)[P(b, d)]^2 = P(b, d)R(f; x, y) + P(b, d) \int_a^b \int_c^d p(s, t) f(s, t) dt \, ds$$

$$+ \sum_{j=0}^N \int_a^b (-1)^{M+1+j} \hat{P}^{(M,j)}(x, s, y) R(f_{(M+1,j)}; s, y) \, ds$$

$$+ \sum_{i=0}^M \int_c^d (-1)^{i+N+1} \tilde{P}^{(i,N)}(x, y, t) R(f_{(i,N+1)}; x, t) \, dt$$

$$+ \sum_{i=0}^M \sum_{j=0}^N \int_a^b \int_a^b \int_c^d (-1)^{M+1+i+j} \hat{P}^{(M,j)}(x, s, y) p(v, t) \frac{(v-x)^i}{i!}$$

$$\times f_{(M+1+i,j)}(s, t) \, dt \, ds \, dv$$

$$+ \sum_{i=0}^M \sum_{j=0}^N \int_c^d \int_a^b \int_c^d (-1)^{i+j+N+1} \tilde{P}^{(i,N)}(x, y, t) p(s, \zeta) \frac{(\zeta-y)^j}{j!}$$

$$\times f_{(i,N+1+j)}(s, t) \, dt \, ds \, d\zeta$$

$$+ 2 \int_a^b \int_c^d \sum_{i=0}^M \sum_{j=0}^N (-1)^{i+j+M+N} \hat{P}^{(M,j)}(x, s, y) \tilde{P}^{(i,N)}(x, y, t)$$

$$\times f_{(M+1+i,N+1+j)}(s, t) \, dt \, ds,$$

$$- P(b, d) \int_a^b \int_c^d (-1)^{M+N} \bar{P}^{(M,N)}(x, s, y, t) f_{(M+1,N+1)}(s, t) \, dt \, ds, \tag{4.3.13}$$

where $\hat{P}^{(M,j)}(x,s,y)$, $\tilde{P}^{(i,N)}(x,y,t)$, $\bar{P}^{(M,N)}(x,s,y,t)$, P and p are as in Theorem 4.3.6.

Proof By Summing (4.3.11) for $j \in \{0, 1, \ldots, N\}$ and (4.3.12) for $i \in \{0, 1, \ldots, M\}$ we obtain respectively.

$$f(x,y)P(b,d) = R(f;x,y) + \sum_{j=0}^{N} \int_a^b \int_c^d (-1)^j p(s,\zeta) \frac{(\zeta-y)^j}{j!} f_{(0,j)}(s,y) \, d\zeta \, ds$$

$$+ \sum_{j=0}^{N} \int_a^b (-1)^{M+1+j} \hat{P}^{(M,j)}(x,s,y) f_{(M+1,j)}(s,y) \, ds \quad (4.3.14)$$

and

$$f(x,y)P(b,d) = R(f;x,y) + \sum_{i=0}^{M} \int_a^b \int_c^d (-1)^i p(v,t) \frac{(v-x)^i}{i!} f_{(i,0)}(x,t) \, dt \, dv$$

$$+ \sum_{i=0}^{M} \int_c^d (-1)^{i+N+1} \tilde{P}^{(i,N)}(x,y,t) f_{(i,N+1)}(x,t) \, dt, \quad (4.3.15)$$

$\forall (x,y) \in I \times J$. Formula (4.3.14) use for partial derivatives $f_{(i,N+1)}$ for $i \in \{0, 1, \ldots, M\}$, so

$$f_{(i,N+1)}(x,t) P(b,d) = R(f_{(i,N+1)}; x, t)$$

$$+ \sum_{j=0}^{N} \int_a^b \int_c^d (-1)^j p(s,\zeta) \frac{(\zeta-t)^j}{j!} f_{(i,N+1+j)}(s,t) \, d\zeta \, ds$$

$$+ \sum_{j=0}^{N} \int_a^b (-1)^{M+1+j} \hat{P}^{(M,j)}(x,s,t) f_{(M+1+i,N+1+j)}(s,t) \, ds. \quad (4.3.16)$$

Formula (4.3.15) use for partial derivatives $f_{(M+1,j)}$ for $j \in \{0, 1, \ldots, N\}$, so

$$f_{(M+1,j)}(s,y) P(b,d) = R(f_{(M+1,j)}; s, y)$$

$$+ \sum_{i=0}^{M} \int_a^b \int_c^d (-1)^i p(v,t) \frac{(v-s)^i}{i!} f_{(M+1+i,j)}(s,t) \, dt \, dv$$

$$+ \sum_{i=0}^{M} \int_c^d (-1)^{i+N+1} \tilde{P}^{(i,N)}(s,y,t) f_{(M+1+i,N+1+j)}(s,t) \, dt. \quad (4.3.17)$$

4.3 Weighted Montgomery Identities for Higher Order Differentiable Function...

Substituting (4.3.16) and (4.3.17) into (4.3.9) and obtain

$$f(x,y)P(b,d) = R(f;x,y) + \int_a^b \int_c^d p(s,t)f(s,t)dt\,ds$$

$$+ \frac{1}{P(b,d)} \sum_{j=0}^{N} \int_a^b (-1)^{M+1+j} \hat{P}^{(M,j)}(x,s,y) \Big[R(f_{(M+1,j)};s,y)$$

$$+ \sum_{i=0}^{M} \int_a^b \int_c^d (-1)^i p(v,t) \frac{(v-s)^i}{i!} f_{(M+1+i,j)}(s,t)\,dt\,dv$$

$$+ \sum_{i=0}^{M} \int_c^d (-1)^{i+N+1} \tilde{P}^{(i,N)}(s,y,t) f_{(M+1+i,N+1+j)}(s,t)\,dt \Big] ds$$

$$+ \frac{1}{P(b,d)} \sum_{i=0}^{M} \int_c^d (-1)^{i+N+1} \tilde{P}^{(i,N)}(x,y,t) \Big[R(f_{(i,N+1)};x,t)$$

$$+ \sum_{j=0}^{N} \int_a^b \int_c^d (-1)^j p(s,\zeta) \frac{(\zeta-t)^j}{j!} f_{(i,N+1+j)}(s,t)\,d\zeta\,ds$$

$$+ \sum_{j=0}^{N} \int_a^b (-1)^{M+1+j} \hat{P}^{(M,j)}(x,s,t) f_{(M+1+i,N+1+j)}(s,t)\,ds \Big] dt$$

$$- \int_a^b \int_c^d (-1)^{M+N} \bar{P}^{(M,N)}(x,s,y,t) f_{(M+1,N+1)}(s,t)\,dt\,ds.$$

We get desired result, after some rearrangements and using theorem of Fubini. □

Remark 4.3.1 For $M = N = 0$, we can get special cases of Theorem 4.3.6, 4.3.7 and 4.3.8 as similar as Theorem 4.3.3, 4.3.4 and 4.3.5 respectively (also see similar case in [156]).

Special Cases

If substitute $p(s,t) = q(s)r(t)$ in (4.3.9), (4.3.10) and (4.3.13) then we obtain the following special cases respectively:

$$f(x,y)P_{a\to b}(q)P_{c\to d}(r) = S(f;x,y) + \int_a^b \int_c^d r(t)q(s)f(s,t)dt\,ds$$

$$+ \sum_{j=0}^{N} \int_a^b (-1)^{M+1+j} \hat{Q}^{(M,j)}(x,s,y) f_{(M+1,j)}(s,y)\,ds$$

$$+ \sum_{i=0}^{M} \int_{c}^{d} (-1)^{i+N+1} \tilde{Q}^{(i,N)}(x,y,t) f_{(i,N+1)}(x,t) \, dt$$

$$- \int_{a}^{b} \int_{c}^{d} \bar{Q}^{(M,N)}(x,s,y,t)(-1)^{M+N} f_{(M+1,N+1)}(s,t) \, dt \, ds,$$

$$f(x,y) P_{a \to b}(q) P_{c \to d}(r) = S(f;x,y) - \int_{a}^{b} \int_{c}^{d} q(s) r(t) f(s,t) \, dt \, ds$$

$$+ \int_{a}^{b} \int_{c}^{d} r(t) q(s) f(s,y) \, dt \, ds + \int_{a}^{b} \int_{c}^{d} r(t) q(s) f(x,t) \, dt \, ds$$

$$+ \sum_{j=1}^{N} \int_{a}^{b} q(s)(-1)^{j} f_{(0,j)}(s,y) \, ds \, S_{c \to d}^{(j)}(r,y)$$

$$+ \sum_{i=1}^{M} S_{a \to b}^{(i)}(q,x) \int_{c}^{d} r(t)(-1)^{i} f_{(i,0)}(x,t) \, dt$$

$$+ \int_{a}^{b} \int_{c}^{d} (-1)^{M+N} \bar{Q}^{(M,N)}(x,s,y,t) f_{(M+1,N+1)}(s,t) \, dt \, ds,$$

$$f(x,y)[P_{a \to b}(q) P_{c \to d}(r)]^{2} = P_{a \to b}(q) P_{c \to d}(r) S(f;x,y)$$

$$+ P_{a \to b}(q) P_{c \to d}(r) \int_{a}^{b} \int_{c}^{d} q(s) r(t) f(s,t) \, dt \, ds$$

$$+ \sum_{j=0}^{N} \int_{a}^{b} (-1)^{M+1+j} \hat{Q}^{(M,j)}(x,s,y) S(f_{(M+1,j)}; s, y) \, ds$$

$$+ \sum_{i=0}^{M} \int_{c}^{d} (-1)^{i+N+1} \tilde{Q}^{(i,N)}(x,y,t) S(f_{(i,M+1)}; x, t) \, dt$$

$$+ \sum_{i=0}^{M} \sum_{j=0}^{N} S_{a \to b}^{(i)}(q,x) \int_{a}^{b} \int_{c}^{d} (-1)^{M+1+i+j} \hat{Q}^{(M,j)}(x,s,y) r(t)$$

$$\times f_{(M+1+i,j)}(s,t) \, dt \, ds$$

$$+ \sum_{i=0}^{M} \sum_{j=0}^{N} S_{c \to d}^{(j)}(r,y) \int_{a}^{b} \int_{c}^{d} (-1)^{i+j+N+1} \tilde{Q}^{(i,N)}(x,y,t) q(s)$$

$$\times f_{(i,N+1+j)}(s,t) \, dt \, ds$$

4.3 Weighted Montgomery Identities for Higher Order Differentiable Function... 289

$$+2\int_a^b \int_c^d \sum_{i=0}^{M} \sum_{j=0}^{N} (-1)^{i+j+M+N} \hat{Q}^{(M,j)}(x,s,y)\tilde{Q}^{(i,N)}(x,y,t)$$

$$\times f_{(M+1+i,N+1+j)}(s,t)\,dt\,ds$$

$$-P_{a\to b}(q)P_{c\to d}(r) \int_a^b \int_c^d (-1)^{N+M} \bar{Q}^{(M,N)}(x,s,y,t) f_{(M+1,N+1)}(s,t)\,dt\,ds,$$

where

$$P_{a\to b}(q) = \int_a^b q(s)\,ds, \qquad S_{a\to b}^{(i)}(q,x) = \int_a^b q(v)\frac{(x-v)^i}{i!}\,dv,$$

$$S_{\binom{a}{c}\Rightarrow\binom{b}{d}}^{(i,j)}(x,y) = S_{a\to b}^{(i)}(q,x) S_{c\to d}^{(j)}(r,y),$$

$$S_{\binom{a}{c}\Rightarrow\binom{b}{d}}^{(0,j)}(y) = P_{a\to b}(q)\, S_{c\to d}^{(j)}(r,y),$$

$$S_{\binom{a}{c}\Rightarrow\binom{b}{d}}^{(i,0)}(x) = S_{a\to b}^{(i)}(q,x)\, P_{c\to d}(r),$$

$$S(f;x,y) = -\sum_{i=1}^{M}\sum_{j=1}^{N} f_{(i,j)}(x,y) S_{\binom{a}{c}\Rightarrow\binom{b}{d}}^{(i,j)}(x,y) - \sum_{j=1}^{N} f_{(0,j)}(x,y) S_{\binom{a}{c}\Rightarrow\binom{b}{d}}^{(0,j)}(y)$$

$$-\sum_{i=1}^{M} f_{(i,0)}(x,y) S_{\binom{a}{c}\Rightarrow\binom{b}{d}}^{(i,0)}(x),$$

$$\hat{Q}^{(M,j)}(x,s,y) = \begin{cases} -S_{\binom{a}{c}\Rightarrow\binom{s}{d}}^{(M,j)}(s,y), & a \le s \le x; \\ +S_{\binom{s}{c}\Rightarrow\binom{b}{d}}^{(M,j)}(s,y), & x < s \le b; \end{cases}$$

$$\tilde{Q}^{(i,N)}(x,y,t) = \begin{cases} -S_{\binom{a}{c}\Rightarrow\binom{b}{t}}^{(i,N)}(x,t), & c \le t \le y; \\ +S_{\binom{a}{t}\Rightarrow\binom{b}{d}}^{(i,N)}(x,t), & y < t \le d; \end{cases}$$

and

$$\bar{Q}^{(M,N)}(x,s,y,t) = \begin{cases} S_{\binom{a}{c}\Rightarrow\binom{s}{t}}^{(M,N)}(s,t), & a \le s \le x,\ c \le t \le y; \\ S_{\binom{b}{c}\Rightarrow\binom{s}{t}}^{(M,N)}(s,t), & x < s \le b,\ c \le t \le y; \\ S_{\binom{a}{d}\Rightarrow\binom{s}{t}}^{(M,N)}(s,t), & a \le s \le x,\ y < t \le d; \\ S_{\binom{b}{d}\Rightarrow\binom{s}{t}}^{(M,N)}(s,t), & x < s \le b,\ y < t \le d. \end{cases}$$

4.3.2 Ostrowski Type Inequalities for Double Weighted Integrals of Higher Order Differentiable Functions

Under this heading we would recall well known Ostrowski inequality (2.0.1) from Chap. 2. This inequality has helped to provide many generalizations see in [156] Pečarić and Vukelić have also provided the generalizations of Ostrowski inequality by applying identities (4.3.2) and (4.3.3). Now in current subsection, by applying identities (4.3.9) and (4.3.10) we would get results for generalization of Ostrowski type inequality for differentiable function of higher order for two variables as below:

Theorem 4.3.9 *Let f be real-valued function in the interval $I \times J$ and $f \in C^{(M+1,N+1)}$ in the same interval, then*

$$\left| f(x,y) - \frac{1}{P(b,d)} \int_a^b \int_c^d p(s,t) f(s,t)\, dt\, ds \right| \leq O(x,y) + \sum_{j=0}^{N} \hat{O}^{(0,j)}(x,y)$$

$$+ \sum_{i=0}^{M} \tilde{O}^{(i,0)}(x,y) + \bar{O}(x,y)$$

holds, $\forall (x,y) \in I \times J$, where

$$O(x,y) = \frac{1}{|P(b,d)|} |R(f;x,y)|,$$

$$\hat{O}^{(0,j)}(x,y) = \frac{1}{|P(b,d)|} \left(\sum_{j=0}^{N} \int_a^b |\hat{P}^{(M,j)}(x,s,y)|^{\hat{q}_j} ds \right)^{1/\hat{q}_j}$$

$$\times \| (-1)^{M+1+j} f_{(M+1,j)} \|_{\hat{p}_j},$$

provided $f_{(M+1,j)} \in L_{\hat{p}_j}(I \times J)$, $1/\hat{p}_j + 1/\hat{q}_j = 1$,

$$\tilde{O}^{(i,0)}(x,y) = \frac{1}{|P(b,d)|} \left(\sum_{i=0}^{M} \int_c^d |\tilde{P}^{(i,N)}(x,y,t)|^{\tilde{q}_i} dt \right)^{1/\tilde{q}_i}$$

$$\times \| (-1)^{i+N+1} f_{(i,N+1)} \|_{\tilde{p}_i},$$

provided $f_{(i,N+1)} \in L_{\tilde{p}_i}(I \times J)$, $1/\tilde{p}_i + 1/\tilde{q}_i = 1$,

$$\bar{O}(x,y) = \frac{1}{|P(b,d)|} \left(\int_a^b \int_c^d |\bar{P}^{(M,N)}(x,s,y,t)|^{\bar{q}} dt\, ds \right)^{1/\bar{q}}$$

$$\times \| (-1)^{M+N} f_{(M+1,N+1)} \|_{\bar{p}},$$

provided that $f_{(M+1,N+1)} \in L_{\bar{p}}(I \times J)$, $1/\bar{p} + 1/\bar{q} = 1$,

4.3 Weighted Montgomery Identities for Higher Order Differentiable Function...

where $\hat{P}^{(M,j)}$, $\tilde{P}^{(i,N)}$, $\bar{P}^{(M,N)}$ are as in Theorem 4.3.6 whereas P and $R(f; x, y)$ are stated in (4.3.1) and (4.3.6) respectively.

Proof Identity (4.3.9) can be written as

$$f(x, y) - \frac{1}{P(b,d)} \int_a^b \int_c^d p(s,t) f(s,t) \, dt \, ds$$

$$= \frac{1}{P(b,d)} \Bigg[R(f; x, y) + \sum_{j=0}^{N} \int_a^b (-1)^{M+1+j} \hat{P}^{(M,j)}(x, s, y) f_{(M+1, j)}(s, y) \, ds$$

$$+ \sum_{i=0}^{M} \int_c^d (-1)^{i+N+1} \tilde{P}^{(i,N)}(x, y, t) f_{(i, N+1)}(x, t) \, dt$$

$$- \int_a^b \int_c^d (-1)^{M+N} \bar{P}^{(M,N)}(x, s, y, t) f_{(M+1, N+1)}(s, t) \, dt \, ds \Bigg].$$

We can easily get desired result by taking absolute value and by using inequality of Hölder for double integrals. □

Remark 4.3.2 For $M = N = 0$, we can get special case of Theorem 4.3.9 as similar as Theorem 4 of [156].

Theorem 4.3.10 *Let f be a real-valued function defined in the interval $I \times J$ such that $f \in C^{(M+1, N+1)}$ in the interval $I \times J$ and $|f_{(M+1,N+1)}|^p$ be integrable function, i.e.,*

$$\left\| (-1)^{N+M} f_{(M+1,N+1)} \right\|_p := \left(\int_a^b \int_c^d |(-1)^{N+M} f_{(M+1,N+1)}(s,t)|^p \, dt \, ds \right)^{1/p} < \infty,$$

where $1/p + 1/q = 1$. Then, it follows

$$\Bigg| \int_a^b \int_c^d p(s,t) f(s,t) \, dt \, ds - \Bigg[R(f; x, y) + \int_a^b \int_c^d p(s,t) f(s, y) \, dt \, ds$$

$$+ \int_a^b \int_c^d p(s,t) f(x, t) \, dt \, ds + \sum_{j=1}^{N} \int_a^b \int_c^d (-1)^j p(s, \zeta) \frac{(\zeta - y)^j}{j!} f_{(0,j)}(s, y) \, d\zeta \, ds$$

$$+ \sum_{i=1}^{M} \int_a^b \int_c^d (-1)^i p(v, t) \frac{(v - x)^i}{i!} f_{(i,0)}(x, t) \, dt \, dv - f(x, y) P(b, d) \Bigg] \Bigg|$$

$$\leq \left(\int_a^b \int_c^d |\bar{P}^{(M,N)}(x, s, y, t)|^p \, dt \, ds \right)^{1/p} \left\| (-1)^{M+N} f_{(M+1, N+1)} \right\|_q,$$

$\forall (x, y) \in I \times J.$

Proof Identity (4.3.10) can be written as

$$\int_a^b \int_c^d f(s,t) p(s,t)\, dt\, ds - \Bigg[R(f;x,y) + \int_a^b \int_c^d p(s,t) f(s,y)\, dt\, ds$$

$$+ \int_a^b \int_c^d p(s,t) f(x,t)\, dt\, ds + \sum_{j=1}^N \int_a^b \int_c^d (-1)^j p(s,\zeta) \frac{(\zeta - y)^j}{j!} f_{(0,j)}(s,y)\, d\zeta\, ds$$

$$+ \sum_{i=1}^M \int_a^b \int_c^d (-1)^i p(v,t) \frac{(v-x)^i}{i!} f_{(i,0)}(x,t)\, dt\, dv - f(x,y) P(b,d) \Bigg]$$

$$= \int_a^b \int_c^d (-1)^{M+N} \bar{P}^{(M,N)}(x,s,y,t) f_{(M+1,N+1)}(s,t)\, dt\, ds.$$

We can easily get desired result by taking absolute value and by using inequality of Hölder for double integrals. □

Remark 4.3.3 For $M = N = 0$, we can obtain special case of Theorem 4.3.10 as similar as Theorem 5 of [156].

4.3.3 Grüss Type Inequalities for Double Weighted Integrals of Higher Order Differentiable Functions

Grüss [65] proved remarkable integral inequality in 1935 which is given in (2.4.1) of Chap. 2.

In [156], by applying identities (4.3.2) and (4.3.3) Pečarić and Vukelić have given new double weighted integrals inequalities for Grüss type. Now by applying differentiable functions of higher order for two variables, we get results for more generalization but for this purpose we use following notations for simplification instead of detailed presentations.

$$B^{(i,j)}(x,y) = p(x,y)[f_{(i,j)}(x,y) g(x,y) + g_{(i,j)}(x,y) f(x,y)] P^{(i,j)}_{\binom{a}{c} \Rightarrow \binom{b}{d}}(x,y), \qquad (4.3.18)$$

$$B(x,y) = p(x,y) \int_a^b \int_c^d p(s,t)[f(s,t) g(x,y) + g(s,t) f(x,y)]\, dt\, ds, \qquad (4.3.19)$$

$$\hat{B}^{(M,j)}(x,y) = p(x,y) \int_a^b [f_{(M+1,j)}(s,y) g(x,y) + g_{(M+1,j)}(s,y) f(x,y)]$$

$$\times \hat{P}^{(M,j)}(x,s,y)\, ds, \qquad (4.3.20)$$

4.3 Weighted Montgomery Identities for Higher Order Differentiable Function...

$$\tilde{B}^{(i,N)}(x,y) = p(x,y)\int_c^d [g(x,y)f_{(i,N+1)}(x,t) + g_{(i,N+1)}(x,t)f(x,y)]$$
$$\times \tilde{P}^{(i,N)}(x,y,t)\,dt, \tag{4.3.21}$$

$$\bar{B}^{(M,N)}(x,y) = p(x,y)\int_a^b \int_c^d [g(x,y)f_{(M+1,N+1)}(s,t) + g_{(M+1,N+1)}(s,t)f(x,y)]$$
$$\times \bar{P}^{(M,N)}(x,s,y,t)\,dt\,ds, \tag{4.3.22}$$

$$C^{(i,j)}(x,y) = |p(x,y)g(x,y)|\,\|f_{(i,j)}(x,y)\|_\infty + |p(x,y)f(x,y)|\,\|g_{(i,j)}(x,y)\|_\infty, \tag{4.3.23}$$

$$D^{(i,j)}(x,y) = \frac{(\max\{b-x, x-a\})^{i+1}}{(i+1)!}\frac{(\max\{d-y, y-c\})^{j+1}}{(j+1)!}$$
$$\times \int_a^b \int_c^d |p(v,\zeta)|\,d\zeta\,dv, \tag{4.3.24}$$

$$D^{(0,j)}(y) = (b-a)\frac{(\max\{d-y, y-c\})^{j+1}}{(j+1)!}\int_a^b \int_c^d |p(v,\zeta)|\,d\zeta\,dv, \tag{4.3.25}$$

$$D^{(i,0)}(x) = (d-c)\frac{(\max\{b-x, x-a\})^{i+1}}{(i+1)!}\int_a^b \int_c^d |p(v,\zeta)|\,d\zeta\,dv, \tag{4.3.26}$$

$$\hat{D}^{(M,j)}(x,y) = \int_a^b |\hat{P}^{(M,j)}(x,s,y)|\,ds, \tag{4.3.27}$$

$$\tilde{D}^{(i,N)}(x,y) = \int_c^d |\tilde{P}^{(i,N)}(x,y,t)|\,dt, \tag{4.3.28}$$

$$\bar{D}^{(M,N)}(x,y) = \int_a^b \int_c^d |\bar{P}^{(M,N)}(x,s,y,t)|\,dt\,ds, \tag{4.3.29}$$

$$G_f(x,y) = R(f;x,y) + \int_a^b \int_c^d p(s,t)f(s,y)\,dt\,ds$$
$$+ \int_a^b \int_c^d p(s,t)f(x,t)\,dt\,ds$$
$$+ \sum_{j=1}^N \int_a^b \int_c^d (-1)^j p(s,\zeta)\frac{(\zeta-y)^j}{j!}f_{(0,j)}(s,y)\,d\zeta\,ds$$
$$+ \sum_{i=1}^M \int_a^b \int_c^d (-1)^i p(v,t)\frac{(v-x)^i}{i!}f_{(i,0)}(x,t)\,dt\,dv, \tag{4.3.30}$$

$$G_g(x,y) = R(g;x,y) + \int_a^b \int_c^d p(s,t)g(s,y)\,dt\,ds$$

$$+ \int_a^b \int_c^d p(s,t) g(x,t) \, dt \, ds$$

$$+ \sum_{j=1}^{N} \int_a^b \int_c^d (-1)^j p(s,\zeta) \frac{(\zeta - y)^j}{j!} g_{(0,j)}(s,y) \, d\zeta \, ds$$

$$+ \sum_{i=1}^{M} \int_a^b \int_c^d (-1)^i p(v,t) \frac{(v - x)^i}{i!} g_{(i,0)}(x,t) \, dt \, dv, \qquad (4.3.31)$$

where $\hat{P}^{(M,j)}(x,s,y)$, $\tilde{P}^{(i,N)}(x,y,t)$, $\bar{P}^{(M,N)}(x,s,y,t)$ are as in Theorem 4.1.12 whereas P and $R(f;x,y)$ are stated in (4.3.1) and (4.3.6) respectively.

Now, we are ready to get our important results of current subsection using notations defined above, which are as follow:

Theorem 4.3.11 *Let $p, f, g : I \times J \to \mathbb{R}$ be three functions such that $g, f \in C^{(M+1,N+1)}$ in the same interval and p is integrable, then*

$$\left| \frac{1}{P(b,d)} \int_a^b \int_c^d p(x,y) f(x,y) g(x,y) \, dy \, dx \right.$$

$$- \left(\frac{1}{P(b,d)} \int_a^b \int_c^d p(x,y) f(x,y) \, dy \, dx \right)$$

$$\left. \times \left(\frac{1}{P(b,d)} \int_a^b \int_c^d p(x,y) g(x,y) \, dy \, dx \right) \right| \leq \frac{1}{2[P(b,d)]^2} \int_a^b \int_c^d \left[\sum_{i=1}^{M} \sum_{j=1}^{N} \right.$$

$$\times C^{(i,j)}(x,y) D^{(i,j)}(x,y) + \sum_{j=1}^{N} C^{(0,j)}(y) D^{(0,j)}(y) + \sum_{i=1}^{M} C^{(i,0)}(x) D^{(i,0)}(x)$$

$$+ C^{(M+1,j)}(x,y) \hat{D}^{(M,j)}(x,y) + C^{(i,N+1)}(x,y) \tilde{D}^{(i,N)}(x,y)$$

$$\left. + C^{(M+1,N+1)}(x,y) \bar{D}^{(M,N)}(x,y) \right] dy \, dx.$$

Proof From (4.3.9) we have following identities

$$f(x,y) P(b,d) = R(f;x,y) + \int_a^b \int_c^d p(s,t) f(s,t) \, dt \, ds$$

$$+ \sum_{j=0}^{N} \int_a^b (-1)^{M+1+j} \hat{P}^{(M,j)}(x,s,y) f_{(M+1,j)}(s,y) \, ds$$

4.3 Weighted Montgomery Identities for Higher Order Differentiable Function...

$$+ \sum_{i=0}^{M} \int_{c}^{d} (-1)^{i+N+1} \tilde{P}^{(i,N)}(x, y, t) f_{(i,N+1)}(x, t) \, dt$$

$$- \int_{a}^{b} \int_{c}^{d} (-1)^{M+N} \bar{P}^{(M,N)}(x, s, y, t) f_{(M+1,N+1)}(s, t) \, dt \, ds \quad (4.3.32)$$

and

$$g(x, y) P(b, d) = R(g; x, y) + \int_{a}^{b} \int_{c}^{d} p(s, t) g(s, t) \, dt \, ds$$

$$+ \sum_{j=0}^{N} \int_{a}^{b} (-1)^{M+1+j} \hat{P}^{(M,j)}(x, s, y) g_{(M+1,j)}(s, y) \, ds$$

$$+ \sum_{i=0}^{M} \int_{c}^{d} (-1)^{i+N+1} \tilde{P}^{(i,N)}(x, y, t) g_{(i,N+1)}(x, t) \, dt$$

$$- \int_{a}^{b} \int_{c}^{d} (-1)^{N+M} \bar{P}^{(M,N)}(x, s, y, t) g_{(M+1,N+1)}(s, t) \, dt \, ds, \quad (4.3.33)$$

for every $(x, y) \in I \times J$. Multiply (4.3.32) by $g(x, y) p(x, y)$ and (4.3.33) by $f(x, y) p(x, y)$ and then sum these identities, we get

$$2P(b,d) p(x,y) f(x,y) g(x,y) = - \sum_{i=1}^{M} \sum_{j=1}^{N} (-1)^{i+j} B^{(i,j)}(x,y)$$

$$- \sum_{j=1}^{N} (-1)^j B^{(0,j)}(y) - \sum_{i=1}^{M} (-1)^i B^{(i,0)}(x) + B(x,y) + (-1)^{M+1+j} \hat{B}^{(M,j)}(x,y)$$

$$+ (-1)^{i+N+1} \tilde{B}^{(i,N)}(x,y) - (-1)^{N+M} \bar{B}^{(M,N)}(x,y). \quad (4.3.34)$$

Integrate above equation to $I \times J$ and divided by $2P(b,d)$ on both sides, and obtain

$$\int_{a}^{b} \int_{c}^{d} f(x,y) g(x,y) p(x,y) \, dy \, dx$$

$$= \frac{1}{2P(b,d)} \int_{a}^{b} \int_{c}^{d} \left[- \sum_{i=1}^{M} \sum_{j=1}^{N} (-1)^{i+j} B^{(i,j)}(x,y) \right.$$

$$\sum_{j=1}^{N}(-1)^j B^{(0,j)}(y) - \sum_{i=1}^{M}(-1)^i B^{(i,0)}(x) + B(x,y) + (-1)^{M+1+j}\hat{B}^{(M,j)}(x,y)$$

$$+ (-1)^{i+N+1}\tilde{B}^{(i,N)}(x,y) - (-1)^{N+M}\bar{B}^{(M,N)}(x,y)\bigg] dy\,dx.$$

It may be written as

$$\frac{1}{P(b,d)}\int_a^b \int_c^d p(x,y)f(x,y)g(x,y)\,dy\,dx$$

$$-\left(\frac{1}{P(b,d)}\int_a^b \int_c^d f(x,y)p(x,y)\,dy\,dx\right)$$

$$\times \left(\frac{1}{P(b,d)}\int_a^b \int_c^d p(x,y)g(x,y)\,dy\,dx\right)$$

$$= \frac{1}{2[P(b,d)]^2}\int_a^b \int_c^d \bigg[-\sum_{i=1}^{M}\sum_{j=1}^{N}(-1)^{i+j} \times B^{(i,j)}(x,y) - \sum_{j=1}^{N}(-1)^j B^{(0,j)}(y)$$

$$-\sum_{i=1}^{M}(-1)^i B^{(i,0)}(x) + (-1)^{M+1+j}\hat{B}^{(M,j)}(x,y)$$

$$+ (-1)^{i+N+1}\tilde{B}^{(i,N)}(x,y) - (-1)^{M+N}\bar{B}^{(M,N)}(x,y)\bigg] dy\,dx. \qquad (4.3.35)$$

Using (4.3.18),...,(4.3.29) we get below inequalities $\forall (x,y) \in I \times J$

$$|(-1)^{i+j}B^{(i,j)}(x,y)| \leq C^{(i,j)}(x,y)\,D^{(i,j)}(x,y),$$
$$|(-1)^j B^{(0,j)}(y)| \leq C^{(0,j)}(y)\,D^{(0,j)}(y),$$
$$|(-1)^i B^{(i,0)}(x)| \leq C^{(i,0)}(x)\,D^{(i,0)}(x),$$
$$|(-1)^{M+1+j}\hat{B}^{(M,j)}(x,y)| \leq C^{(M+1,j)}(x,y)\,\hat{D}^{(M,j)}(x,y),$$
$$|(-1)^{i+N+1}\tilde{B}^{(i,N)}(x,y)| \leq C^{(i,N+1)}(x,y)\,\tilde{D}^{(i,N)}(x,y),$$
$$|(-1)^{M+N}\bar{B}^{(M,N)}(x,y)| \leq C^{(M+1,N+1)}(x,y)\,\bar{D}^{(M,N)}(x,y).$$

We can easily get desired result by taking absolute value in (4.3.35) on both sides and by using above these inequalities in it. □

4.3 Weighted Montgomery Identities for Higher Order Differentiable Function...

Theorem 4.3.12 Let $p, f, g : I \times J \to \mathbb{R}$ be three functions such that $g, f \in C^{(M+1,N+1)}$ in the same interval and p is integrable, then

$$\left| \frac{1}{P(b,d)} \int_a^b \int_c^d p(x,y) f(x,y) g(x,y) \, dy \, dx \right.$$
$$+ \left(\frac{1}{P(b,d)} \int_a^b \int_c^d p(x,y) f(x,y) \, dy \, dx \right)$$
$$\times \left(\frac{1}{P(b,d)} \int_a^b \int_c^d p(x,y) g(x,y) \, dy \, dx \right)$$
$$\left. - \frac{1}{2[P(b,d)]^2} \int_a^b \int_c^d p(x,y) \left[g(x,y) G_f(x,y) + f(x,y) G_g(x,y) \right] dy \, dx \right|$$
$$\leq \frac{1}{2[P(b,d)]^2} \int_a^b \int_c^d C^{(M+1,N+1)}(x,y) \, \bar{D}^{(M,N)}(x,y) \, dy \, dx.$$

Proof From (4.3.10) we have below identities

$$f(x,y) P(b,d) = G_f(x,y) - \int_a^b \int_c^d p(s,t) f(s,t) \, dt \, ds$$
$$+ \int_a^b \int_c^d (-1)^{N+M} \bar{P}^{(M,N)}(x,s,y,t) f_{(M+1,N+1)}(s,t) \, dt \, ds,$$
(4.3.36)

$$g(x,y) P(b,d) = G_g(x,y) - \int_a^b \int_c^d p(s,t) g(s,t) \, dt \, ds$$
$$+ \int_a^b \int_c^d (-1)^{N+M} \bar{P}^{(M,N)}(x,s,y,t) g_{(M+1,N+1)}(s,t) \, dt \, ds,$$
(4.3.37)

for $(x,y) \in I \times J$. Multiply (4.3.36) by $g(x,y) p(x,y)$ and (4.3.37) by $f(x,y) p(x,y)$ and then sum these identities, we get

$$2 P(b,d) p(x,y) f(x,y) g(x,y) = G_f(x,y) p(x,y) g(x,y)$$
$$+ G_g(x,y) p(x,y) f(x,y)$$
$$- B(x,y) + (-1)^{M+N} \bar{B}^{(M,N)}(x,y).$$
(4.3.38)

Integrate above equation to $I \times J$ and divided by $2P(b,d)$ on both sides, and obtain

$$\int_a^b \int_c^d p(x,y) f(x,y) g(x,y) \, dy \, dx$$
$$= \frac{1}{2P(b,d)} \int_a^b \int_c^d p(x,y) [G_f(x,y) g(x,y) + G_g(x,y) f(x,y)] \, dy \, dx$$
$$- \frac{1}{P(b,d)} \left(\int_a^b \int_c^d f(x,y) p(x,y) \, dy \, dx \right) \left(\int_a^b \int_c^d g(x,y) p(x,y) \, dy \, dx \right)$$
$$+ \frac{1}{2P(b,d)} \int_a^b \int_c^d (-1)^{N+M} \bar{B}^{(M,N)}(x,y) \, dy \, dx, \tag{4.3.39}$$

since we have

$$|(-1)^{M+N} \bar{B}^{(M,N)}(x,y)| \leq C^{(M+1,N+1)}(x,y) \, \bar{D}^{(M,N)}(x,y). \tag{4.3.40}$$

From (4.3.39) and (4.3.40) get desired inequality. □

Remark 4.3.4 For $N = M = 0$, we can obtain special cases of Theorems 4.3.11 and 4.3.12 similar as Theorems 6 and 7 of [156] respectively, we may also obtain as similar results as in [66].

Bibliography

1. R.P. Agarwal, P.J.Y. Wong, *Error Inequalities in Polynomial Interpolation and Their Applications* (Kluwer Academic Publishers, Dordrecht, 1993)
2. F. Ahmad, M.A. Rana, *Elements of Numerical Analysis* (National Book Foundation, Allah Wala Printers, Lahore, 1995)
3. N.I. Akhiezer, *The Classical Moment Problem and Some Related Questions in Analysis* (Translated from Russian by N. Kemmer) (Hafner Publishing, New York, 1965)
4. A.A. Aljinović, J. Pečarić, A. Vukelić, On some Ostrowski type inequalities via Montgomery identity and Taylor's formula II. Tamkang J. Math. **36**(4), 279–301 (2005)
5. A.A. Aljinović, A.R. Khan, J.E. Pečarić, Weighted majorization theorems via generalization of Taylor's formula. J. Inequal. Appl. **196**, 1–22 (2015)
6. M.W. Alomari, A companion of Ostrowski's inequality for mappings whose first derivatives are bounded and applications in numerical integration. Kragujevac J. Math. **36**(1), 77–82 (2012)
7. M.W. Alomari, A companion of Dragomir's generalization of the Ostrowski inequality and applications to numerical integration. Ukrainian M. J. **64**(4), 491–510 (2012)
8. M.W. Alomari, Two point Ostrowski's inequality. Result. Math. **72**(3), 1499–1523 (2017)
9. M.W. Alomari, M. Darus, Some Ostrowski type inequalities for convex functions with applications. RGMIA **13**(1), 3 (2010)
10. M.W. Alomari, M. Darus, Some Ostrowski type inequalities for Quasi-convex functions with applications to special means. RGMIA **13**(2), 3 (2010)
11. G. Anastassiou, M.R. Hooshmandasl, A. Ghasemi, F. Muftakharzadeh, Montgomery identities for fractional integrals and related fractional inequalities. J. Inequal. Pure Appl. Math. **10**(4), Art. 97, 6pp. (2009)
12. T.W. Anderson, *An Introduction to Multivariate Statistical Analysis*, 3rd edn. (Wiley, Hoboken, 2003)
13. M. Anwar, J. Jakšetić, J.E. Pečarić, Atiq ur Rehman, Exponential convexity, positive semi-definite matrices and fundamental inequalities. J. Math. Inequal. **4**(2), 171–189 (2010)
14. L. Avazpour, Fractional Ostrowski type inequalities for functions whose derivatives are prequasinvex. J. Inequal. Spec. Funct. **9**(2), 15–29 (2018)
15. I.A. Baloch, J. Pečarić, M. Praljak, Generalization of Levinson's inequality. J. Math. Inequal. **9**(2), 571–586 (2015)
16. N.S. Barnett, S.S. Dragomir, An Ostrowski type inequality for double integrals and applications for cubature formulae. Soochow J. Math. **27**(1), 1–10 (2001)
17. N.S. Barnett, S.S. Dragomir, On the weighted Ostrowski inequality. J. Inequal. Pure Appl. Math. **8**(4), 1–10 (2007)

18. N.S. Barnett, P. Cerone, S.S. Dragomir, J. Roumeliotis, Some inequalities for the dispersion of a random variable whose PDF is defined on a finite interval. J. Inequal. Pure Appl. Math. **2**(1), 1, 18pp. (2001)
19. N.S. Barnett, S.S. Dragomir, A. Sofo, Better bounds for an inequality of Ostrowski type with applications. Demonstratio Math. **34**(3), 533–542 (2001)
20. N.S. Barnett, P. Cerone, S.S. Dragomir, *Inequalities for Random Variable Over a Finite Interval* (Nova Science Publishers, Hauppauge, 2004)
21. E.F. Beckenbach, R. Bellman, *Inequalities* (Springer, Berlin, 1961)
22. S.N. Bernstein, Sur les fonctions absolument monotones. Acta Math. **52**(1), 1–66 (1929)
23. M. Bohner, A.R. Khan, M. Khan, F. Mehmood, M.A. Shaikh, Generalized perturbed Ostrowski-type inequalities. Ann. Univ. Mariae Curie-Sk lodowska Sect. A (to appear)
24. K. Boukerrioua, A. Guezane-Lakoud, On generalization of Čebyšev type inequalities. J. Inequal. Pure Appl. Math. **8**(2), Art. 55, 9pp. (2007)
25. H.D. Brunk, Integral inequalities for functions with nondecreasing increments. Pac. J. Math. **14**, 783–793 (1964)
26. P.S. Bullen, *Handbook of Means and Their Inequalities* (Kluwer Academic Publishers, Dordrecht, 2003)
27. P.S. Bullen, D.S. Mitrinović, P.M. Vasić, *Means and Their Inequalities* (Kluwer Academic Publishers, Dordrecht, 1988)
28. C. Buse, P. Cerone, S.S. Dragomir, J. Roumeliotis, A refinement of Grüss type inequality for the Bochner integral of vector-valued functions in hilbert spaces and applications. J. Korean Math. Soc. **43**(5), 911–929 (2006)
29. S.I. Butt, N. Mehmood, J.E. Pečarić, New generalizations of Popoviciu type inequalities via new Green functions and Fink's identity. Trans. A. Razmadze Math. Inst. **171**(3), 293–303 (2017)
30. P.L. Čebyšev, Sur les Expressions Approximative des Intégrales Définies par les Autres Prises Entre les Mêmes Limites. Proc. Math. Soc. Charkov **2**, 93–98 (1882)
31. P.L. Čebyšev, O. približennyh vyraženijah odnih integralov čerez drugie. Soobščenija i protokoly zasedaniĭ Matematičeskogo občestva pri Imperatorskom Har'kouskom Universitete, No. 2, 93 – 98; Polnoe sobranie sočineniĭ P. L. Čebyševa. Moskva-Leningrad **1882**, 128–131 (1948)
32. P.L. Čebyšev, Ob odnom rjade, dostavljajuscem predel'nye veličiny integralov pri razloženii podintegral'noĭ funkcii na množeteli. Priloženi k 57 tomu Zapisok Imp. Akad. Nauk, No. 4; Polnoe sobranie sočineniĭ P. L. Čebyševa., Moskva-Leningrad **1883**, 157–169 (1948)
33. P. Cerone, On Chebyshev functional bounds, in *Proceedings of the Conference on Differential and Difference Equations and Applications* (Hindawi Publishing Corporation, London, 2006), pp. 267–277
34. P. Cerone, S.S. Dragomir, A refinement of the Grüss inequality and applications. Tamkang J. Math. Tamkang J. Math. **38**(1), 37–49 (2007)
35. P. Cerone, S.S. Dragomir, Some new Ostrowski-type bounds for Čebyšev functional and applications. J. Math. Inequal. **8**(1), 159–170 (2014)
36. P. Cerone, S.S. Dragomir, J. Roumeliotis, An inequality of Ostrowski type for mappings whose second derivatives are bounded and applications. East Asian Math. J. **15**(1), 4, 9pp. (1999)
37. P. Cerone, S.S. Dragomir, C.E.M. Pearce, A generalized trapezoid inequality for functions of bounded variation. Turk. J. Math. **24**(2), 147–163 (2000)
38. Z. Ciesielski, A note on some inequalities of Jensen's type. Ann. Polonici Math. **4**, 269–274 (1957/1958)
39. P.J. Davis, *Interpolation and Approximation* (Blaisdell, Boston, 1961)
40. L. Dedic, M. Matic, J. Pečarić, On some generalizations of Ostrowski's inequality for Lipschitz functions and functions of bounded variation. Math. Inequal. Appl. **3**(1), 1–14 (2000)

41. C.B. Dolićanin, V.B. Nikolić, D. Ć. Dolićanin, Application of finite difference method to study of the phenomenon in the theory of thin plates. Ser. A: Appl. Math. Inform. and Mech. **2**(1), 29–43 (2010)
42. S.S. Dragomir, Some integral inequalities of Grüss type. J. Pure Appl. Math. **31**(4), 397–415 (2000)
43. S.S. Dragomir, On the Ostrowski's integral inequality for mappings with bounded variation and applications. Math. Inequal. Appl. **4**(1), 59–66 (2001)
44. S.S. Dragomir, Refinements of the generalised trapozoid and Ostrowski inequalities for functions of bounded variation. Arch. Math. **91**(5), 450–460 (2008)
45. S.S. Dragomir, New Grüss type inequalities for functions of bounded variation and applications. Appl. Math. Lett. **25**(10), 1475–1479 (2012)
46. S.S. Dragomir, A companion of Ostrowski's inequality for functions of bounded variation and applications. Int. J. Nonlinear Anal. Appl. **5**, 89–97 (2014)
47. S.S. Dragomir, N.S. Barnett, An Ostrowski type inequality for mappings whose second derivatives are bounded and applications. J. Ind. Math. Soc. **66**(1–4), 237–245 (1999)
48. S.S. Dragomir, T.M. Rassias, *Ostrowski Type Inequalities and Applications in Numerical Integration* (Kluwer Academics Publishers, Dordrecht, 2002)
49. S.S. Dragomir, A. Sofo, An integral inequality for twice differentiable mappings and applications. Tamkang J. Math. **31**(4), 257–266 (2000)
50. S.S. Dragomir, S. Wang, An inequality of Ostrowski-Grüss type and its applications to the estimation of error bounds for some special means and for some numerical quadrature rules. Comput. Math. Appl. **33**(11), 15–20 (1997)
51. S.S. Dragomir, S. Wang, A new inequality Ostrowski's type in L_p norm and applications to some special means and some numerical quadrature rules. Tamkang J. Math. **28**, 239–244 (1997)
52. S.S. Dragomir, P. Cerone, N.S. Barnett, J. Roumeliotis, An inequality of the Ostrowski type for double integrals and applications for cubature formulae. Tamsui Oxf. J. Math. Sci. **16**(1), 1–16 (2000)
53. S.S. Dragomir, P. Cerone, J. Roumeliotis, A new generalization of Ostrowski integral inequality for mappings whose derivatives are bounded and applications in numerical integration and for special means. Appl. Math. Lett. **13**, 19–25 (2000)
54. S.S. Dragomir, N.S. Barnett, P. Cerone, An Ostrowski type inequality for double integrals in terms of L_p norms and application in numerical integration. Rev. Anal. Numér. Théor. Approx. **32**(2), 161–169 (2003)
55. S.S. Dragomir, N. Irshad, A.R. Khan, Generalization of weighted Ostrowski Grüss type inequality by using Korkine's identity. Stud. Univ. Babeş-Bolyai Math. **65**(2), 183–198 (2020)
56. S.S. Dragomir, A.R. Khan, M. Khan, F. Mehmood, M.A. Shaikh, A new integral version of generalised Ostrowski-Grüss type inequality with applications (submitted)
57. E. Eban, R. Rothschild, A. Mizrahi, I. Nelken, G. Elidan, C. Carvalho, P. Ravikumar, (eds.), Dynamic copula networks for modeling real-valued time series. J. Mach. Learn. Res. **31**, 247–255 (2013)
58. K. Fan, Problem 4471. Am. Math. Month. **60**, 195–197 (1953)
59. G. Fasshauer, *Meshfree Approximation Methods with MATLAB* (World Scientific Publishing, Hackensack, 2007)
60. A.M. Fink, Bounds on the deviation of the function from its averages. Czech. Math. J. **42**(117), 289–310 (1998)
61. A.M. Fink, An essay on the history of inequalities. J. Math. Anal. Appl. **249**, 118–134 (2000)
62. D. Gleich, Finite Calculus: A tutorial for solving nasty sums (Stanford University, 2005)
63. V.L. Gontscharoff, *Theory of Interpolation and Approximation of Functions* (Gostekhizdat, Moscow, 1954)
64. R. Gorenflo, F. Mainardi, *Fractional Calculus: Integral and Differential Equations of Fractional Order* (Springer, New York, 1997)
65. G. Grüss, Über das Maximum des absoluten Betrages von. Math. Z. **39**(1), 215–226 (1935)

66. A. Guezane-Lakoud, F. Aissaoui, New Čebyšev type inequalities for double integrals. J. Math. Inequal. **5**(4), 453–462 (2011)
67. J. Hadamard, Étude sur les propriétés des fonctions entrières et en particulier d'une fonction considérée par Riemann. J. Math. Pures Appl. **58**, 171–216 (1893)
68. D. Halliday, R. Resnick, *Physics Parts I and II*. Wiley International Edition (Wiley/Toppan Company, Tokyo, 1966)
69. G.H. Hardy, J.E. Littlewood, G. Pólya, *Inequalities* (Cambridge University Press, Cambridge, 1934). (second edition: 1952)
70. G.H. Hardy, J.E. Littlewood, G. Pólya, *Inequalities* (Cambridge University Press, Cambridge, 1978)
71. F. Hausdorff, Moment probleme fur̈ ein endliches Intervall. Math. Z. **16**, 220–248 (1923)
72. C. Hermite, Sur deux limites d'une intégrale définie. Mathesis **3**, 82 (1883)
73. M.O. Hölder, Über einen Mittelwertsatz, Nachr. Ges. Wiss. Göttingen 38–47 (1889)
74. M. Imtiaz, N. Irshad, A.R. Khan, Generalization of weighted Ostrowski integral inequality for twice differentiable mappings. Adv. Inequal. Appl. **2016**, Art. 20, 17pp. (2016)
75. N. Irshad, Generalization of Ostrowski type inequalities with applications. Unpublished doctoral dissertation, University of Karachi, Karachi (2019)
76. N. Irshad, A.R. Khan, Some applications of quadrature rules for mappings on $L_p[u, v]$ space via Ostrowski type inequality. J. Numer. Anal. Approx. Theory **46**, 141–149 (2017)
77. N. Irshad, A.R. Khan, Generalization of Ostrowski inequality for differentiable functions and its applications to numerical quadrature rules. J. Math. Anal. **8**(1), 79–102 (2017)
78. N. Irshad, A.R. Khan, On weighted Ostrowski-Grüss inequality with applications. Transylv. J. Math. Mech. **10**(1), 15–20 (2018)
79. N. Irshad, A.R. Khan, H. Musharraf, Generalized Ostrowski inequalities and computational integration. J. Numer. Anal. Approx. Theory **49**(2), 155–176 (2020)
80. N. Irshad, A.R. Khan, H. Musharraf, Generalized fractional Ostrowski type inequality. J. Inequal. Spec. Funct. **11**(4), 16–26 (2020)
81. J. Jakšetić, J. Pečarić, Exponential convexity method. J. Convex Anal. **20**(1), 181–197 (2013)
82. J.L.W.V. Jensen, Om konvexe funktioner og uligheder mellem Middelvaerdier. Nyt Tidsskr. Math. **16B**, 49–69 (1905)
83. J.L.W.V. Jensen, Sur les fonctions convexes et les inégalités entre les valeurs moyennes. Acta Math. **30**, 175–193 (1906)
84. J.-F. Jouanin, G. Riboulet, T. Roncalli, Financial Applications of Copula Functions. Par Giorgio Szego, edn. (Wiley, Hoboken, 2004)
85. A.R. Khan, General inequalities for generalized convex functions. Unpublished doctoral dissertation. Abdus Salam School of Mathematical Sciences, GC University, Lahore (2014)
86. A.R. Khan, F. Mehmood, Double weighted integrals identities of Montgomery for differentiable function of higher order. J. Math. Stat. **15**(1), 112–121 (2019)
87. A.R. Khan, F. Mehmood, Some remarks on functions with non-decreasing increments. J. Math. Anal. **11**(1), 1–16 (2020)
88. A.R. Khan, F. Mehmood, Positivity of sums for higher order $\nabla-$convex sequences and functions. Global J. Pure Appl. Math. **16**(1), 93–105 (2020)
89. A.R. Khan, J.E. Pečarić, Positivity of general linear inequalities for n-convex functions via the Taylor formula using new Green functions. Commun. Optim. Theory **2019**, Article ID 5, 10pp. (2019)
90. A.R. Khan, J.E. Pečarić, Hermite interpolation with Green functions and positivity of general linear inequalities for $n-$convex functions. J. Math. Anal. **11**(4), 1–15 (2020)
91. A.R. Khan, J.E. Pečarić, Positivity of sums and integrals for n-convex functions via Abel-Gontscharoff's interpolating polynomial and Green functions. Southeast Asian Bull. Math. **45**(2), 197–216 (2021)
92. A.R. Khan , J.E. Pečarić, Positivity of sums and integrals for n-convex functions via the Fink identity and new Green functions. Stud. Univ. Babes-Bolyai Math. **66**(4), 613–627 (2021)
93. A.R. Khan, J.E. Pečarić, Positivity of sums and integrals for $n-$convex functions via extension of Montgomery identity using new Green functions. Trans. Math. Comp. Sci. (to appear)

94. A.R. Khan, S. Saadi, Generalized Jensen-Mercer inequality for functions with nondecreasing increments. Abs. Appl. Anal. **2016**, 5231476, 1–12 (2016)
95. A.R. Khan, N. Latif, J.E. Pečarić, Exponential convexity for majorization. J. Inequal. Appl. **2012**, 105 (2012)
96. A.R. Khan, J.E. Pečarić, S. Varošanec, On some inequalities for functions with nondecreasing increments of higher order. J. Inequal. Appl. **8**, 1–14 (2013)
97. A.R. Khan, J.E. Pečarić, S. Varošanec, Popoviciu type characterization of positivity of sums and integrals for convex functions of higher order. J. Math. Inequal. **7**(2), 195–212 (2013)
98. A.R. Khan, J. Pečarić, M. Praljak, Popoviciu type inequalities for n convex functions via extension of Montgomery identity. An. Ştiinţ. Univ. "Ovidius" Constanţa Ser. Mat. **24**(3), 161–188 (2016)
99. A.R. Khan, J.E. Pečarić, M. Praljak, S. Varošanec, Positivity of sums for n-convex functions via Taylor's formula and Green function. Adv. Stud. Contemp. Math. **27**(4), 515–537 (2017)
100. A.R. Khan, J.E. Pečarić, M. Praljak, A note on generalized Mercer's inequality. Bull. Malays. Math. Sci. Soc. **2**(40), 881–889 (2017)
101. A.R. Khan, J.E. Pečarić, M. Praljak, S. Varošanec, *General Linear Inequalities and Positivity/higher Order Convexity*. Monographs in Inequalities, vol. 12 (Element, Zagreb, 2017)
102. A.R. Khan, J.E. Pečarić, M. Praljak, S. Varošanec, Positivity of sums of $n-$convex functions via Taylor's formula and Green function. Adv. Stud. Contemp. Math. **27**(4), 515–537 (2017)
103. A.R. Khan, J.E. Pečarić , M. Praljak, Weighted averages of $n-$convex functions via extension of Montgomery's identity. Arab. J. Math. **2020**(9), 381–392 (2020)
104. C.H. Kimberling, A probabilistic interpretation of complete monotonicity. Aequationes Math. **10**, 152–164 (1974)
105. I.B. Lacković, On convexity of arithmetic integral mean. Elektrotehn. Fak. Ser. Mat. Fiz. **381–409**, 117–120 (1972). [Univ. Beograd. Publ.]
106. H.P. Langtangen, A. Logg, *Solving PDEs in Python*. The FEniCS Tutorial I, Simula Springer Briefs on Computing, vol. 3 (Springer, Berlin, 2016), pp. 11–35
107. P. Laux, S. Wagner, A. Wagner, J. Jacobeit, A. Bárdossy, H. Kunstmann, Modelling daily precipitation features in the Volta Basin of West Africa. Int. J. Climatol. **29**(7), 937–954 (2009)
108. B.-L. Lin, Every waking moment Ky Fan (1914–2010). Not. Am. Math. Soc. **57**(11), 1444–1447 (2010)
109. Z. Liu, A generalization of two sharp inequalities in two independent variables. Comp. Math. App. **59**, 2809–2814 (2010)
110. W. Liu, New bounds for the companions of Ostrowski's inequality and applications. Filomat **28**(1), 167–178 (2014)
111. W. Liu, Y. Sun, A refinement of the companion of Ostrowski inequality for functions of bounded variation and applications (2012). arXiv:1207.3861v1
112. F. Longin, B. Solnik, Extreme correlation of international equity markets. J. Finance **56**(2), 649–676 (2001)
113. A. Mahajan, D.K. Ross, A note on completely and absolutely monotone functions. Canad. Math. Bull. **25**(2), 143–148 (1982)
114. A.W. Marshall, I. Olkin, B.C. Arnold, *Inequalities: Theory of Majorization and Its Applications*, 2nd edn. (Springer, New York, 2011)
115. M. Masjed-Jamei, S.S. Dragomir, An analogue of the Ostrowski inequality and applications. Filomat **28**(2), 373–381 (2014)
116. M. Masjed-Jamei, S.S. Dragomir, Generalization of Ostrowski-Grüss inequality. Anal. Appl. **12**(2), 117–130 (2014)
117. M. Matić, J.E. Pečarić, N. Ujević, Improvement and further generalization of inequalities of Ostrowski-Grüss type. Comput. Math. Appl. **39**, 161–175 (2000)
118. F. Mehmood, On functions with nondecreasing increments. Unpublished doctoral dissertation, Department of Mathemtics, University of Karachi, Karachi (2019)

119. F. Mehmood, *Functions with Non-decreasing Increments and Popoviciu Type Identities and Inequalities for Sums and Integrals*, 1st edn. (Book Publisher International, India and UK, 2021)
120. F. Mehmood, A.R. Khan, M. Khan, M.A. Shaikh, Generalization of some weighted Čebyšev-type inequalities. J. Mech. Cont. Math. Sci. **15**(4), 13–20 (2020)
121. F. Mehmood, A.R. Khan, M.A.U. Siddique, Concave and concavifiable functions and some related results. J. Mech. Cont. Math. Sci. **15**(6), 268–279 (2020)
122. F. Mehmood, G.M. Khan, K. Saleem, F. Nawaz, Z.A. Naveed, A. Rahman, Majorization theorem for concavifiable functions. Global J. Pure Appl. Math. **16**(4), 569–575 (2020)
123. F. Mehmood, K. Saleem, Z.A. Naveed, G.M. Khan, A. Rahman, A companion of weighted Ostrowski's type inequality and applications to numeriacl integration. Global J. Pure Appl. Math. **16**(4), 577–586 (2020)
124. F. Mehmood, A.R. Khan, M.A.U. Siddique, Some results related to convexifiable functions. J. Mech. Cont. Math. Sci. **15**(12), 36–45 (2020)
125. F. Mehmood, A.R. Khan, F. Nawaz, A. Nazir, Some remarks on results related to ∇−convex function. J. Math. Fund. Sci. **53**(1), 67–85 (2021)
126. F. Mehmood, A.R. Khan, M. Adnan, Positivity of integrals for higher order ∇−convex and completely monotonic functions. Sahand Commun. Math. Anal. (to appear)
127. F. Mehmood, A.R. Khan, Generalized identities and inequalities of Čebyšev and Ky Fan for ∇−convex function. J. Prime Res. Math. (to appear)
128. M. Merkle, Completetly monotone functions: a digest. Anal. Number Theory Approx. Theory Spec. Funct. **(2014)**, 347–364 (2014)
129. G.V. Milovanović, A.S. Cvetkovic, Nonstandard Gaussian quadrature formulae based on operator values. Adv. Comput. Math. **32**, 431–486 (2010)
130. I.Ž. Milovanović, J.E. Pečarić, On some inequalities for ∇−convex sequences of higher order. Period. Math. Hungar. **17**(1), 21–24 (1986)
131. D.S. Mitrinović, *Analytic Inequalities* (Springer, New York, 1970)
132. D.S. Mitrinović, J.E. Pečarić, History, variations and generalizations of the Čebyšev inequality and question of some priorities II. Rad Jugoslav. Akad. Znan. Umjet. No. 450, 139–156 (1990)
133. D.S. Mitrinović, P.M. Vasić, History, variations and generalizations of the Čebyšev inequality and the question of some priorities. Univ. Beograd. Publ. Elektrotehn. Fak. Ser. Mat. Fiz No. **461–497**, 1–30 (1974)
134. D.S. Mitrinović, J.E. Pečarić, A.M. Fink, *Inequalities Involving Functions and Their Integrals and Derivatives* (Kluwer Academic, Dordrecht, 1991)
135. D.S. Mitrinović, J.E. Pečarić, A.M. Fink, *Classical and New Inequalities in Analysis* (Kluwer Academic Publishers, Dordrecht, 1993)
136. P.M. Morillas, A characterization of absolutely monotonic (Δ) functions of a fixed order. Publ. DeĹ Inst. Math. **78**(92), 93–105 (2005)
137. J. Munkhammar, J. Widén, A copula method for simulating correlated instantaneous solar irradiance in spatial networks. Solar Energy **143**, 10–21 (2017)
138. F. Nawaz, Z.A. Naveed, F. Mehmood, G.M. Khan, K. Saleem, A companion of weighted Ostrowski's type inequality for functions whose 1st derivatives are bounded with applications. Global J. Pure Appl. Math. **16**(4), 515–522 (2020)
139. C.P. Niculescu, L.E. Persson, *Convex Functions and Their Applications. A Contemporary Approach* (Springer, New York, 2006)
140. A. Onken, S. Grünewälder, M.H. Munk, K. Obermayer, Analyzing short-term noise dependencies of spike-counts in macaque prefrontal cortex using copulas and the flashlight transformation. PLoS Comput. Biol. **5**(11), e1000577 (2009)
141. A.M. Ostrowski, Über die Absolutabweichung einer Differentiebaren Funktion von Ihren Integralmittelwert. Comment. Math. Helv. **10**, 226–227 (1937)
142. A.M. Ostrowski, On an integral inequality. Aequationes Math. **4**, 358–373 (1970)
143. B.G. Pachpatte, On Čebyšev type inequalities involving functions whose derivatives belong to L_p spaces. J. Inequal. Pure Appl. Math. **7**(2), 58 (2006)

144. B.G. Pachpatte, On Čebyšev-Grüss type inequalities via Pečarić's extention of the Montgomery identity. J. Inequal. Pure Appl. Math. **7**(1), 11 (2006)
145. B.G. Pachpatte, New inequalities of Čebyšev type for double integrals. Demonstratio Math. **XI**(1), 43–50 (2007)
146. J.E. Pečarić, On the Čebyšev inequality. Bul. Inst. Politehn. Timişoara **25**(39), 5–9 (1980)
147. J.E. Pečarić, An inequality for m-convex sequences. Mat. Vesnik **5**(2), 201–203 (1981)
148. J.E. Pečarić, On some inequalities for functions with nondecreasing increments. J. Math. Anal. Appl. **98**(1), 188–197 (1984)
149. J.E. Pečarić, On the Ostrowski's generalization of Čebyšev's inequality. J. Math. Anal. Appl. **102**(2), 479–487 (1984)
150. J.E. Pečarić, Some further remarks on the Ostrowski generalization of Čebyšev's inequality. J. Math. Anal. Appl. **123**(1), 18–33 (1987)
151. J.E. Pečarić, *Konveksne funkcije: nejednakosti* (Naučna knjiga, Beograd, 1987)
152. J.E. Pečarić, On Jessen's inequality for convex functions, III. J. Math. Anal. Appl. **156**, 231–239 (1991)
153. J.E. Pečarić, J. Perić, Improvements of Giaccardi and Petrović inequality and related Stolarsky type means. An. Univ. Craiova Ser. Mat. Inform. **39**(1), 65–75 (2012)
154. J.E. Pečarić, B. Savić, O Novom Postupku Razvijanja Funkcija u Red i Nekim Primjenama. Zbor. Rad. AKoV (Beograd) **9**, 171–202 (1983)
155. J.E. Pečarić, S. Varosanec, A note on Simpson's inequality for functions of bounded variation. Tamkang J. Math. **31**(3), 239–242 (2000)
156. J.E. Pečarić, A. Vukelic, Montgomery's identities for functions of two variables. J. Math. Anal. Appl. **332**(1), 617–630 (2007)
157. J.E. Pečarić, B.A. Mesihović, I.Ž. Milovanović, N. Stojanović, On some inequalities for convex and ∇−convex sequences of higher order II. Period. Math. Hungar. **17**(4), 313–320 (1986)
158. J.E. Pečarić, F. Proschan, Y.L. Tong, *Convex Functions, Partial Orderings and Statistical Applications* (Academic Press, New York, 1992)
159. J.E. Pečarić, A. Perušić, K. Smoljak, Generalizations of Steffensen's inequality by Abel-Gontscharoff polynomial. Khayyam J. Math. **1**(1), 45–61 (2015)
160. J.E. Pečarić, M. Praljak, A. Witkowski, Linear operator inequality for convex functions of order n at a point. Math. Ineq. Appl. **18**, 1201–1217 (2015)
161. H. Pham, *Handbook of Reliability Engineering* (Springer, Berlin, 2003)
162. T. Popoviciu, Notes sur les fonctions convexes d'orde superieur III. Mathematica (Cluj) **16**, 74–86 (1940)
163. T. Popoviciu, Notes sur les fonctions convexes d'orde superieur IV. Disquisitiones Math. **1**, 163–171 (1940)
164. T. Popoviciu, *Les fonctions convexes* (Herman and Cie, Editeurs, Paris, 1944)
165. A.W. Robert, D.E. Varberg, *Convex Functions* (Academic Press, New York, 1973)
166. R.T. Rockafellar, *Convex analysis* (Princeton University Press, Princeton, 1970)
167. P.K. Sahoo, T. Riedel, *Mean Value Theorems and Functional Equations* (World Scientific Publishing, River Edge, 1998)
168. S. Saminger-Platz, J.D.J. Arias-Garcia, R. Mesiar, E.P. Klement, Characterizations of bivariate conic, extreme value, and Archimax copulas. De Gruyter Open **5**, 45–58 (2017)
169. M.Z. Sarikaya, H. Ogunmez, On new inequalities via Riemann Liouville fractional integration. Abst. Appl. Anal., **2012**, 428983, 10pp. (2012)
170. M.Z. Sarikaya, H. Yaldiz, E. Set, On fractional inequalities via Montgomery identities. Int. J. Open Problems Complex Anal. **6**(2), 36–43 (2014)
171. M.Z. Sarikaya, H. Yaldiz, N. Basak, New fractional inequalities of Ostrowski-Grüss type. Le Matematiche **69**(1), 227–235 (2014)
172. M.Z. Sarikaya, H. Filiz, M.E. Kiris, On some generalized integral inequalities for Riemann-Liouville fractional integrals. Filomat **29**(6), 1307–1314 (2015)

173. C. Scholzel, P. Friederichs, Multivariate non-normally distributed random variables in climate research– introduction to the copula approach. Nonlinear Process. Geophys. **15**(5), 761–772 (2008)
174. M. Sen, *Introduction to Fractional-Order Operators and Their Engineering Applications* (University of Notre Dame, Notre Dame, 2014)
175. M.A. Shaikh, A.R. Khan, F. Mehmood, Estimates for weighted Ostrowski-Grüss type inequalities with applications (submitted)
176. M.A. Shaikh, F. Mehmood, A.R. Khan, Two-point Ostrowski type inequality with parmeter (submitted)
177. M. Shaked, J.G. Shanthikumar, Parametric stochastic convexity and concavity of stochastic processes. Ann. Inst. Stat. Math. **42**, 509–531 (1990)
178. B. Simon, *Convexity: An Analytic Viewpoint* (Cambridge University Press, Cambridge, 2011)
179. M.R. Spiegel, *Theory and Problems of Vector Analysis and an Introduction to Tensor Analysis*, SI (Metric) Edition, Schaum's Outline Series (McGraw-Hill, New York, 1974)
180. J. Šremr, Absolutely continuous functions of two variables in the sense of Carathéodory. Electron. J. Differ. Equ. **2010**(154), 1–11 (2010)
181. O. Stolz, *Grundzüge der differential-und integralrechnung I* (Teubner, Leipzig, 1893)
182. D. Thompson, R. Kilgore, Estimating joint flow probabilities at stream confluences using copulas. Transp. Res. Rec. **2262**, 200–206 (2011)
183. F. Tong, L. Guan, A simple proof of the generalized Ostrowski-Grüss type integral inequality. Int. J. Math. Anal. **2**(18), 889–892 (2008)
184. N. Ujević, A generalization of Ostrowski's inequality and applications in numerical integration. Appl. Math. Lett. **17**, 133–137 (2004)
185. J.M. Whittaker, *Interpolation Function Theory* (Cambridge University Press, Cambridge, 1935)
186. D.V. Widder, Necessary and sufficient conditions for the representation of a function by a doubly infinite Laplace integral. Bull. Am. Math. Soc. **40**(4), 321–326 (1934)
187. D.V. Widder, Completely convex function and Lidstone series. Trans. Am. Math. Soc. **51**, 387–398 (1942)
188. D.V. Widder, *The Laplace Transforms* (Princeton University Press, Princeton, 1946)
189. S. Wu, Construction of asymmetric copulas and its application in two-dimensional reliability modelling. Eur. J. Oper. Res. **238**(2), 476–485 (2014)
190. X.J. Yang, Refinement of Hölder inequality and application to Ostrowski inequality. Appl. Math. Comput. **138**, 455–461 (2003)
191. S.C. Yang, T.J. Liu, H.P. Hong, Reliability of tower and tower-line systems under spatiotemporally varying wind or earthquake loads. J. Struct. Eng. **143**, 04017137 (2017)
192. F. Zafar, Some generalizations of Ostrowski inequalities and their applications to numerical integration and special means (Unpublished doctoral dissertation), Bahauddin Zakariya University, Multan (2010)
193. F. Zafar, N.A. Mir, A generalized integral inequality for twice differentiable mappings. Kragujevac J. Math. **32**, 81–96 (2009)
194. F. Zafar, N.A. Mir, A generalization of Ostrowski-Grüss type inequality for first differentiable mappings. Tamsui Oxford J. Math. Sci. **26**(1), 61–76 (2010)
195. Y. Zhang, M. Beer, S.T. Quek, Long-term performance assessment and design of offshore structures. Comput. Struct. **154**, 101–115 (2015)

Index

Symbols

(m, n)-completely monotonic function, 215
(m, n)-convex function, 215
(m, n)-divided difference of a function, 214
(m, n)-finite difference of a function, 214
$(m, n) - \nabla$-convex function, 215
log-convex, 36
log$-$convex in $J-$sense, 36
∇-convex function, 203
m-∇-convex, 195
m-completely monotonic function, 215
n-convex functions at a point, 30
$n-$exponentially convex, 37
$n-$exponentially convex in the $J-$sense, 37
Čebyšev functional, 18

A

Abel-Gontscharoff interpolating polynomial for two-points, 71
Absolutely monotonic function, 204
Arithmetic integral mean, 199

B

Bernstein polynomial, 50
Bounded variation, 122

C

Completely monotonic function, 204
Convexity at a point, 13
Convex sequence, 3
Copula, 198

D

Discrete (m, n)-convex function, 215
Discrete m-convex function, 214
Discrete Čebyšev-Popoviciu type identity, 257
Discrete Čebyšev-Popoviciu type inequality, 262
Discrete Čebyšev type identity and inequality, 253

E

Exponentially convex function, 206
Exponentially convex functions in the $J-$sense, 37
Extension of Montgomery identity, 6

F

Finite difference and Divided difference of order (m, n), of a sequence, 215
Finite difference of a function of order m, 195
Fink identity, 63
Fubini's theorem, 10
Function with nondecreasing increments, 194
Function with nondecreasing increments of mth order, 197

G

Generalized Ostrowski fractional integral inequality, 167
Grüss inequality, 146
Green's function, 7

H
Hölder's Integral Inequality, 19
Hermite-Hadamard Inequality, 89
Hermite interpolating polynomial, 45
Hermite interpolation of a function, 44

I
Integral Čebyšev-Popoviciu type identity, 263
Integral Čebyšev-Popoviciu type inequality, 269
Integral Ky Fan type identity, 273

J
Jensen-Boas inequality for function with nondecreasing increments, 207
Jensen m-convex function, 196
Jensen-Mercer inequality, 211

K
K–monotonic function, 205

L
Lagrange and of Cauchy type Mean Value Theorems, 22
Laplace transform, 205
Laplace transformation of Borel measure, 205
Levinson's type inequality, 208

M
Montgomery identity, 6

N
New Green functions, 24
Non-standard quadrature formulae, 89
Numerical quadrature rules, 115, 125, 140, 159

O
Ostrowski-Grüss inequality, 147
Ostrowski inequality, 87
Ostrowski inequality for bounded above differentiable functions, 111
Ostrowski inequality for bounded below differentiable functions, 108
Ostrowski inequality for bounded differentiable functions, 91
Ostrowski inequality with parameters of double integrals for L_p space, 179

P
Peano kernel, 90
Popoviciu type discrete identity for function, 220
Popoviciu type discrete identity for sequence, 217
Popoviciu type integral identity for function, 226
Pre-Grüss inequality, 20
Probability density functions, 157

S
Standard quadrature formulae, 89

T
Taylor's theorem, 24
Taylor's two point formula, 53
Total variation, 122
Two-dimensional Induction, 228

U
Ultramodular function, 197

W
Weighted expectation, 158
Weighted Grüss inequality, 146
Weighted Korkine's identity, 148
Weighted Ostrowski integral inequality, 132
Wright–convex function, 200

CPSIA information can be obtained
at www.ICGtesting.com
Printed in the USA
LVHW081258190323
741962LV00004B/292